# WHAT EVERY ENGINEER SHOULD KNOW ABOUT

# RELIABILITY AND RISK ANALYSIS

## WHAT EVERY ENGINEER SHOULD KNOW

**A Series**

*Editor*

**William H. Middendorf**

*Department of Electrical and Computer Engineering*
*University of Cincinnati*
*Cincinnati, Ohio*

1. What Every Engineer Should Know About Patents, *William G. Konold, Bruce Tittel, Donald F. Frei, and David S. Stallard*
2. What Every Engineer Should Know About Product Liability, *James F. Thorpe and William H. Middendorf*
3. What Every Engineer Should Know About Microcomputers: Hardware/Software Design, A Step-by-Step Example, *William S. Bennett and Carl F. Evert, Jr.*
4. What Every Engineer Should Know About Economic Decision Analysis, *Dean S. Shupe*
5. What Every Engineer Should Know About Human Resources Management, *Desmond D. Martin and Richard L. Shell*
6. What Every Engineer Should Know About Manufacturing Cost Estimating, *Eric M. Malstrom*
7. What Every Engineer Should Know About Inventing, *William H. Middendorf*
8. What Every Engineer Should Know About Technology Transfer and Innovation, *Louis N. Mogavero and Robert S. Shane*
9. What Every Engineer Should Know About Project Management, *Arnold M. Ruskin and W. Eugene Estes*
10. What Every Engineer Should Know About Computer-Aided Design and Computer-Aided Manufacturing: The CAD/CAM Revolution, *John K. Krouse*
11. What Every Engineer Should Know About Robots, *Maurice I. Zeldman*
12. What Every Engineer Should Know About Microcomputer Systems Design and Debugging, *Bill Wray and Bill Crawford*
13. What Every Engineer Should Know About Engineering Information Resources, *Margaret T. Schenk and James K. Webster*
14. What Every Engineer Should Know About Microcomputer Program Design, *Keith R. Wehmeyer*
15. What Every Engineer Should Know About Computer Modeling and Simulation, *Don M. Ingels*

16. What Every Engineer Should Know About Engineering Workstations, *Justin E. Harlow III*
17. What Every Engineer Should Know About Practical CAD/CAM Applications, *John Stark*
18. What Every Engineer Should Know About Threaded Fasteners: Materials and Design, *Alexander Blake*
19. What Every Engineer Should Know About Data Communications, *Carl Stephen Clifton*
20. What Every Engineer Should Know About Material and Component Failure, Failure Analysis, and Litigation, *Lawrence E. Murr*
21. What Every Engineer Should Know About Corrosion, *Philip Schweitzer*
22. What Every Engineer Should Know About Lasers, *D. C. Winburn*
23. What Every Engineer Should Know About Finite Element Analysis, *edited by John R. Brauer*
24. What Every Engineer Should Know About Patents: Second Edition, *William G. Konold, Bruce Tittel, Donald F. Frei, and David S. Stallard*
25. What Every Engineer Should Know About Electronic Communications Systems, *L. R. McKay*
26. What Every Engineer Should Know About Quality Control, *Thomas Pyzdek*
27. What Every Engineer Should Know About Microcomputers: Hardware/Software Design, A Step-by-Step Example. Second Edition, Revised and Expanded, *William S. Bennett, Carl F. Evert, and Leslie C. Lander*
28. What Every Engineer Should Know About Ceramics, *Solomon Musikant*
29. What Every Engineer Should Know About Developing Plastics Products, *Bruce C. Wendle*
30. What Every Engineer Should Know About Reliability and Risk Analysis, *M. Modarres*

ADDITIONAL VOLUMES IN PREPARATION

# WHAT EVERY ENGINEER SHOULD KNOW ABOUT

# RELIABILITY AND RISK ANALYSIS

**M. Modarres**

*Center for Reliability Engineering*
*University of Maryland*
*College Park, Maryland*

MARCEL DEKKER, INC. NEW YORK • BASEL

**Library of Congress Cataloging-in-Publication Data**

Modarres, M. (Mohammad)
What every engineer should know about reliability and risk analysis / M. Modarres
p. cm. -- (What every engineer should know about ; v. 30)
Includes bibliographical references and index.
ISBN 0-8247-8958-X
1. Reliability (Engineering) 2. Risk assessment. I. Title.
II. Title: Reliability and risk analysis. III. Series.
TA169.M63 1993
620'.00452--dc20 92-32998
CIP

This book is printed on acid-free paper.



MARCEL DEKKER, INC.
270 Madison Avenue, New York, New York 10016

Current printing (last digit):
10 9 8 7 6 5 4

PRINTED IN THE UNITED STATES OF AMERICA

To my wife, Susan,
for her patience and understanding.
And to my son, Ceena.

*The quest for certainty blocks the search for meaning. Uncertainty is the very condition to impel man to unfold his powers.*

Erich Fromm, *Man for Himself* (1947)

# Preface

This book provides an introduction to reliability and risk analysis, both for engineering students at the undergraduate and graduate levels, and for practicing engineers. Since reliability analysis is a multidisciplinary subject, the scope is not limited to any one engineering discipline; rather, the material is applicable to most engineering disciplines. I developed the contents of this book from material I presented over the last 10 years in undergraduate and graduate-level courses in Reliability Analysis and Risk Assessment that I have been teaching at the University of Maryland. The book presents basic and advanced methods in reliability analysis that are commonly used in practice. The book presents these methods along with a number of examples.

The emphasis of the book is the introduction and explanation of the practical methods used in reliability, and risk studies, and discussion of their use and limitations. These methods cover a wide range of topics that are used in routine engineering activities. The book assumes that the readers have little or no background in probability and statistics. Thus, an introductory chapter (Chapter 1) defines reliability, availability and risk analysis, and Chapter 2 provides review of probability and statistics essential to understanding of the reliability methods discussed in the book.

I have structured the book so that basic reliability methods are described first in Chapter 3 in the context of a basic engineering unit (i.e., a component). Next, in Chapter 4 these analytical methods are described in the context of a more complex engineering unit (i.e., a system containing many interacting components). The material in Chapters 1 through 4 are more appropriate for an undergraduate course in reliability engineering.

The availability concept and reliability considerations for repairable systems are discussed in Chapter 5. This chapter also explains the corresponding use of the analytical methods discussed in the earlier chapters when performing availability analysis of components and engineering systems. In chapter 6, I discuss a number of important methods frequently used in reliability, availability, and risk studies. For example, in Section 6.2, I discuss the concept of uncertainty, şources of uncertainty, parameter and model uncertainty, and probabilistic methods for quantifying and propagating parameter uncertainties in engineering systems (or models). Examples clarifying the uses of these methods and their shortcomings are also presented.

In Chapter 7, I discuss the method of risk assessment. A number of the analytical methods explained in the preceding chapters are integrated. Probabilistic risk assessment has been a major topic of interest in light of hazards imposed by many engineering designs and processes. Interest on risk analysis has increased due to accidents that have recently resulted in significant public attention, such as the nuclear accident at Three Mile Island; the accident at a chemical plant in Bhopal, India; and the Challenger disaster. The book could have not been materialized without the help and corrections from some of my students and colleagues. It would be difficult to name all, but some names to mention includes: L. Chen, Y. Guan, K. Hsueh, Y.S. Hu, D. Koo, A. Mosleh, M. Roush and J.N. Wang. Also the editorial help of D. Grimsman and typing and graphical help of Ms. M. Tedijanto and Ms.Y. Zhang are highly appreciated.

M. Modarres

# Contents

# WHAT EVERY ENGINEER SHOULD KNOW ABOUT

# RELIABILITY AND RISK ANALYSIS

# 1
# Reliability, Availability and Risk Analysis in Perspective

Overall performance of an item (component, product, subsystem or system) results from implementation of various programs that ultimately improve the performance of the item. Historically, these programs have been installed through a trial-and-error approach. For example, they are sometimes established based on empirical evidence gathered during investigation of failures. An example of such programs is a root-cause failure analysis program. It is worthwhile, at this point, to understand why a well established reliability analysis and engineering program can influence the performance of today's items. For this reason, let us first define what constitutes the performance of an item.

The performance of an item can be described by four elements:

- Capability or the item's ability to satisfy functional needs;
- Efficiency or the item's ability to effectively utilize the energy supplied;
- Reliability or the item's ability to start or continue to operate;
- Maintainability or the item's ability to quickly start following its failure.

It is evident that the first two measures are influenced by the design, construction, production or manufacturing of the item. Capability and efficiency reflect the levels to which the item is designed and built. For example, the designer ensures that design levels are adequate to meet the functional requirements of a product. On the other hand, reliability is an operations related issue and is influenced by the item's potential to remain operational. In a repairable item, the ease with which the item is repaired, maintained and returned to operation is measured by its maintainability. Based on the above definitions it would be conceivable to have an item that is highly reliable, but does not achieve a high performance. For example, when an item does not meet its stated design objectives. Clearly humans play a major role in the design, construction, production, operation, and maintenance of the item. This common role can significantly influence the values of the four performance measures. The role of humans is often determined by various programs and activities that support the four elements of performance, proper implementation of which leads to a **quality** item.

To put all of these factors in perspective, let's consider the development of a high performance product in an integrated framework. For this purpose, let's consider the so-called diamond tree conceptually shown in Fig. 1.1. In this tree, the top goal of "High performance" during the life cycle of an items is hierarchically decomposed into various goals and subgoals. By looking downward from the top of this structure, one can describe **how** various goals and subgoals are achieved, and by looking upward, one can identify **why** a goal needs to be achieved. Fig. 1.1 shows only typical goals, but also reflects the general goals involved in designing and operating a high performance item. For more detailed description of the diamond tree the readers are referred to Hunt and Modarres(1985).

The role of reliability, quality and maintainability in the overall framework shown in Fig. 1.1 is clear. A more detailed look at the goals of improving reliability, in an integrated manner, would yield a better perspective on the role of reliability analysis as shown by the hierarchy depicted in Fig. 1.2. From this, one can put into a proper context the role of reliability and availability analysis.

Clearly, reliability is an important element of achieving high performance since it directly and significantly influences the item's performance and ultimately its life-cycle cost and economics. Poor reliability in design, manufacturing, construction, and operation would directly cause increased warranty costs, liabilities, recalls, and repair costs. Poor quality would also lead to poor performance. Therefore, a high quality design, production, manufacturing and operation program leads to low failures, effective maintenance and repair, and ultimately high performance.

## 1.1 Definition of Reliability

Reliability has two connotations. One is probabilistic in nature; the other is deterministic. In this book, we generally deal with the probabilistic aspect. Let us first define what we mean by reliability. The most widely accepted definition of reliability is **the ability of an item (product, system, ... etc.) to operate under desig-**

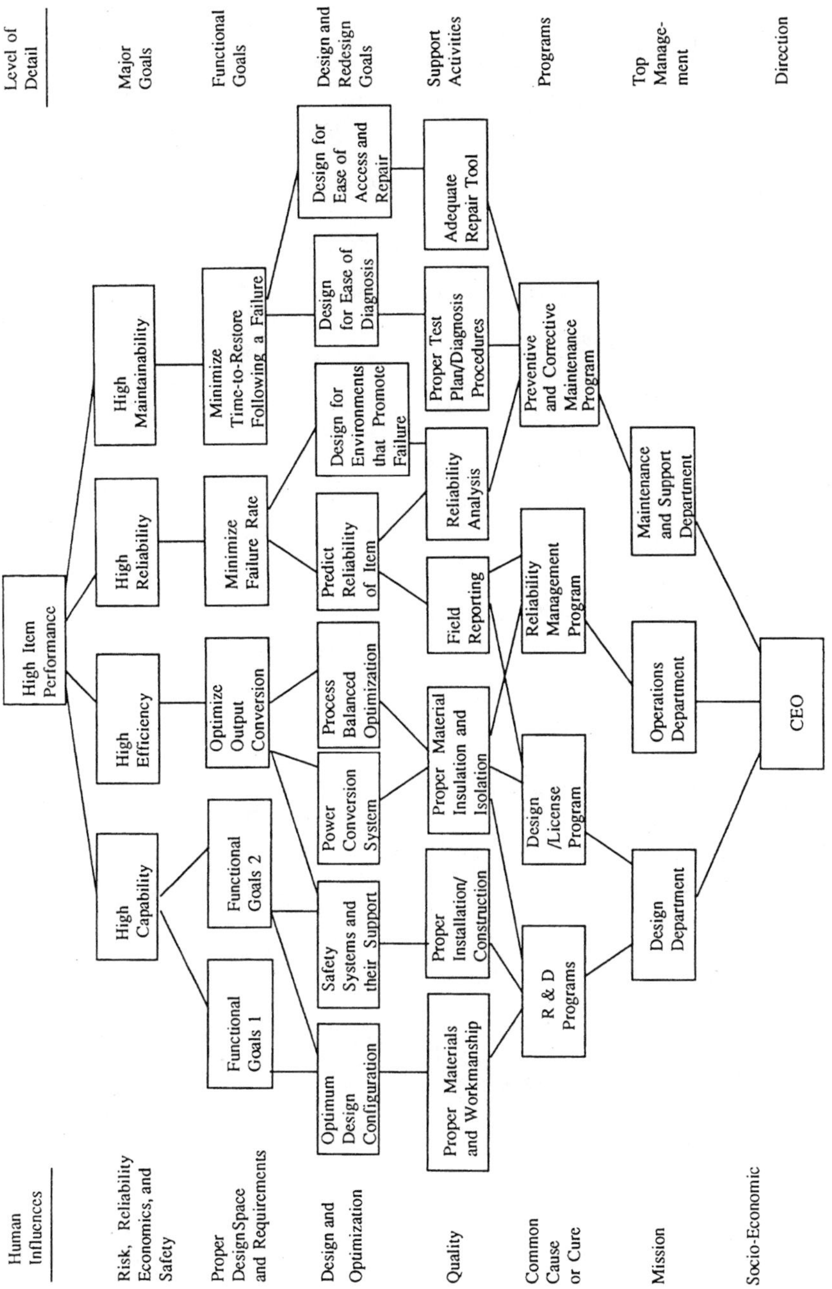


*Fig. 1.1 A Conceptual Diamond Tree Representation for Achieving High Performance*

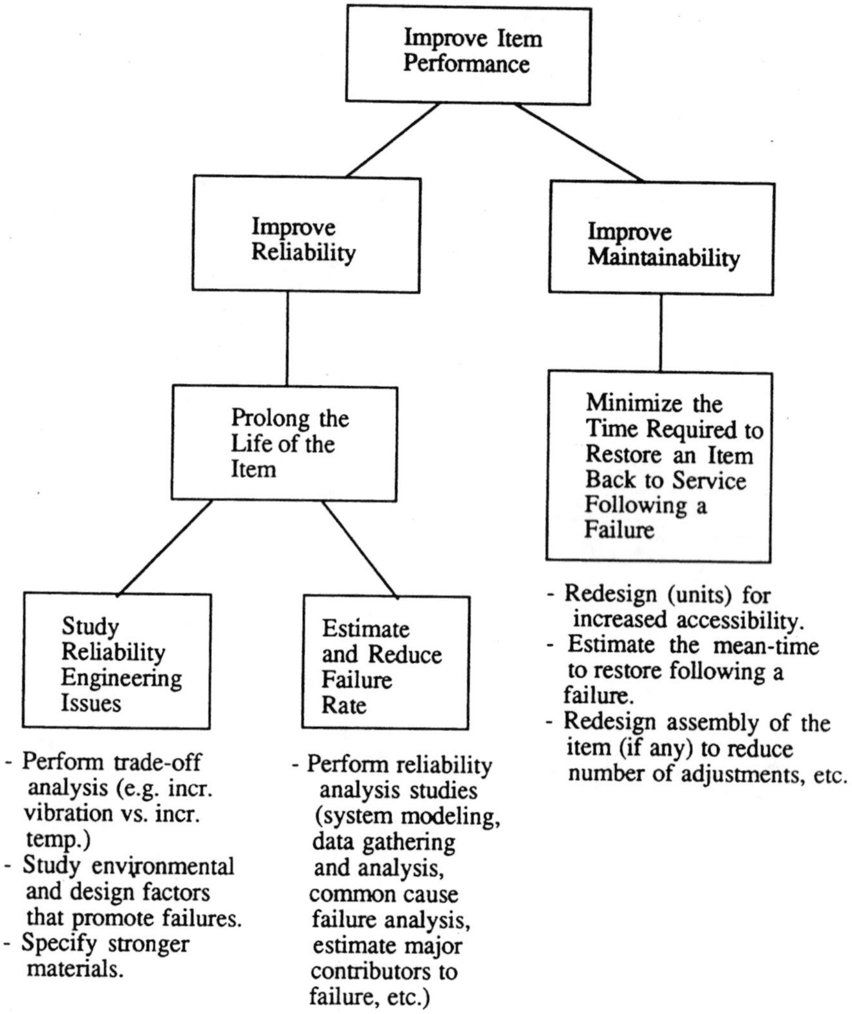


*Fig. 1.2 A Conceptual Hierarchy for Improving Performance*

**nated operating conditions for a designated period of time or number of cycles.** The **ability** of an item can be designated through a probability (the probabilistic connotation), or can be designated deterministically. The deterministic approach, in essence, deals with understanding how and why an item fails, and how it can be designed and tested to prevent such failures from occurrence or recurrence. This includes such analyses as deterministic analysis and review of field failure reports, understanding physics of failure, the role and degree of test and inspection, performing redesign, or performing reconfiguration. In practice, this is an important aspect of reliability analysis.

The probabilistic treatment of an item's reliability according to the definition above can be summarized by

$$R(t) = \Pr(T \geq t | c_1, c_2, \ldots),. \tag{1.1}$$

where,

$t =$ the designated period of time or cycles for the item's operation (mission time),
$T =$ time to failure or cycle to failure of the item,
$R(t) =$ reliability of the item, and
$c_1, c_2, \ldots =$ designated conditions, such as environmental conditions.

Often, in practice, $c_1, c_2, \ldots$ are implicitly considered in the probabilistic reliability analysis and thus (1.1) reduces to

$$R(t) = \Pr(T \geq t). \tag{1.2}$$

Expressions (1.1) and (1.2) are discussed further in Chapter 3.

## 1.2 Definition of Availability

Availability analysis is performed to verify that an item has a satisfactory probability of being operational so it can achieve its intended objective. In Fig. 1.2, an item's availability can be considered as combination of its reliability and maintainability. Accordingly, when no maintenance or repair is performed (e.g., in nonrepairable items), reliability can be considered as instantaneous availability.

Mathematically, the availability of an item is a measure of the fraction of time that the item is in operating condition in relation to total or calendar time. There are several measures of availability, namely, inherent availability, achieved availability, and operational availability. For further definition of these availability measures, see

Ireson and Coombs (1988). Here, we describe inherent availability, which is the most common definition used in the literature.

A more formal definition of availability is **the probability that an item, when used under stated conditions in an ideal support environment (i.e., ideal spare parts, personnel, diagnosis equipment, procedures, etc.), will be operational at a given time**. Based on this definition, the average availability of an item during an interval of time $T$ can be expressed by

$$A = \frac{u}{u + d}, \tag{1.3}$$

where

$u$= uptime during time T.
$d$= downtime during time T.
T= u+d.

Time-dependent expressions of availability and measures of availability for different types of equipment are discussed in more detail in Chapter 5. The mathematics and methods for reliability analysis discussed in this book are also equally applicable to availability analysis.

## 1.3 Definition of Risk

Risk can be viewed both qualitatively and quantitatively. Qualitatively speaking, when there is a source of danger (hazard), and when there are no safeguards against exposure of the hazard, then there is a possibility of loss or injury. This possibility is referred to as risk. The loss or injury could result from business, social, or military activities; operation of equipment; investment; etc. Risk can be formally defined as **the potential of loss or injury resulting from exposure to a hazard**.

In complex engineering systems, there are often safeguards against exposure of hazards. The higher the level of safeguards, the lower the risk. This also underlines the importance of highly reliable safeguard systems and shows the roles of and relationship between reliability analysis and risk analysis.

In this book, we are concerned with quantitative risk analysis. Since quantitative risk analysis involves estimation of the degree or probability of loss, risk analysis is fundamentally intertwined with the concept of probability of occurrence of hazards. Risk analysis consists of answers to the following questions[see Kaplan and Garrick (1981)]:

1) What can go wrong that could lead to an outcome of hazard exposure?
2) How likely is this to happen?
3) If it happens, what consequences are expected?

To answer question 1, a list of outcomes (or scenarios of events leading to the outcome) should be defined. The likelihood of these scenarios should be estimated (answer to question 2), and the consequence of each scenario should be described (answer to question 3). Therefore, risk can be defined, quantitatively, as the following set of triplets:

$$R = < S_i, P_i, C_i >, \qquad i = 1, 2, ..., n, \tag{1.4}$$

where

$S_i$ is a scenario of events that lead to hazard exposure,
$P_i$ is the likelihood of scenario i, and
$C_i$ is the consequence (or evaluation measure) of scenario i e.g., a measure of the degree of damage or loss.

Since (1.4) involves estimation of the likelihood of occurrence of events (e.g., failure of safeguard systems), most of the methods described in Chapters 2 through 6 become relevant. However, we have specifically devoted Chapter 7 to a detailed, quantitative description of these methods as applied to risk analysis.

## BIBLIOGRAPHY

1. Hunt, R.N., and M. Modarres (1985). "Use of Goal Tree Methodology to Evaluate Institutional Practices and Their Effect on Power Plant Hardware Performance," American Nuclear Society Topical Meeting on Probabilistic Safety Methods and Applications, San Fransisco.
2. Ireson, W.G., and C.F. Coombs eds.(1988). *Handbook of Reliability Engineering and Management*, McGraw-Hill, New York.
3. Kaplan, S., and J. Garrick (1981). "On the Quantitative Definition of Risk," *Risk Analysis*, vol. 1, No.1.

# 2
# Basic Mathematics

## 2.1 Introduction

In this chapter, we discuss the elements of mathematical reliability that are relevant to the study of reliability of items. Since reliability is defined as a conditional probability in Chapter 1, we begin with a presentation of basic concepts of probability. We then discuss fundamental concepts of statistics that are used in reliability analysis.

## 2.2 Elements of Probability

Probability is a concept that people use formally and casually every day. The weather forecasts are associated with probability. People use probability in their casual conversations to show their perception of the likely occurrence or nonoccurrence of particular events. Odds are given for the outcome of sporting events, and are used in gambling.

Formal use of probability concepts is widespread in science, for example, astronomy, biology, and engineering. In this chapter, we discuss the formal application of probability in the field of reliability engineering.

### *2.2.1 Sets and Boolean Algebra*

To present operations associated with probability, it is often necessary to use sets. A set is a collection of items or elements, each with some specific characteristics. A set that includes all items of interest is referred to as a universal set, denoted by $\Omega$. A subset refers to a collection of items that belong to a universal set. For example, if set $\Omega$ represents the collection of all pumps in a power plant, then the collection of electrically driven pumps is a subset $E$ of $\Omega$. Graphically, the relationship between subsets and sets can be illustrated by using Venn diagrams. The Venn diagram in Fig. 2.1 shows the universal set $\Omega$ by a rectangle, and subsets $E_1$ and $E_2$ by circles. It can also be seen that $E_2$ is a subset of $E_1$. The relationship

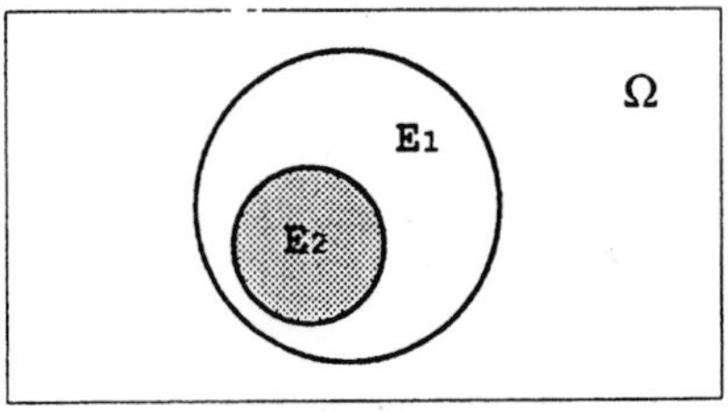


*Fig. 2.1 Venn Diagram*

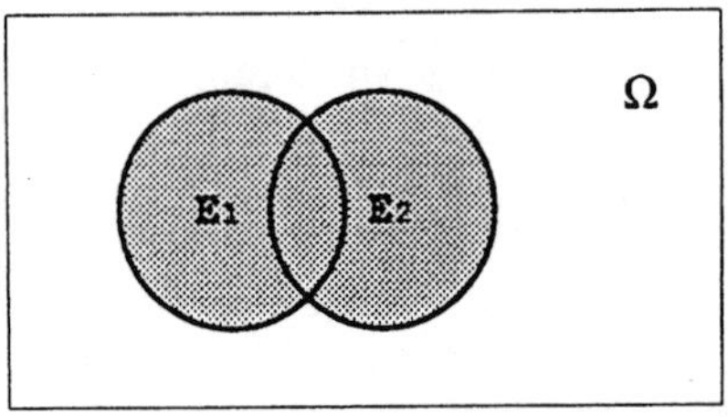


*Fig. 2.2 Union of Two Sets, $E_1$ and $E_2$*

between subsets $E_1$ and $E_2$ and the universal set can be symbolized by $E_2 \subset E_1 \subset \Omega$.

The **complement** of a set $E$, denoted by $\bar{E}$, and called "not E," is the set of all items in the universal set that do not belong to set E. In Fig. 2.1, the nonshaded area represents $\bar{E}_2$. It is clear that sets $E_2$ and $\bar{E}_2$ together comprise $\Omega$.

The **union** of two sets, $E_1$ and $E_2$, is a set that contains all items that belong to $E_1$ or $E_2$. The union is symbolized either by $E_1 \cup E_2$ or $E_1 + E_2$, and is read "$E_1$ **or** $E_2$." That is, the set $E_1 \cup E_2$ represents all elements that are in $E_1$, $E_2$, or both $E_1$ and $E_2$. The shaded area in Fig. 2.2 shows the union of sets $E_1$ and $E_2$. Suppose $E_1$ and $E_2$ represent positive odd and even numbers between 1 and 10, respectively. Then

$$E_1 = \{1, 3, 5, 7, 9\}$$
$$E_2 = \{2, 4, 6, 8, 10\}$$

The union of these two sets is:

$$E_1 \cup E_2 = \{1, 2, 3, 4, 5, 6, 7, 8, 9, 10\},$$

or, if $E_1 = \{x, y, z\}$ and $E_2 = \{x, t, z\}$, then $E_1 \cup E_2 = \{x, y, z, t\}$.

The **intersection** of two sets, $E_1$ and $E_2$, is the set of items that are common to both $E_1$ and $E_2$. This set is symbolized by $E_1 \cap E_2$ or $E_1 \cdot E_2$, and is read "$E_1$ **and** $E_2$." In Fig. 2.3, the shaded area

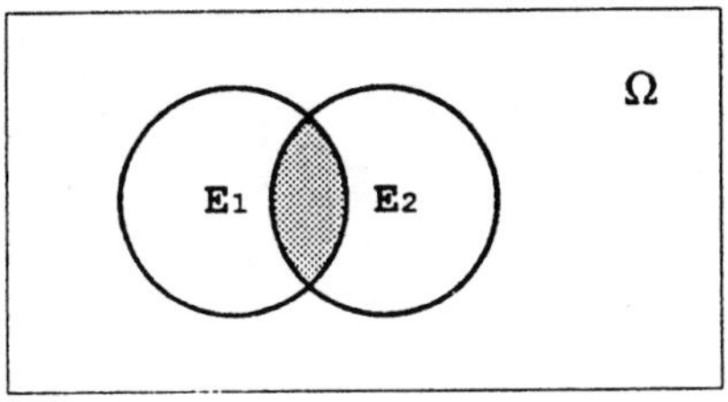


*Fig. 2.3 Intersection of Two Sets, $E_1$ and $E_2$*

represents the intersection of $E_1$ and $E_2$.

Suppose $E_1$ is a set of manufactured devices that operate for $t > 0$ but fail prior to 1000 hours of operation. If set $E_2$ represents a set of devices that operate between 500 and 2000 hours, then $E_1 \cap E_2$ can be obtained as follows:

$$E_1 = \{t \mid 0 < t < 1000\},$$
$$E_2 = \{t \mid 500 < t < 2000\},$$
$$E_1 \cap E_2 = \{t \mid 500 < t < 1000\}.$$

A **null** or **empty set**, $\phi$, refers to a set that contains no items. One can immediately infer that the complement of a universal set is a null set, and vice versa. That is,

$$\bar{\Omega} = \phi, \tag{2.1}$$
$$\Omega = \bar{\phi}.$$

Two sets, $E_1$ and $E_2$, are termed **mutually exclusive** or **disjoint** when $E_1 \cap E_2 = \phi$. In this case, there are no elements common to $E_1$ and $E_2$. Two mutually exclusive sets are illustrated in Fig. 2.4.

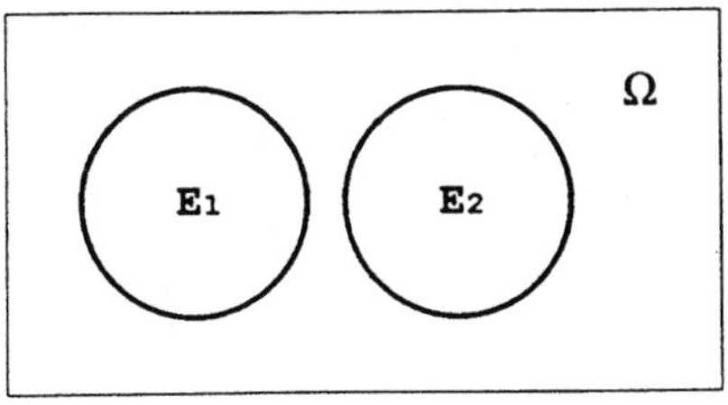


*Fig. 2.4 Mutually Exclusive Sets, $E_1$ and $E_2$*

From the discussions thus far, or from examination of a Venn diagram, the following conclusions can be easily drawn:

- The intersection of a set $E$ and a null set is a null set:

$$E \cap \phi = \phi. \tag{2.2}$$

- The union of a set $E$ and a null set is $E$:

$$E \cup \phi = E. \tag{2.3}$$

- The intersection of a set $E$ and the complement of $E$ is a null set:

$$E \cap \bar{E} = \phi. \tag{2.4}$$

- The intersection of a set E and a universal set is E:

$$E \cap \Omega = E. \tag{2.5}$$

- The union of a set E and a universal set is the universal set:

$$E \cup \Omega = \Omega. \tag{2.6}$$

- The complement of the complement of a set $E$ is $E$:

$$(\bar{\bar{E}}) = E. \tag{2.7}$$

- The union of two identical sets E is E:

$$E \cup E = E. \tag{2.8}$$

- The intersection of two identical sets $E$ is $E$:

$$E \cap E = E. \tag{2.9}$$

---

*Example 2.1*

Simplify the following expressions:

$$E_1 \cap E_1 \cup \phi.$$

*Solution:*

Since $E_1 \cap E_1 = E_1$, then the expression reduces to $E_1 \cup \phi = E_1$.

---

Boolean algebra provides a means of evaluating sets. The rules are fairly simple. The sets of axioms in Table 2.1 provide all the major relations of interest in Boolean algebra including some of the expressions discussed in equations (2.1) through (2.9).

*Table 2.1 Laws of Boolean Algebra*

| | |
|---|---|
| $X \cap Y = Y \cap X$<br>$X \cup Y = Y \cup X$ | Commutative Law |
| $X \cap (Y \cap Z) = (X \cap Y) \cap Z$<br>$X \cup (Y \cup Z) = (X \cup Y) \cup Z$ | Associative Law |
| $X \cap (Y \cup Z) = (X \cap Y) \cup (X \cap Z)$<br>$X \cup (Y \cap Z) = (X \cup Y) \cap (X \cup Z)$ | Distributive Law |
| $X \cap X = X$<br>$X \cup X = X$ | Idempotent Law |
| $X \cap (X \cup Y) = X$<br>$X \cup (X \cap Y) = X$ | Absorption Law |
| $X \cap \bar{X} = \phi$<br>$X \cup \bar{X} = \Omega$<br>$(\bar{\bar{X}}) = X$ | Complementation Law |
| $(\overline{X \cap Y}) = \bar{X} \cup \bar{Y}$<br>$(\overline{X \cup Y}) = \bar{X} \cap \bar{Y}$ | deMorgan's Theorem |

---

*Example 2.2*

Simplify the Boolean expression:

$$\overline{[(A \cap B) \cup (A \cap \bar{B}) \cup (\bar{A} \cap \bar{B})]}$$

| | |
|---|---|
| $= (\overline{A \cap B}) \cap (\overline{A \cap \bar{B}}) \cap (\overline{\bar{A} \cap \bar{B}})$ | deMorgan's Theorem |
| $= (\bar{A} \cup \bar{B}) \cap (\bar{A} \cup \bar{\bar{B}}) \cap (\bar{\bar{A}} \cup \bar{\bar{B}})$ | deMorgan's Theorem |
| $= (\bar{A} \cup \bar{B}) \cap (\bar{A} \cup B) \cap (A \cup B)$ | complementation Law |
| $= [\bar{A} \cup (\bar{A} \cap \bar{B}) \cup (B \cap \bar{A}) \cup (\bar{B} \cap B)] \cap (A \cup B)$ | Absorption Law |
| $= [(\bar{A} \cup \bar{B} \cap \bar{A}) \cup (\bar{B} \cap B)] \cap (A \cup B)$ | Absorption Law |
| $= (\bar{A} \cup \bar{B} \cap \bar{A}) \cap (A \cup B)$ | Distribution Law |
| $= \bar{A} \cap (A \cup B)$ | Distributive Law |
| $= (\bar{A} \cap A) \cup (\bar{A} \cap B)$ | Complementation Law |
| $= \bar{A} \cap B.$ | |

---

*2.2.2 Basic Laws of Probability*

In probability theory, the elements that comprise a set are outcomes of an experiment. Thus, the universal set $\Omega$ represents the mutually exclusive listing of all possible outcomes of the experiment and is referred to as the **sample space** of the experiment. In examining the outcomes of rolling a die, the sample space $S = \{1, 2, 3, 4, 5, 6\}$. This sample space consists of six items (elements) or **sample points**. In probability concepts, a combination of several sample points is called an **event**. An event is, therefore, a subset of the sample space. For example, the event of an odd outcome when rolling a die represents a subset containing sample points 1, 3, and 5.

Associated with any event $E$ of a sample space $S$ is a probability (Pr) shown by $\Pr(E)$ and obtained from the following equation:

$$\Pr(E) = \frac{m(E)}{m(S)}, \tag{2.10}$$

where $m(.)$ denotes the number of elements in the set (.) and refers to the **size of the set**.

The probability of getting an odd number when tossing a die is determined by using m(odd outcomes)=3 and m(sample space)=6. In this case Pr(odd outcomes)=3/6 or 0.5.

Note that (2.10) represents a comparison of the relative size of the subset represented by the event $E$ to the sample space $S$. This is true when all sample points are equally likely to be the outcome. When all sample points are not equally likely to be the outcome, the sample points may be weighted according to their relative frequency of occurrence over many trials or according to expert judgment.

It is important that the readers appreciate some intuitive differences between three major conceptual interpretations of probability described below.

### Classical Interpretation of Probability (Equally Likely Concept)

In this interpretation, the probability of an event $E$ can be obtained from (2.10), provided that the sample space contains $N$ equally likely and different outcomes (i.e., $m(S) = N$), $n$ of which have an outcome (event) $E$ (i.e., $m(E) = n$). Thus $\Pr(E) = n/N$. This definition is frequently inadequate for engineering applications. For example, if failures of a pump after a start in a process plant are observed, it is often not known whether all failures to start are equally likely. Nor is it clear whether or not the whole spectrum of possible events is observed. That case is not similar to rolling a perfect die, with each side having an equal probability of 1/6 at any time in future.

### Frequency Interpretation of Probability (Empirical Concept)

In this interpretation, the limitation on the lack of knowledge about the overall sample space is remedied by defining the probability as the limit of $n/N$ as $N$ becomes large. Therefore, $\Pr(E) = \lim_{N\to\infty} n/N$. Thus if we have observed 2,000 starts of a pump in which 20 failed, and if we assume that 2,000 is a large number, then the probability of the pump failure to start is $20/2000 = 0.01$.

The frequency interpretation is the most widely used definition today. However, some argue that because it does not cover cases in which little or no experience (or evidence) is available, nor cases where estimates concerning our observations are principally intuitive, a broader definition is required. This has led to the third interpretation of probability.

### Subjective Interpretation of Probability

In this interpretation, $\Pr(E)$ is a measure of the degree of belief one holds in a specified event $E$. To better understand this interpretation, consider the probability of improving a system by making a design change. The designer believes that such a change results in a performance improvement in one out of three missions in which the system is used. It would be difficult to describe this problem through the first two interpretations. That is, the classical interpretation is inadequate since there is no reason to believe that performance is as likely to improve as to not improve. The frequency interpretation is not applicable because no historical data exist to show how often a design change resulted in improving the system. Thus the subjective interpretation will provide a broad definition of the probability concept.

### Calculus of Probability

The basic rules used to combine and treat the probability of an event are not affected by the interpretations discussed above; we can proceed without adopting any of them. (There is much dispute among probability scholars regarding these interpretations. Readers are referred to Cox (1946) for further discussions of this subject.)

In general, the axioms of probability can be defined for a sample space $S$ as follows:

1) $\Pr(E) \geq 0$, for every event $E$ such that $E \subset S$,
2) $\Pr(E_1 \cup E_2 \cup \ldots \cup E_n) = \Pr(E_1) + \Pr(E_2) + \ldots + \Pr(E_n)$,

where the events $E_1, E_2, \ldots, E_n$ are such that no two have a point in common,

3) $\Pr(S) = 1$.

It is important to understand the concept of independent events before attempting to multiply and add probabilities. Two events are **independent** if the occurrence or nonoccurrence of one does not depend on or change the probability of the occurrence of the other. Mathematically, this can be shown by

$$\Pr(E_1 \mid E_2) = \Pr(E_1), \tag{2.11}$$

where $\Pr(E_1 \mid E_2)$ is the probability of $E_1$, given that $E_2$ has occurred. To better illustrate, let's consider the result of a test on 200 manufactured parts. It is observed that 23 parts fail to meet the length limitation imposed by the designer, and 18 fail to meet the height limitation. Additionally, 7 parts fail to meet both length and height limitations. Therefore, 152 parts meet both of the specified requirements. Let $E_1$ represent the event that a part does not meet the specified length, and $E_2$ represent the event that the part does not meet the specified height. According to (2.10) $\Pr(E_1) = (7 + 23)/200 = 0.15$, and $\Pr(E_2) = (18 + 7)/200 = 0.125$. Furthermore, among 25 parts (7+18) that have at least event $E_2$, 7 parts also have event $E_1$. Thus, $\Pr(E_1 \mid E_2) = 7/25 = 0.28$. Since $\Pr(E_1 \mid E_2) \neq \Pr(E_1)$, event $E_1$ and $E_2$ are dependent.

We shall now discuss the rules for evaluating the probability of simultaneous occurrence of two or more events, that is, $\Pr(E_1 \cap E_2)$. For this purpose, we recognize two facts: First, when $E_1$ and $E_2$ are independent, the probability that both $E_1$ and $E_2$ occur simultaneously is simply the multiplication of the probability that $E_1$ and $E_2$ occur individually. That is, $\Pr(E_1 \cap E_2) = \Pr(E_1) \cdot \Pr(E_2)$. Second, when $E_1$ and $E_2$ are dependent, the probability that both $E_1$ and $E_2$ occur simultaneously is obtained from the following expressions:

$$\Pr(E_1 \cap E_2) = \Pr(E_1) \cdot \Pr(E_2 \mid E_1). \tag{2.12}$$

We will elaborate further on (2.12) when we discuss Bayes' Theorem. It is easy to see that when $E_1$ and $E_2$ are independent, and (2.11) is applied, (2.12) reduces to

$$\Pr(E_1 \cap E_2) = \Pr(E_1) \cdot \Pr(E_2).$$

In general, the probability of joint occurrence of $n$ independent events $E_1, E_2, \ldots, E_n$ is the product of their individual probabilities.

That is,

$$\Pr(E_1 \cap E_2 \cap \ldots \cap E_n) = \Pr(E_1) \cdot \Pr(E_2) \ldots \Pr(E_n) = \prod_{i=1}^{n} \Pr(E_i). \quad (2.13)$$

Similarly, the probability of joint occurrence of $n$ dependent events $E_1, E_2, \ldots, E_n$ is obtained from

$$\begin{aligned}\Pr(E_1 \cap E_2 \cap \ldots \cap E_n) = \Pr(E_1) \cdot \Pr(E_2 \mid E_1) \cdot \Pr(E_3 \mid E_1 \cap E_2) \ldots \\ \ldots \Pr(E_n \mid E_1 \cap E_2 \cap \ldots \cap E_{n-1}),\end{aligned} \quad (2.14)$$

where $\Pr(E_3 \mid E_1 \cap E_2 \cap \ldots)$ denotes the conditional probability of $E_3$ given the occurrence of both $E_1$ and $E_2$, and so on.

---

*Example 2.3*

Suppose that Vendor 1 provides 40% and Vendor 2 provides 60% of electronic devices used in a computer. It is further known that 2.5% of Vendor 1's supplies are defective, and only 1% of Vendor 2's supplies are defective. What is the probability that a unit is both defective and supplied by Vendor 1? What is the same probability for Vendor 2?

*Solution:*

$$\begin{aligned}E_1 &= \text{the event that a device is from Vendor 1,}\\ E_2 &= \text{the event that a device is from Vendor 2,}\\ D &= \text{the event that a device is defective,}\\ D \mid E_1 &= \text{the event that a device is known to be from Vendor 1 and defective,}\\ D \mid E_2 &= \text{the event that a device is known to be from Vendor 2 and defective.}\end{aligned}$$

Then,

$$\Pr(E_1) = 0.40,\ \Pr(E_2) = 0.60, \Pr(D \mid E_1) = 0.025,\ \text{and} \Pr(D \mid E_2) = 0.01.$$

From (2.14), the probability that a defective device is from Vendor 1 is

$$\Pr(E_1 \cap D) = \Pr(E_1) \cap \Pr(D \mid E_1) = (0.4) \cdot (0.025) = 0.01.$$

Similarly,

$$\Pr(E_2 \cap D) = 0.006.$$

---

The evaluation of the probability of union of two events depends on whether or not these events are mutually exclusive. To illustrate this point, let's consider the 200 electronic parts that we discussed earlier. The union of two events $E_1$ and $E_2$ includes those parts that do not meet the length requirement, or the height requirement, or both. That is, a total of $23 + 18 + 7 = 48$. Thus $\Pr(E_1 \cup E_2) = 48/200 = 0.24$. In other words, 24% of the parts do not meet one or both of the requirements. We can easily see that $\Pr(E_1 \cup E_2) \neq \Pr(E_1) + \Pr(E_2)$, since $0.24 \neq 0.125 + 0.15$. The reason for this inequality is the fact that the two events $E_1$ and $E_2$ are not mutually exclusive. In turn, $\Pr(E_1)$ will include the probability of inclusive events $E_1 \cap E_2$, and $\Pr(E_2)$ will also include events $E_1 \cap E_2$. Thus, joint events are counted twice in the expression $\Pr(E_1) + \Pr(E_2)$. Therefore, $\Pr(E_1 \cap E_2)$ must be subtracted from this expression. This description, which can also be seen in a Venn diagram, leads to the following expression for evaluating the probability of the union of two events that are not mutually exclusive:

$$\Pr(E_1 \cup E_2) = \Pr(E_1) + \Pr(E_2) - \Pr(E_1 \cap E_2). \tag{2.15}$$

Since $\Pr(E_1 \cap E_2) = 7/200 = 0.035$, then $\Pr(E_1 \cup E_2) = 0.125 + 0.15 - 0.035 = 0.24$, which is what we expect to get. From (2.15) one can easily infer that if $E_1$ and $E_2$ are mutually exclusive, then $\Pr(E_1 \cup E_2) = \Pr(E_1) + \Pr(E_2)$. If events $E_1$ and $E_2$ are dependent, then by using (2.12), we can write (2.15) in the following form:

$$\Pr(E_1 \cup E_2) = \Pr(E_1) + \Pr(E_2) - \Pr(E_1) \cdot \Pr(E_2 \mid E_1). \tag{2.16}$$

Equation (2.15) for two events can be logically extended to $n$ events.

$$\begin{aligned} &\Pr(E_1 \cup E_2 \cup \ldots \cup E_n) \\ &= [\Pr(E_1) + \Pr(E_2) + \ldots + \Pr(E_n)] \\ &- [\Pr(E_1 \cap E_2) + \Pr(E_1 \cap E_3) + \ldots + \Pr(E_{n-1} \cap E_n)] \\ &+ [\Pr(E_1 \cap E_2 \cap E_3) + \Pr(E_1 \cap E_2 \cap E_4) + \ldots \\ &+ \Pr(E_{n-2} \cap E_{n-1} \cap E_n)] \\ &- \ldots (-1)^{n+1} [E_1 \cap E_2 \cap \ldots \cap E_n]. \end{aligned} \tag{2.17}$$

Equation (2.17) consists of $2^n - 1$ terms. If events $E_1, E_2, \ldots, E_n$ are mutually exclusive, then

$$\Pr(E_1 \cup E_2 \cup \ldots \cup E_n) = \Pr(E_1) + \Pr(E_2) + \ldots + \Pr(E_n). \tag{2.18}$$

When events $E_1, E_2, \ldots, E_n$ are not mutually exclusive, a useful approximation known as a **rare event approximation** can be used. In this approximation, (2.18) is used if all $\Pr(E_i)$ are small (e.g., $< 1/50n$).

---

*Example 2.4*

Determine the maximum error in the right hand side of (2.17) if (2.18) is used instead of (2.17). Find this error for $n = 2, 3, 4$, and assume $\Pr(E_i) \leq 1/50n$.

*Solution:*

For maximum error assume $\Pr(E_i) = \frac{1}{50n}$

For $n = 2$, using (2.17),

$$\Pr(E_1 \cup E_2) = \frac{2}{50 \times 2} - (\frac{1}{50 \times 2})^2 = 0.01990.$$

Using (2.18),

$$\Pr(E_1 \cup E_2) = \frac{2}{50 \times 2} = 0.02000,$$

$$|\max \% \text{ Error}| = |\frac{0.01990 - 0.02000}{0.01990} \times 100| = 0.50\%.$$

For $n = 3$, using (2.18),

$$\Pr(E_1 \cup E_2 \cup E_3) = \frac{3}{50 \times 3} - 3\left(\frac{1}{50 \times 3}\right)^2 + \left(\frac{1}{50 \times 3}\right)^3 = 0.01987,$$

$$\Pr(E_1 \cup E_2 \cup E_3) = \frac{3}{50 \times 3} = 0.02000,$$

$$|\max \% \text{ error}| = 0.65\%.$$

Similarly for $n = 4$,

$$|\max \% \text{ Error}| = 0.76\%.$$

---

For dependent events, (2.17) can also be expanded to the form of equation (2.16) by using (2.14). If all events are independent, then according to (2.13), (2.15) can be further simplified to

$$\Pr(E_1 \cup E_2) = \Pr(E_1) + \Pr(E_2) - \Pr(E_1) \cdot \Pr(E_2). \tag{2.19}$$

Equation (2.19) can be algebraically reformatted to the easier form of

$$\Pr(E_1 \cup E_2) = 1 - (1 - \Pr(E_1)) \cdot (1 - \Pr(E_2)). \tag{2.20}$$

Equation (2.19) can be expanded in the case of $n$ independent events to

$$\Pr(E_1 \cup E_2 \cup \ldots \cup E_n) = 1 - (1 - \Pr(E_1)) \cdot (1 - \Pr(E_2)) \ldots (1 - \Pr(E_n)). \tag{2.21}$$

---

*Example 2.5*

A particular type of a computer chip is manufactured by three different suppliers. It is known that 5% of chips from Supplier 1, 3% from Supplier 2, and 8% from Supplier 3 are defective. If one chip is selected from each supplier, what is the probability that at least one of the chips is defective?

*Solution:*

$D_1 =$ the event that a chip from Supplier 1 is defective,
$D_2 =$ the event that a chip from Supplier 2 is defective,
$D_3 =$ the event that a chip from Supplier 3 is defective.

$D_1 \cup D_2 \cup D_3$ is the event that at least one chip from Supplier 1, Supplier 2, or Supplier 3 is defective. Since the occurrence of events $D_1, D_2$, and $D_3$ are independent, we can use (2.21) to determine the probability of $D_1 \cup D_2 \cup D_3$. Thus,

$$\Pr(D_1 \cup D_2 \cup D_3) = 1 - (1 - 0.05) \cdot (1 - 0.03) \cdot (1 - 0.08) = 0.152.$$

---

In probability evaluations, it is sometimes necessary to evaluate the probability of the complement of an event, that is, $\Pr(\bar{E})$. To obtain this value, let's begin with (2.10) and recognize that the probability of the sample space $S$ is 1. The sample space can also be expressed by event $E$ and its complement $\bar{E}$. That is

$$\Pr(S) = 1 = \Pr(E \cup \bar{E}).$$

Since $E$ and $\bar{E}$ are mutually exclusive, $\Pr(E \cup \bar{E}) = \Pr(E) + \Pr(\bar{E})$. Thus, $\Pr(E) + \Pr(\bar{E}) = 1$. By rearrangement, it follows that

$$\Pr(\bar{E}) = 1 - \Pr(E). \tag{2.22}$$

It is important to emphasize the difference between independent events and mutually exclusive events. These two concepts are often confused. In fact, two events that are mutually exclusive are not independent. Since two mutually exclusive events $E_1$ and $E_2$ have no intersection, that is, $E_1 \cap E_2 = \phi$, then $\Pr(E_1 \cap E_2) = \Pr(E_1) \cdot \Pr(E_2 \mid E_1) = 0$. This means that $\Pr(E_2 \mid E_1) = 0$, since $\Pr(E_1) \neq 0$. For two independent events, we expect to have $\Pr(E_2 \mid E_1) = \Pr(E_2)$, which is not zero except for the trivial case of $\Pr(E_2) = 0$. This indicates that two mutually exclusive events are indeed dependent.

### *2.2.3 Bayes' Theorem*

An important law known as **Bayes' Theorem** follows directly from the concept of conditional probability, a form of which is described in (2.12). For example, three forms of equation (2.12) for events A and E are

$$\Pr(A \cap E) = \Pr(A) \cdot \Pr(E \mid A), \quad \text{and}$$
$$\Pr(A \cap E) = \Pr(E) \cdot \Pr(A \mid E), \quad \text{or}$$
$$\Pr(A) \cdot \Pr(E \mid A) = \Pr(E) \cdot \Pr(A \mid E).$$

By solving for $\Pr(A \mid E)$, it follows that

$$\Pr(A \mid E) = \frac{\Pr(A) \cdot \Pr(E \mid A)}{\Pr(E)}. \tag{2.23}$$

This equation is known as Bayes' Theorem.

It is easy to prove that if event $E$ depends on some previous events that can occur in one of the $n$ different ways $A_1, A_2, \ldots, A_n$, then (2.23) can be generalized to

$$\Pr(A_j \mid E) = \frac{\Pr(A_j)\Pr(E \mid A_j)}{\sum_{i=1}^{n} \Pr(A_i)\Pr(E \mid A_i)}. \tag{2.24}$$

The right-hand side of the Bayes' equation consists of two terms: $\Pr(A_j)$, called the **prior probability**, and $\Pr(E \mid A_j)/\sum_{i=1}^{n} \Pr(E \mid A_i)\Pr(A_i)$, the **relative likelihood** or the factor by which the prior probability is revised based on evidential observations (e.g., limited failure observations). $\Pr(A_j \mid E)$ is called the **posterior probability**, that is, given event $E$, the probability of event $A_i$ can be updated [from prior probability $\Pr(A_j)$]. Clearly when more evidence (in the form of events $E$) becomes available, $\Pr(A_j \mid E)$ can be further updated. Bayes' Theorem provides a means of changing one's knowledge about an event in light of new evidence related to the event. Because of the updating capability, Bayes' Theorem is also useful for failure data analysis. We return to this topic and its application in failure data evaluation in Chapter 3. For further studies about Bayes' Theorem, refer to Lindley (1965).

---

*Example 2.6*

Suppose that 70% of an inventory of memory chips used by a computer manufacturer comes from Vendor 1 and 30% from Vendor 2, and that 99% of chips from Vendor 1 and 88% of chips from Vendor

2 are not defective. If a chip from the manufacturer's inventory is selected and is defective, what is the probability that the chip was made by Vendor 1. What is the probability of selecting a defective chip (irrespective of the vendor)?

*Solution:*
Let

$$\begin{aligned} A_1 &= \text{event that a chip is supplied by Vendor 1,} \\ A_2 &= \text{event that a chip is supplied by Vendor 2,} \\ E &= \text{event that a chip is defective,} \\ E \mid A_1 &= \text{event that a chip known to be made by Vendor 1 is defective,} \\ A_1 \mid E &= \text{event that a chip known to be defective is made by Vendor 1.} \end{aligned}$$

Thus,

$$\Pr(A_1) = 0.7, \quad \Pr(A_2) = 0.3, \quad \Pr(E \mid A_1) = 1 - 0.99 = 0.01,$$
$$\Pr(E \mid A_2) = 1 - 0.88 = 0.12.$$

Using (2.24),

$$\Pr(A_1 \mid E) = \frac{0.7 \cdot 0.01}{0.7 \cdot 0.01 + 0.3 \cdot 0.12} = 0.163.$$

Thus, the prior probability that Vendor 1 was the supplier 0.7, is changed to a posterior probability of 0.163 in light of evidence that the chosen unit is defective. From the denominator of (2.24),

$$\Pr(E) = \sum_{i=1}^{n} \Pr(A_i) \Pr(E \mid A_i) = 0.043.$$

---

## 2.3 Probability Distributions

In this section, we concentrate on probability distributions that are used to describe the populations from which samples represented in sample space are obtained. In Chapter 3, we also discuss application of certain probability distributions in reliability analysis. A fundamental aspect in describing probability distributions is the concept of random variable (r.v.). We begin with this concept and then continue with the basics of probability distributions useful to reliability analysis.

### *2.3.1 Random Variable*

Let's consider an experiment with a number of possible outcomes. If the occurrence of each outcome is governed by chance (random outcome), then possible outcomes may be assigned a numerical value. The process of assigning a number to each outcome is achieved by using a random variable.

An upper case letter (e.g., $X, Y$) is used to represent a r.v., and a lower case letter is used to determine the numerical value that the r.v. can take. For example, if r.v. $X$ represents the number of system breakdowns in a process plant, then $x_i$ shows the actual number of observed breakdowns during a given period of time $i$ (e.g., number of breakdowns per year). In this case, the sample space of $X$ is $S = \{x_1, x_2, \ldots, x_k\}$ where $k$ denotes the number of observed periods or intervals.

Random variables can be divided into two classes, namely, **discrete** and **continuous**. A r.v. is said to be discrete if its sample space is finite and countable, such as the number of system breakdowns in a given period of time. A r.v. is said to be continuous if it can only take a continuum of values. That is, it takes on values in an interval(s) as opposed to a specific countable number. Continuous r.v.s are a result of measured variables as opposed to counted data. For example, the operation of several light bulbs can be shown by a r.v. $T$, which takes a continuous survival time $t$ for each light bulb. This may be shown by the sample $S = \{t \mid t > 0\}$, which reads "the set of all values $t$ such that $t$ is greater than zero."

### *2.3.2 Some Basic Discrete Distributions*

In the remainder of this section, we briefly discuss certain probability distributions that are fundamental to reliability evaluations. A probability distribution model of an experiment is a function that assigns a probability to each possible value that the random variable of that experiment takes on, such that the total probability assigned is unity. For a discrete random variable, the probability distribution is denoted by the symbol $\Pr(x_i)$, where $x_i$ is one of the values that r.v. $X$ takes on. Consider a r.v. $X$ and the sample space $S$ designating the countable realizations of $X$. $S = \{x_1, x_2, \ldots, x_k\}$ where $k$ is a finite or infinite number. The discrete probability distribution of this space is then a function $\Pr(x_i)$, such that

1. $\Pr(x_i) \geq 0, \quad i = 1, 2, \ldots, k,$ and
2. $\sum_{i=1}^{k} \Pr(x_i) = 1.$

(2.25)

## (a) Uniform Distribution

Suppose that all outcomes of an experiment are equally likely. Thus, in the sample space $S = \{x_1, x_2, \ldots, x_k\}$,

$$\Pr(x_i) = p = 1/k, \quad \text{for all } i = 1, 2, \ldots, k. \tag{2.26}$$

An example of this distribution is rolling a die. If r.v. $X$ describes the upturned faces, then the discrete number of outcomes is $k = 6$. Thus, $\Pr(x_i) = 1/6, \quad x_i = 1, 2, \ldots, 6$.

## (b) Binomial Distribution

Consider a series of $n$ independent trials in an experiment. R.v. $X$ denotes the total number of independent occurrences of a known event. Since the number of occurrences is a real integer, the sample space is $S = \{0, 1, 2, \ldots, n\}$, and the binomial distribution is defined as

$$\Pr(x) = \binom{n}{x} p^x (1-p)^{n-x}, \qquad x = 0, 1, 2, \ldots, n, \tag{2.27}$$

and shows the probability that a known event or outcome occurs exactly $x$ times out of $n$ trials. In (2.27), $x$ is the number of times that a given outcome has occurred. The parameter $p$ indicates the probability that a given outcome will occur. The symbol $\binom{n}{x}$ is the total number of ways that a given outcome can occur without regard to the order of occurrence. By definition,

$$\binom{n}{x} = \frac{n!}{x!(n-x)!}, \quad \text{where } n! = n(n-1)(n-2)\ldots 1, \quad \text{and } 0! = 1. \tag{2.28}$$

---

*Example 2.7*

A random sample of 15 valves is observed. From past experience, it is known that the probability of a given failure within 500 hours following maintenance is 0.18. Calculate the probability that these valves will experience $0, 1, 2, \ldots, 15$, independent failures within 500 hours following their maintenance.

*Solution:*

Here the r.v. $X$ designates the failure of a valve that can take on values of $0, 1, 2, \ldots, 15$, and $p = 0.18$. Using (2.27),

| $x$ | $\Pr(x_i)$ | $x$ | $\Pr(x_i)$ |
|---|---|---|---|
| 0 | 0.051 | 8 | $2.90 \times 10^{-3}$ |
| 1 | 0.168 | 9 | $4.61 \times 10^{-4}$ |
| 2 | 0.258 | 10 | $5.66 \times 10^{-5}$ |
| 3 | 0.245 | 11 | $5.26 \times 10^{-6}$ |
| 4 | 0.161 | 12 | $3.15 \times 10^{-7}$ |
| 5 | 0.078 | 13 | $1.60 \times 10^{-8}$ |
| 6 | 0.054 | 14 | $4.95 \times 10^{-10}$ |
| 7 | 0.014 | 15 | $6.75 \times 10^{-12}$ |
| | | | $\sum \Pr(x_i) = 1.00$ |

*Example 2.8*

In a process plant, there are two identical diesel generators for emergency a.c. needs. One of these diesels is sufficient to provide the needed emergency a.c. Operational history indicates that, on average, there is one failure in 100 demands for one of these diesels.

a) What is the probability that at a given time of demand both diesel generators will fail?

b) If, on average, there are 12 demands per year for emergency a.c., what is the probability of at least one failure for diesel A in a given year? (Assume diesel A is demanded first.)

c) What is the probability that for the case described in (b), both diesels A and B will fail simultaneously at least one time in a given year?

d) What is the probability in (c) for exactly one simultaneous failure in a given year?

*Solution:*

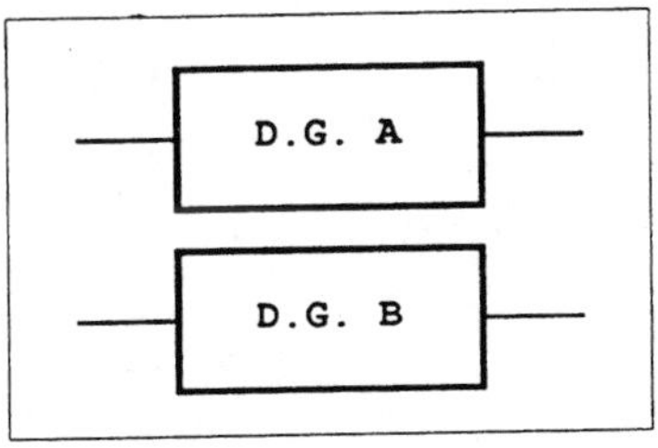


a) $q = 1/100 = 0.01 \quad p = 99/100 = 0.99$

Assume A and B are independent. (See Chapter 6 for dependent treatments.)

$\Pr(A \cap B) = \Pr(A) \times \Pr(B) = 0.01 \times 0.01 = 1 \times 10^{-4}$.
That is, there is a 1/10000 chance that both A and B will fail at a given demand.

b) $\Pr(X = 12) = \binom{12}{12}(0.99)^{12}(0.01)^0 = 0.886$,
This is the probability of exactly 12 successes of A. Therefore, Pr(at least 1 failure in one year of A ) $= 1 - 0.886 = 0.114$.

c)

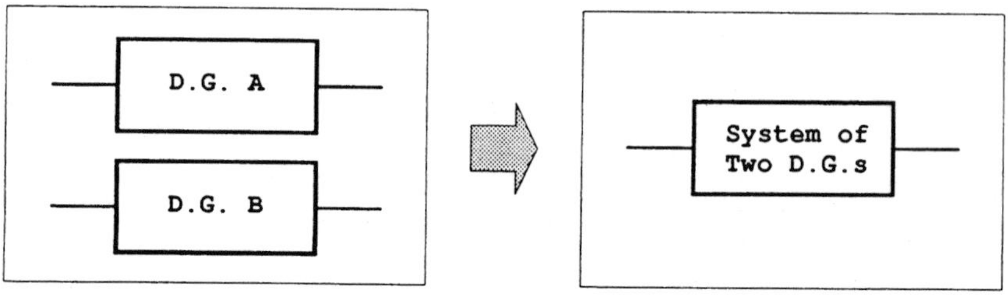


$$q = 0.0001$$

$$p = 0.9999$$

$\Pr(X = 12) = \binom{12}{12}(0.9999)^{12}(0.0001)^0 = 0.9988$,
That is the probability of 12 success of both A **and** B.
Pr(at least 1 failure of both A and B in one year) $= 1 - 0.9988 = 0.0012$.

d) For exactly **one simultaneous** failure in a given year,
$\Pr(X = 11) = \binom{12}{11}(0.9999)^{11}(0.0001)^1 = 0.001196$.

---

## (c) Hypergeometric Distribution

A r.v. $x$ is said to have a hypergeometric distribution if

$$\Pr(x) = \frac{\binom{Np}{x}\binom{N-Np}{n-x}}{\binom{N}{n}}, \quad \max(0, n + Np - N) < x < \min(Np, n), \qquad (2.29)$$

where $N$ is the total population, $n$ is the sample size (taken from the total population), and $p$ is the parameter of the distribution.

The hypergeometric distribution is very useful to certain quality control problems and certain acceptance sampling problems. In the acceptance sampling problems, $Np$ can be replaced by the parameter

$D$ (for sampling with replacement) to reflect the number of units with a given characteristic (e.g., defective, high powered, etc.). Therefore, (2.29) can be written as

$$\Pr(x) = \frac{\binom{D}{x}\binom{N-D}{n-x}}{\binom{N}{n}}, \quad \max(0, n+D-N) \le x \le \min(D, n). \tag{2.30}$$

---

*Example 2.9*

A manufacturer has a stockpile of 286 computer units. It is known that 121 of the units have a higher reliability than the other units. If a sample of four computer units is selected without replacement, what is the probability that no units, two units and all four units are from high reliability units?

*Solution:*

In this problem, using (2.30),

$$\begin{aligned} x &= \text{number of high reliability units in the sample,} \\ n &= \text{number of units in the selected sample,} \\ n-x &= \text{number of non-high reliability units in the sample,} \\ N &= \text{number of units in the stockpile,} \\ D &= \text{number of high reliability units in the stockpile,} \\ N-D &= \text{number of non-high reliability units in the stockpile.} \end{aligned}$$

Possible values of $x$ are $0 \le x \le 4$.

| $x$ | $\Pr(x)$ |
|---|---|
| 0 | $\binom{121}{0}\binom{286-121}{4-0}/\binom{286}{4} = 0.109$ |
| 2 | $\binom{121}{2}\binom{286-121}{4-2}/\binom{286}{4} = 0.360$ |
| 4 | $\binom{121}{4}\binom{286-121}{4-4}/\binom{286}{4} = 0.031$ |

---

## (d) Poisson Distribution

If the average number of events in an experiment in a given interval of time or in a specified region is constant, then the r.v. $X$ representing the number of occurrences of these events follows a Poisson distribution. For example, r.v. $X$ can represent the number of failures of a process plant per year, or the number of buses arriving at a given station per hour. Also, it can represent the number of cracks per unit area of a metal sheet. A r.v. $X$ that follows a Poisson distribution must possess the following properties:

1. The number of events occurring in one time interval or specified region are independent of those occurring in other time intervals or regions.
2. The probability of a single event occurring in a short time interval or in a small region is directly proportional to the length of the time interval or the size of the region.

When r.v. $X$ has a Poisson distribution, then

$$\Pr(x) = \frac{\rho^x \exp(-\rho)}{x!}, \quad \text{for } \rho > 0, \quad x = 0, 1, 2, \ldots \tag{2.31}$$
$$= 0 \text{ otherwise.}$$

$\rho = \lambda t$,
$\lambda =$ Poisson event occurrence intensity per unit of time or per unit of region,
$t =$ total time or region of interest.

---

*Example 2.10*

A nuclear plant receives its electric power from a utility grid outside of the plant. From past experience, it is known that loss of grid power occurs at a rate of once a year. What is the probability that over a period of 3 years no power outage will occur? That at least two, power outages will occur?

*Solution:*

Here,
$\lambda =$ 1/year,
$t =$ 3 years,
$X = 0, \quad \text{and } X \geq 2,$
$\rho = 1 \cdot 3 = 3.$

$$\Pr(X = 0) = \frac{3^0 \cdot \exp(-3)}{0!} = 0.05.$$

$$\Pr(X = 1) = \frac{3^1 \exp(-3)}{1!} = 0.149.$$

$$\Pr(X \geq 2) = 1 - \Pr(x \leq 1) = 1 - \Pr(X = 0) - \Pr(X = 1)$$
$$= 1 - 0.05 - 0.149 = 0.80.$$

---

*Example 2.11*

Inspection of a high pressure pipe reveals that on average, two pits per meter of pipe has occurred during its service. If the hypothesis that this pitting intensity is representative of all similar pipes is true, what is the probability that there are fewer than five pits in a 10-meter long pipe branch? What is the probability that there are five or more pits?

*Solution:*

Here,

$$\lambda = 2, \quad t = 10, \quad \rho = 2 \cdot 10 = 20$$

$$\Pr(X < 5) = \Pr(x = 4) + \Pr(X = 3) + \Pr(X = 2) + \Pr(X = 1) + \Pr(X = 0),$$

$$\Pr(X < 5) = \frac{20^4 \exp(-20)}{4!} + \frac{20^3 \exp(-20)}{3!} + \ldots = 1.69 \times 10^{-5},$$

$$\Pr(X \geq 5) = 1 - 1.4 \times 10^{-5} = 0.99983.$$

---

Poisson distribution can be used instead of a binomial distribution when the $p$ parameter of the binomial distribution is small (e.g., when $p \leq 0.1$) and when parameter $n$ is large. In this case, the $\rho$ parameter of the Poisson distribution is obtained by substituting $np$ for $\rho$ in (2.31).

---

*Example 2.12*

A radar system uses 650 similar electronic devices. Each device has a probability of 0.00015 per month of failing. If all these devices operate independently, what is the probability that there are no failures in a given year?

*Solution:*

Let $x$ denote the number of failures per year. Then

$$p = 0.00015 \times 12 = 0.0018 \text{ failures/year,} \quad \text{and}$$

$$\rho = np = 0.0018 \times 650 = 1.17.$$

From (2.31)

$$\Pr(X = 0) = \frac{1.17^0 \exp(-1.17)}{0!} = 0.31.$$

---

### (e) Geometric Distribution

If repeated independent trials result in an outcome with a probability $p$, then the probability distribution of the r.v.$X$, which represents the number of trials following which the desired outcome occurs for the first time is said to be geometrically distributed. This special form of binomial distribution is given by

$$\Pr(X) = p(1-p)^{x-1}, \quad x = 1, 2, \ldots \tag{2.32}$$

The term $(1-p)^{x-1}$ denotes the probability that the given outcome will not occur in the first $x-1$ trials, and when multiplied by $p$, accounts for the occurrence of the outcome in the $x^{th}$ trial.

---

*Example 2.13*

In a nuclear power plant diesel, generators are used to provide emergency electric power to the safety systems. It is known that, on average, 1 in every 52 tests performed on a diesel generator results in diesel generator failure. What is the probability that 10 consecutive tests are performed before a failure is observed? (The failure occurs at the 10 th test.)

*Solution:*

Using (2.32) with $X = 10$ and $p = 1/52 = 0.0192$, yields

$$\Pr(X = 10) = (0.0192)(1 - 0.0192)^9 = 0.016.$$

---

The books by Johnson and Kotz (1969) and Hahn and Shapiro (1967) are good references for other discrete probability distributions.

*2.3.3 Some Basic Continuous Distributions*

In this section, we present certain continuous probability distribution functions that are fundamental to reliability analysis. A continuous r.v. $X$ has a probability of zero of assuming exactly any of its possible values. For example, if r.v. $T$ represents the time within which a given emergency action is performed by a pilot, then the probability that a given pilot will perform this emergency action in exactly 2 minutes, for examples, is extremely remote. In this situation, it is appropriate to present the probability associated with a small range of values that the r.v. can possess. For example, one can determine $\Pr(t_1 < T < t_2)$, or the probability that the pilot would perform the emergency action sometime between 1 and 2 minutes.

We shall designate the probability distribution of a continuous r.v. $T$ by the functional symbol $f(t)$, called **probability density function** (pdf). Since $T$ is continuous, the graph of $f(t)$ will also be continuous over the possible range of $t$. A pdf is constructed such that the area under its curve bounded by all possible ranges of $t$ is equal to 1. The probability that $T$ assumes a value between $t_1$ and $t_2$ can mathematically be determined from the integral calculus given by

$$\Pr(t_1 < T < t_2) = \int_{t_1}^{t_2} f(t)dt. \tag{2.33}$$

In general, the pdf $f(t)$, representing a continuous sample space, must have the following properties:

1. $f(t) \geq 0$ for all t,
2. $\int_{-\infty}^{\infty} f(t)dt = 1$, and
3. $\int_{t_1}^{t_2} f(t)dt = \Pr(t_1 < T < t_2)$.

---

*Example 2.14*

Let the r.v. $T$ have the pdf

$$\begin{aligned} f(t) &= t^2/a, \quad 0 < t < 6 \\ &= 0, \text{ otherwise.} \end{aligned}$$

What is the value of a? Find $\Pr(1 < T < 3)$.

*Solution:*

$$\int_{-\infty}^{\infty} f(t)dt = \int_{t_1}^{t_2} f(t)dt = 1.$$

$$\begin{aligned} \int_{t_1}^{t_2} f(t)\,dt = \int_0^6 (t^2/a)\,dt = \left.\frac{t^3}{3a}\right|_0^6 &= \frac{216}{3a} - 0 \\ &= 1, \quad \text{then } a = \frac{216}{3} = 72. \end{aligned}$$

$$\Pr(1 < T < 3) = \int_1^3 \frac{t^2}{72}dt = \left.\frac{t^3}{216}\right|_1^3 = \frac{27}{216} - \frac{1}{216} = 0.12.$$

---

## (a) Normal Distribution

Perhaps the most well known and important continuous probability distribution is the normal distribution. A normal pdf (sometimes called Gaussian distribution) has the symmetric bell-shaped curve shown in Fig. 2.5, called the normal curve. In 1733, DeMoivre developed the mathematical representation of the normal pdf, as follows:

$$f(t) = \frac{1}{\sigma\sqrt{2\pi}} \exp(-1/2[(t-\mu)/\sigma]^2), \quad -\infty \leq t \leq \infty, \quad -\infty < \mu < \infty, \tag{2.34}$$

where $\mu$ and $\sigma$ are the two parameters of the distribution, and $\pi = 3.14159\ldots$, $e = 2.71828\ldots$, and $\sigma > 0$.

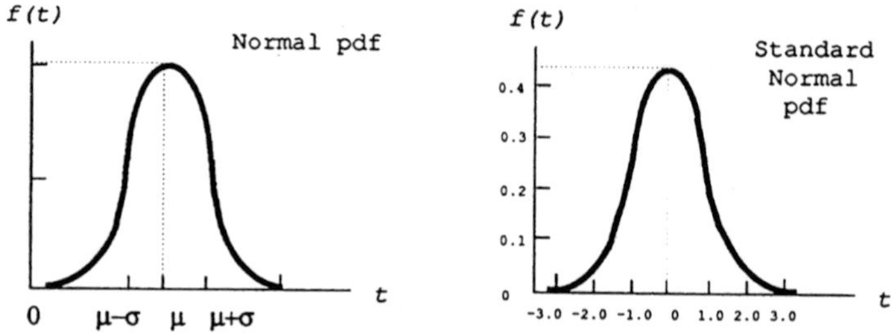


*Fig. 2.5 Normal and Standard Normal Distributions*

It is evident from (2.34) that once $\mu$ and $\sigma$ are specified, the normal curve can be determined. We will see later in Chapter 3 that the parameter $\mu$, which is referred to as the **mean**, and the parameter $\sigma$, which is referred to as the **standard deviation** of the pdf have special statistical significance.

According to (2.34), the probability that the random variable $T$ assumes a value between ordinates $t = t_1$ and $t = t_2$ is obtained from

$$\Pr(t_1 < T < t_2) = \frac{1}{\sigma\sqrt{2\pi}} \int_{t_1}^{t_2} \exp(-1/2[(t-\mu)/\sigma]^2)dt. \tag{2.35}$$

The difficulty in evaluating the integral function of a normal pdf necessitates the numerical integration and tabulation of normal curve areas. However, it would be an impossible task to provide a separate table for every conceivable value of $\mu$ and $\sigma$. One way to get around this difficulty is the transformation of the normal **pdf** to the so-called **standard normal pdf** which has a mean of zero ($\mu = 0$) and a standard deviation of 1 ($\sigma = 1$). This can be achieved by means of the r.v. transformation $Z$, such that

$$Z = \frac{T-\mu}{\sigma}. \tag{2.36}$$

That is, whenever r.v. $T$ assumes a value $t$, the corresponding value of r.v. $Z$ is given by $z = (t-\mu)/\sigma$. Therefore, if $T$ takes on values $t = t_1$ and $t = t_2$, the r.v. $Z$ takes on values $z_1 = (t_1 - \mu)\sigma$, and $z_2 = (t_2 - \mu)/\sigma$. Based on this transformation, we can write

$$\Pr(t_1 < T < t_2) = \frac{1}{\sigma\sqrt{2\pi}} \int_{t_1}^{t_2} \exp(-1/2[(t-\mu)/\sigma]^2)dt,$$
$$= \frac{1}{\sqrt{2\pi}} \int_{z_1}^{z_2} \exp(-Z^2/2)dZ = \Pr(z_1 < Z < z_2), \quad (2.37)$$

where $Z$, as seen from (2.37) is a r.v. represented by a normal pdf with a mean of zero and a standard deviation of 1. Since the standard normal pdf is characterized by a constant mean and standard deviation, only one table is necessary to provide the areas under the normal curves. Table A.1 presents the area under the standard normal curve corresponding to $\Pr(-\infty < Z < a)$.

---

*Example 2.15*

A manufacturer says his light bulbs have a mean life of 1700 hours and a standard deviation of 280 hours. Assuming the light bulb lives are normally distributed, calculate the probability that a given light bulb will last less than 1000 hours.

*Solution:*

First, the corresponding Z value is obtained.

$$Z = \frac{1000 - 1700}{280} = -2.5.$$

Notice that the lower tail of a normal pdf can extend to $-\infty$. The lower limit on this example is zero. However, the area under the normal curve representing this problem from $-\infty$ to 0 is negligible. Thus, from Table A.1, we have

$$\Pr(0 < T < 1000) \approx \Pr(-\infty < T < 1000) = \Pr(-\infty < Z < -2.5) = 0.0062.$$

---

## (b) Lognormal Distribution

A random variable is said to be lognormally distributed if its logarithm is normally distributed. The lognormal distribution has considerable applications in engineering. One major application of this distribution is to present random variables that are the result of the product of many independent random variables.

The transformation $T = \ln Y$ or $Y = \exp(T)$ transfers the normal pdf representing r.v. $T$ with mean $\mu_t$ and standard deviation $\sigma_t$ to a lognormal pdf.

$$f(y) = \frac{1}{\sigma_t y\sqrt{2\pi}} \exp\left[\frac{-1}{2\sigma_t^2}(\ln y - \mu_t)^2\right], \tag{2.38}$$

$$y \geq 0, \quad -\infty < \mu_t < -\infty, \quad \sigma_t > 0.$$

Fig. 2.6 shows the pdf of the lognormal distribution for different values of $\mu_t$ and $\sigma_t$.

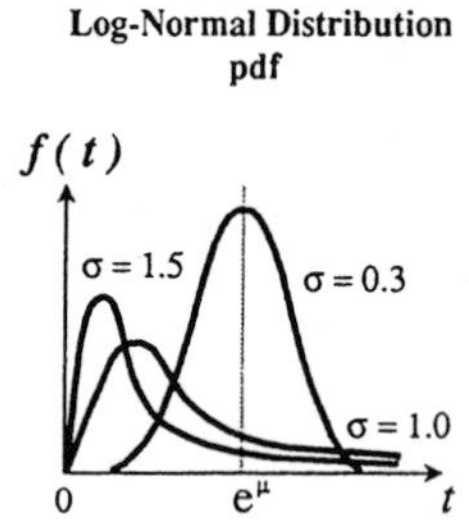


*Fig. 2.6 lognormal Distribution*

The area under the curves $f(y)$ between two points, $y_1$ and $y_2$, which is equal to the probability that r.v. $Y$ takes a value between $y_1$ and $y_2$, can be determined using a procedure similar to that outlined for the normal distribution. Here, since $\ln y$ is normally distributed, a standard normal distribution with

$$z_1 = \frac{\ln y_1 - \mu_t}{\sigma_t}, \qquad \text{and}$$

$$z_2 = \frac{\ln y_2 - \mu_t}{\sigma_t}, \tag{2.39}$$

provides the necessary transformation to calculate the probabilities. In this case,

$$\begin{aligned} \Pr(y_1 < Y < y_2) &= \Pr(\ln y_1 < \ln Y < \ln y_2) \\ &= \Pr(\ln y_1 < T < \ln y_2) = \Pr(z_1 < Z < z_2). \end{aligned}$$

If $\mu_t$ and $\sigma_t$ are not known and instead $\mu_y$ and $\sigma_y$ are known, the following equations can be used to obtain $\mu_t$ and $\sigma_t$:

$$\mu_t = \ln\left[\mu_y/(1 + \sigma_y^2/\mu_y^2)^{1/2}\right] \tag{2.40}$$

$$\sigma_t = \left[\ln(1 + \sigma_y^2/\mu_y^2)\right]^{1/2}. \tag{2.41}$$

From (2.40) and (2.41), $\mu_y$ and $\sigma_y$ can also be determined in terms of $\mu_t$ and $\sigma_t$.

$$\mu_y = \exp(\mu_t + \sigma_t^2/2) \tag{2.42}$$

$$\sigma_y = \left[\exp(\sigma_t^2) - 1\right]^{1/2} \mu_y. \tag{2.43}$$

## (c) Exponential Distribution

This distribution is widely used in reliability evaluation to model a r.v. representing time-to-failure of a device (often a device composed of several independent units). The distribution is a one-parameter pdf defined by

$$\begin{aligned} f(t) &= \lambda \exp(-\lambda t) \quad \lambda, t > 0 \\ &= 0, \quad t \leq 0. \end{aligned} \tag{2.44}$$

It is easy to show that requirements $\int_{\text{all } t} f(t)dt = 1$ and $f(t) > 0$ for a valid pdf are met for this distribution. In reliability engineering applications, the parameter $\lambda$ represents the probability per unit of time that a device fails (referred to as a failure rate of the device). We show in Chapter 3 that in reliability applications, the exponential pdf can be considered as a special form of a Poisson distribution. Fig. 2.7 illustrates the exponential distribution for a value of $\lambda$.

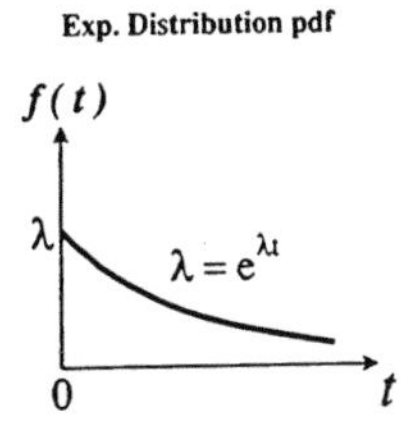


*Fig. 2.7 Exponential Distribution*

---

*Example 2.16*

A system has a constant failure rate of $10^{-3}$/hour. What is the probability that this system will fail before $t = 1000$ hours? Determine the probability that it works for at least 1000 hours.

*Solution:*

$$\Pr(t < 1000) = \int_0^{1000} \lambda \exp(-\lambda t)dt = -\exp(-\lambda t)\Big|_0^{1000} = 1 - \exp(-1) = 0.632.$$

$$\Pr(t > 1000) = 1 - \Pr(t < 1000) = 0.368.$$

---

## (d) Weibull Distribution

This distribution is widely used to represent the **time to failure** or **life length** of the components in a system, measured from a start time to the time that a component fails. The continuous r.v. $T$ representing the time to failure follows a Weibull distribution if

$$f(t) = \frac{\beta t^{\beta-1}}{\alpha^\beta} \exp[-(t/\alpha)^\beta], \quad t, \alpha, \beta > 0,$$
$$= 0, \quad \text{otherwise.} \tag{2.45}$$

Fig. 2.8 shows the Weibull distribution for various parameters of $\alpha$ and $\beta$. A careful inspection of these graphs reveals that the parameter $\beta$ has a considerable effect on the shape of the distribution. Therefore, $\beta$ is referred to as the **shape parameter**. The parameter $\alpha$, on the other hand controls the scales of the distribution. For this reason, $\alpha$ is referred to as the **scale parameter**. If we let $\beta = 1$, the Weibull distribution reduces to an exponential distribution with $\lambda = 1/\alpha$. For values of $\beta > 1$, the distribution becomes bell-shaped with some skew. We will further elaborate on this distribution and its use in reliability analysis in Chapter 3.

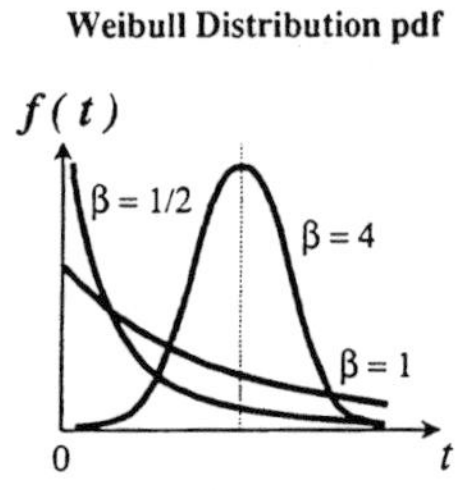


*Fig. 2.8 Weibull Distribution*

## (e) Gamma Distribution

The gamma distribution can be thought of as an extension of the exponential distribution. For example, if the time $T_i$ between successive failures of a system has an exponential distribution, then a r.v.$T$ such that $T = T_1 + T_2 + \ldots + T_n$ follows a gamma distribution. In this context, $T$ represents the cumulative time to the $n^{th}$ failure.

An alternative way to interpret this distribution is to consider a situation in which a system is subject to shocks of identifiable magnitude with the shocks occurring according to the Poisson process (with parameter $\lambda$). If the system fails after receiving $n$ shocks, then the time to failure of such a system follows a gamma distribution.

The pdf of a gamma distribution with parameters $\beta$ and $\alpha$ are given by

$$f(t) = \frac{1}{\beta^{\alpha}\Gamma(\alpha)} t^{\alpha-1} \exp(-t/\beta) \qquad \alpha, \beta, t \geq 0, \tag{2.46}$$

where $\Gamma(\alpha)$ denotes a gamma function evaluated at $\alpha$. The parameter $\alpha$ is not necessarily an integer. The parameter $\beta$ is referred to as the shape parameter and the parameter $\alpha$ is referred to as the scale parameter. As expected, if $\alpha = 1$, (2.46) reduces to an exponential distribution. Chapter 3 provides more discussions regarding the application of this distribution in reliability analysis. Fig. 2.9 shows the gamma distribution for some values of $\alpha$ and $\beta$.

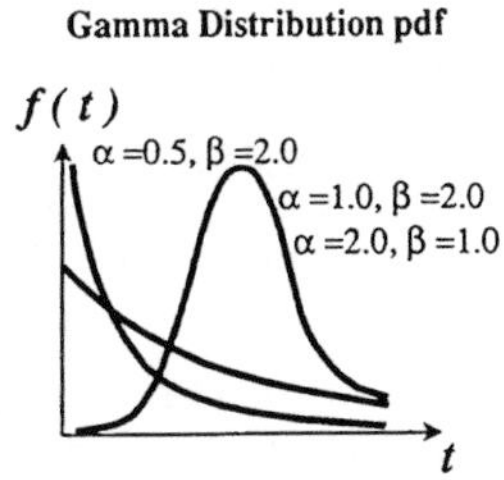


*Fig. 2.9 Gamma Distribution*

## 2.4 Empirical Distribution

In some cases of reliability evaluation, we are faced with a collection of observed statistics (e.g., time-to-failure of a group of tested devices) obtained through sampling. These observed values may not follow any known and well-established probability distributions. In these cases, the observed data can be easily represented through empirical distributions. To better understand this process, consider the observed time-to-failure data of 100 identical electronic devices that are placed on a life test. The observed time-to-failure data are tabulated in Table 2.2.

The measure of interest here is the probability associated with each interval of time to failure of the device. This can be obtained using (2.10) i.e., dividing each interval frequency by the total number of devices tested (100). Sometimes it is important in reliability evaluation to indicate how well a set of observed data fits a known

*Table 2.2 Class Interval*

| Class Interval (hrs) | Frequency |
|---|---|
| 0– 100 | 35 |
| 100– 200 | 26 |
| 200– 300 | 11 |
| 300– 400 | 12 |
| 400– 500 | 6 |
| 500– 600 | 3 |
| 600– 700 | 4 |
| 700– 800 | 2 |
| 800– 900 | 0 |
| 900–1000 | 1 |
| Total | 100 |

distribution, to determine whether a hypothesis that the data originate from a known distribution is true. For this purpose, it is necessary to calculate the expected frequencies of device failures from the known distribution and compare them with the observed frequencies. Several methods exist to determine the adequacy of such a fit. We discuss these methods further in Section 2.9 of this chapter as well as in Chapter 3.

To understand how the expected frequencies are determined, let's consider the time-to-failure data for the electronic device example discussed above, and assume that it is believed the time-to-failure data for this device exhibit an exponential distribution with the parameter $\lambda = 5 \times 10^{-3}(\text{hr}^{-1})$. The probability that a r.v.$T$ takes values between 0 and 100 hours according to (2.33) is

$$\Pr(0 < T < 100) = \int_0^{100} 5 \times 10^{-3} \exp(-5 \times 10^{-3} t)dt$$

$$= [1 - \exp(-5 \times 10^{-3} t)]\Big|_0^{100} = 0.393.$$

By multiplying this probability by the total number of devices observed (100), we will be able to determine the expected frequency. The expected frequency here would be $0.393 \times 100 = 39.3$ for the 0-100 interval. The results for the rest of the intervals are shown in the following table.

| Class Interval | Observed Frequency | Expected Frequency |
|---|---|---|
| 0- 100 | 35 | 39.3 |
| 100- 200 | 26 | 23.8 |
| 200- 300 | 11 | 14.5 |
| 300- 400 | 12 | 8.8 |
| 400- 500 | 6 | 5.3 |
| 500- 600 | 3 | 3.2 |
| 600- 700 | 4 | 2.0 |
| 700- 800 | 2 | 1.2 |
| 800- 900 | 0 | 0.7 |
| 900-1000 | 1 | 0.4 |

A comparison of the observed and expected frequencies would reveal differences as great as 4.6. Fig. 2.10 illustrates the graphic representation of this empirical distribution and its comparison to the exponential distribution with $\lambda = 1 \times 10^{-3}$. The graphic representation of empirical data is commonly known as a histogram.

## 2.5 Joint and Marginal Distributions

Thus far, we have discussed distribution functions that are restricted to one-dimensional sample spaces. There are, however, situations in which more than one r.v. is simultaneously measured and recorded in multidimensional sample spaces. For example, in a study of human reliability in a control room situation, one can simultaneously measure the r.v. $T$ representing time that various operators spend to fulfill an emergency task, and measure the r.v.$E$ representing the level of training that these various operators have had for performing these emergency tasks. Since one expects $E$ and $T$ to have some relationships, a joint distribution of both r.v. $T$ and r.v. $E$ can be used to show variations in the r.v.'s.

Let $X$ and $Y$ be two r.v.'s ( not necessarily independent ). The probability distribution for their simultaneous occurrence is denoted by $f(x, y)$ and is called a **joint probability density function** of $X$ and $Y$. If r.v. $X$ and $Y$ are discrete, the joint probability distribution function can be denoted by $\Pr(X = x,\ Y = y)$, or simply $\Pr(x, y)$. That is, $\Pr(x, y)$ gives the probability that the outcomes $x$ and $y$ occur simultaneously. For example, if r.v. $X$ represents the number of a given circuit in a process plant, and Y represents the number of failures of the circuit in the most recent year, then $\Pr(7, 1)$ is the probability that a given process plant has seven circuits and has

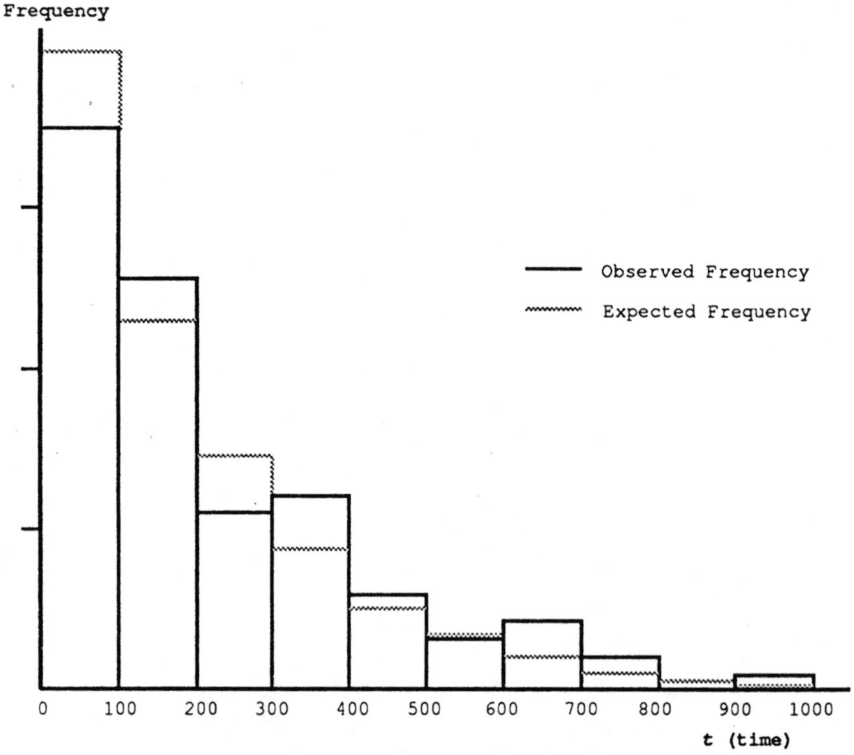


*Fig. 2.10 Histogram of the Comparison of Expected and Observed Data*

failed one time in the most recent year. The function $f(x,y)$ is a joint probability density function of continuous r.v. $X$ and $Y$ if

1. $f(x,y) \geq 0, \qquad -\infty \leq x, y \leq \infty$, and

2. $$\int_{-\infty}^{\infty}\int_{-\infty}^{\infty} f(x,y)\,dx\,dy = 1. \tag{2.47}$$

Similarly, the function $\Pr(x,y)$ is a joint probability function of the discrete random variable $X$ and $Y$ if

1. $\Pr(x,y) \geq 0$ for all values of $x$ and $y$, and

2. $$\sum_x \sum_y \Pr(x,y) = 1. \tag{2.48}$$

The probability that two or more joint r.v.'s fall within a specified subset of the sample space is given by

$$\Pr(x_1 < X < x_2, y_1 < Y < y_2) = \int_{x_1}^{x_2} \int_{y_1}^{y_2} f(x,y)\,dx\,dy \quad \text{for continuous r.v.'s,}$$

and by

$$\Pr(X = x, Y = y) = \sum_{x_1,\ldots} \sum_{y_1,\ldots} \Pr(x,y) \quad \text{for discrete r.v.'s.} \tag{2.49}$$

The **marginal probability** density function of $X$ and $Y$ is defined respectively by

$$g(x) = \int_{-\infty}^{\infty} f(x,y)\,dy, \quad \text{and} \tag{2.50}$$
$$h(y) = \int_{-\infty}^{\infty} f(x,y)\,dx,$$

for continuous r.v.'s, and by

$$\Pr(x) = \sum_{y} \Pr(x,y), \quad \text{and} \tag{2.51}$$
$$\Pr(y) = \sum_{x} \Pr(x,y),$$

for the discrete r.v.'s.
Using (2.12), the conditional probability of an event $y$, given event $x$, is

$$\Pr(Y = y \mid X = x) = \frac{\Pr(X = x \cap Y = y)}{\Pr(X = x)} = \frac{\Pr(x,y)}{\Pr(x)}, \quad \Pr(x) > 0 \tag{2.52}$$

when $X$ and $Y$ are discrete r.v.s.

Similarly, one can extend the same concept to continuous r.v.s $X$ and $Y$ and write

$$f(x \mid y) = \frac{f(x,y)}{h(y)}, \quad h(\eta) > 0, \quad \text{or} \tag{2.53}$$
$$f(y \mid x) = \frac{f(x,y)}{g(x)}, \quad g(x) > 0,$$

where (2.52) and (2.53) are called **conditional probability density functions** of discrete and continuous r.v.s. The conditional probability density functions have the same properties as any other distribution. Similar to (2.11), if r.v. $X$ and r.v. $Y$ are independent, then $f(x \mid y) = f(x)$ for continuous r.v.'s, or $\Pr(x \mid y) = \Pr(x)$ for discrete r.v.s. This would lead to the conclusion that for independent r.v.s $X$ and $Y$,

$$f(x,y) = g(x) \cdot h(y), \text{ if X and Y are continuous, and}$$
$$\Pr(x,y) = \Pr(x) \cdot \Pr(y), \text{ if X and Y are discrete.} \quad (2.54)$$

Equation (2.54) can be expanded to a more general form of

$$f(x_1, x_2, \ldots, x_n) = f_1(x_1) \cdot f_2(x_2), \ldots, f_n(x_n), \quad (2.55)$$

where $f(x_1, x_2, \ldots, x_n)$ is a joint probability density function of r.v.'s $X_1, X_2, \ldots, X_n$, and $f_1(x_1), f_2(x_2), \ldots, f_n(x_n)$ are marginal probability density functions of $X_1, X_2, \ldots, X_n$ respectively.

---

*Example 2.17*

Let r.v. $T_1$ represent the time (in minutes) that a nuclear power plant operator spends to locate and mitigate a small loss of coolant accident, and let r.v. $T_2$ represent the length of time (in minutes) that they need to spend reading procedures related to this accident. If r.v.s $T_1$ and $T_2$ are represented by the joint probability function

$$f(t_1, t_2) = c(t_1^{1/3} + t_2^{1/5}), \quad \text{when } 60 > t_1 > 0,\ 10 > t_2 > 0, \quad \text{and}$$
$$= 0 \quad \text{otherwise.}$$

Find the following:

1. The value of c.
2. The probability that an operator will be able to handle a small loss-of-coolant accident in less than 10 minutes. Assume the operator in this accident should take less than 2 minutes to read the necessary procedures.
3. Whether r.v.$X$ and r.v.$Y$ are independent.

*Solution:*

1. $f(t_1 < 60, t_2 < 10) = \int_0^{t_1=60} \int_0^{t_2=10} c(t_1^{1/3} + t_2^{1/5}) dt_1 dt_2 = 1, \quad c = 3.92 \times 10^{-4}$.
2. $\Pr(t_1 < 10, t_2 < 2) = \int_0^{t_1=10} \int_0^{t_2=2} 3.92 \times 10^{-4}(t_1^{1/3} + t_2^{1/5}) = 0.02$.
3. $f(t_2) = \int_0^{60} f(t_1, t_2) dt_1 = 3.92 \times 10^{-4}(176.16 + 60t_2^{1/5})$.
   Similarly,

$f(t_1) = \int_0^{10} f(t_1, t_2)dt_2 = 3.92 \times 10^{-4}(10t_1^{1/3} + 13.2)$.
Since $f(t_1, t_2) \neq f(t_1) \cdot f(t_2)$, $t_1$ and $t_2$ are not independent.

## 2.6 Cumulative Distribution Functions

The discrete and continuous distributions that we discussed in Sections 2.3.2 and 2.3.3 provide a measure of point probabilities in the case of discrete r.v.'s, and point densities in the case of continuous r.v.'s. A useful measure when applying these distributions in reliability analysis problems is the **cumulative distribution function** (cdf). This function, denoted by F(.), is defined as follows:

$$F(x) = \Pr(X \leq x) = \sum_{X \leq x} \Pr(x), \text{ for a discrete r.v. } X,$$

$$F(t) = \int_{-\infty}^{t} f(t)dt, \text{ for a continuous r.v. } T. \quad (2.56)$$

*Example 2.18*

For the Poisson distribution model with parameter $\rho$, find the cdf.

*Solution:*

$$F(x) = \Pr(X \leq x)$$
$$= \sum_{k=0}^{x} \frac{\rho^k \exp(-\rho)}{k!} = \frac{\rho^0 \exp(-\rho)}{0!} + \frac{\rho^1 \exp(-\rho)}{1!} + \ldots + \frac{\rho^x \exp(-\rho)}{x!}.$$

*Example 2.19*

For the Weibull distribution model with parameter $\beta$, find the cdf.

*Solution:*

$$F(t) = \Pr(T \leq t) = \int_0^t \beta/\alpha(t/\alpha)^{\beta-1} \exp(-t/\alpha)dt = 1 - \exp(-t/\alpha)^{\beta}.$$

## 2.7 Elements of Statistics

In this section, we introduce basic concepts of statistics useful to reliability engineering and analysis. We review several statistical techniques related to the evaluation of sample data, and draw inferences to the population from which the sample data were obtained.

The expectation value of r.v. $X$ is a characteristic of $X$ that applies to continuous and discrete r.v.'s. Consider a discrete r.v. $X$ that takes values $x_i$ with corresponding probabilities $\Pr(x_i)$. The **expected value** of $X$ denoted by $E(X)$ is defined by

$$E(X) = \sum_i x_i \Pr(x_i), \tag{2.57}$$

if the right-hand side of the equation converges.

Accordingly, if $T$ is a continuous r.v. with a pdf of $f(t)$, then the expected value of $T$ is defined by the integral

$$E(T) = \int_{-\infty}^{\infty} t f(t)\, dt, \tag{2.58}$$

if the integral converges.

$E(X)[orE(T)]$ is a widely used concept in statistics known as the **mean**, or in mechanics known as the center of gravity, and sometimes denoted by $\mu$. $E(X)$ is also referred to as the **first moment about the origin**. In general, the $k^{th}$ moment about the origin (ordinary moment) is denoted as $E(X_k)$ for all $k \geq 1$.

In general, one can obtain the expected value of any real-value function of a r.v. In the case of a discrete distribution, $Pr(x_i)$, the expected value of function $g(X)$ is defined as

$$E[g(X)] = \sum_{i=1}^{k} g(x_i)Pr(x_i). \tag{2.59}$$

Similarly, for a continuous r.v. $T$, the expected value of $g(T)$ is defined as:

$$E[g(T)] = \int_{-\infty}^{\infty} g(t)f(t)dt. \tag{2.60}$$

---

*Example 2.20*

Determine the first and second moments about origin for the Poisson distribution.

*Solution:*

a)

$$E(X) = \sum_{x=0}^{\infty} \frac{x \exp(-\rho)\rho^x}{x!} = \sum_{x=1}^{\infty} \frac{x \exp(-\rho)\rho^x}{x!}$$
$$= \rho \sum_{x=1}^{\infty} \frac{\exp(-\rho)\rho^{x-1}}{(x-1)!}.$$

Let $y = x - 1$. Then

$$E(X) = \rho \sum_{y=0}^{\infty} \frac{\exp(-\rho)\rho^y}{y!}.$$

According to (2.25)

$$\sum_{y=0}^{\infty} \frac{\exp(-\rho)\rho^y}{y!} = 1.$$

Thus $E(X) = \rho$.

b) Using (2.59), we can write

$$E(X^2) = \sum_{x=0}^{\infty} \frac{x^2 \exp(-\rho)\rho^x}{x!} = \sum_{x=1}^{\infty} \frac{x^2 \exp(-\rho)\rho^x}{x!}$$
$$= \rho \sum_{x=1}^{\infty} \frac{x \exp(-\rho)\rho^{x-1}}{(x-1)!}.$$

Let $y = x - 1$. Then

$$E(X^2) = \rho \sum_{y=0}^{\infty} (y+1) \frac{\exp(-\rho)\rho^y}{y!}$$
$$= \rho \sum_{y=0}^{\infty} \frac{y \exp(-\rho)\rho^y}{y!} + \rho \sum_{y=0}^{\infty} \frac{\exp(-\rho)\rho^y}{y!}.$$

Since $\sum_{y=0}^{\infty} \frac{y \exp(-\rho)\rho^y}{y!} = \rho$, and $\sum_{y=0}^{\infty} \frac{\exp(-\rho)\rho^y}{y!} = 1$,

$$E(X^2) = \rho^2 + \rho.$$

---

*Example 2.21*

Determine $E(T_1)$ and $E(T_2)$ if $T$ is an exponential r.v. with parameter $\lambda$.

*Solution:*

By using (2.58) we get

$$E(T) = \lambda \int_0^\infty t \exp(-\lambda t)\, dt$$
$$= -t \exp(-\lambda t)\Big|_0^\infty + \int_0^\infty \exp(-\lambda t)\ dt$$
$$= 0 - \frac{\exp(-\lambda t)}{\lambda}\Big|_0^\infty = 1/\lambda.$$

Similarly, by using (2.60)

$$E(T^2) = \int_0^\infty t^2 \lambda \exp(-\lambda t)\, dt = 2/\lambda^2.$$

---

A measure of dispersion or variation of r.v. $X$ about its mean is called variance, and is denoted by $v(X)$ or $\sigma^2$. The variance is also referred to as the **second moment about the mean** (or the moment of inertia in mechanics), sometimes referred to as the central moment, and is defined as

$$v(X) = \sigma^2 = E[(X - \mu)^2], \tag{2.61}$$

where $\sigma$ is known as the **standard deviation**, $v$ is known as the **variance**. In general, the $k^{th}$ moment about the mean is denoted as

$$g(X) = E[(X - \mu)^k], \quad \text{for all } k \geq 0. \tag{2.62}$$

Table 2.3 presents some useful laws or theorems that permit us to simplify complex expectations. The laws shown in Table 2.3 apply both to discrete and continuous r.v.'s.

*Table 2.3 Expectation Mathematics*

1. $E(aX) = aE(X)$
2. $E(a) = a$
3. $E[g(X) \pm h(X)] = E[g(X)] \pm E[h(X)]$
4. $E[X \pm Y] = E[X] \pm E[Y]$
5. $E[X \cdot Y] = E[X] \cdot E[Y]$ if $X$ and $Y$ are independent

One useful method of determining the moments about the origin of a distribution is the use of Laplace transforms. Suppose Laplace transform of pdf $f(t)$ is $F(S)$, then

$$F(S) = \int_0^\infty f(t)\exp(-St)\,dt, \quad \text{and} \tag{2.63}$$

$$\frac{-dF(S)}{dS} = \int_0^\infty t f(t)\exp(-St)\,dt. \tag{2.64}$$

Since for $S = 0$ the left-hand side of (2.64) reduces to $E(t)$, then

$$E(t) = -\left[\frac{dF(S)}{dS}\right]_{S=0}.$$

In general, it is possible to show that

$$E(t^k) = \left[(-1)^k \frac{d^k F(S)}{dS^k}\right]_{S=0}. \tag{2.65}$$

---

*Example 2.22*

Using the results of Example 2.18, find the variance $v(X)$.

*Solution:*

From Table 2.3 and (2.61),

$$v(T) = E(T-\mu)^2 = E(T^2 + \mu^2 - 2\mu T) = E(T^2) + E(\mu^2) - E(2\mu T).$$

Since $E(T^2) = 2/\lambda^2$, and $E(\mu^2) = \mu^2 = [E(T)]^2 = 1/\lambda^2$, then

$$E(2\mu T) = 2\mu E(T) = 2\mu^2 = 2/\lambda^2, \text{ and}$$

$$v(T) = 2/\lambda^2 + 1/\lambda^2 - 2/\lambda^2 = 1/\lambda^2.$$

---

The concept of expectation value equally applies to joint probability distribution. The expectation of a real-value function $h$ of discrete r.v.s $X_1, X_2, \ldots, X_n$ is

$$E[h(X_1, X_2, \ldots, X_n)] = \sum_{x_1}\sum_{x_2}\cdots\sum_{x_n} h(x_1, x_2, \ldots, x_n) \Pr(x_1, x_2, \ldots, x_n), \tag{2.66}$$

where $\Pr(x_1, x_2, \ldots, x_n)$ is the discrete joint pdf of r.v.'s $X_i$. When dealing with continuous r.v.'s, the summation term in (2.66) replaces with integrals

$$E[h(X_1, X_2, \ldots, X_n)] = \int_{-\infty}^{\infty}\int_{-\infty}^{\infty}\cdots\int_{-\infty}^{\infty} h(x_1, x_2, \ldots, x_n)$$
$$f(x_1, x_2, \ldots, x_n)\, dx_1, dx_2 \ldots dx_n \tag{2.67}$$

where $f(x_1, x_2, \ldots, x_n)$ is the continuous joint pdf of r.v.'s $X_i$. In the case of a bivariate distribution with two r.v.'s $X_1$ and $X_2$, the expectation of the function

$$h(X_1, X_2) = [X_1 - E(X_1)][X_2 - E(X_2)]$$

is called the **covariance** of r.v.'s $X_1$ and $X_2$, and is denoted by $Cov(X_1, X_2)$. Using Table 2.3, it is easy to show that

$$Cov(X_1, X_2) = E(X_1 \cdot X_2) - E(X_1)E(X_2). \tag{2.68}$$

A common measure of determining the relation between two r.v.s is a dimensionless parameter known as **linear correlation coefficient**, which carries information about two aspects of the relationship:

1. Strength, measured on a scale from 0 to 1; and
2. Direction, indicated by the plus or minus sign. Denoted by $\rho(X_1, X_2)$, the linear correlation coefficient between r.v.'s $X_1$ and $X_2$ is defined as

$$\rho(X_1, X_2) = \frac{Cov(X_1, X_2)}{[v(X_1)v(X_2)]^{1/2}} \tag{2.69}$$

Clearly, if $X_1$ and $X_2$ are independent, then from (2.68), $Cov(X_1, X_2) = 0$, and from (2.69), $\rho(x_1, x_2) = 0$. In the application of statistical methods, sums of r.v.'s may be used. In these cases,

$$E(\sum_{i=1}^{n} a_i X_i) = \sum_{i=1}^{n} a_i E(X_i), \tag{2.70}$$

$$v(\sum_{i=1}^{n} a_i X_i) = \sum_{i=1}^{n} a_i^2 v(X_i) + 2\sum_{i=1}^{n-1}\sum_{j=i+1}^{n} a_i a_j Cov(X_i, X_j). \tag{2.71}$$

In cases where r.v.'s are independent, (2.71) becomes simplified to

$$v(\sum_{i-1}^{n} a_i X_i) = \sum_{i=1}^{n} a_i^2 v(X_i). \tag{2.72}$$

See Section 6.2 for additional discussion on using (2.66) and (2.67) as well as expectation values for other types of functions.

---

*Example 2.23*

Find the linear correlation coefficient of r.v. $T_1$ and $T_2$ (in example 2.17).

$$f(t_1,t_2) = 3.92 \times 10^{-4}(t_1^{1/3} + t_2^{1/5}), \quad 60 > t_1 > 0 \quad 10 > t_2 > 0$$
$$= 0 \quad \text{otherwise.}$$

*Solution:*

From Example 2.17, part 3,

$$f(t_1) = 3.92 \times 10^{-4}[10t_1^{1/3} + 13.2],$$
$$f(t_2) = 3.92 \times 10^{-4}[176.17 + 60t_2^{1/5}],$$
$$E(t_1) = \int_0^{60} t_1 f(t_1)dt_1$$
$$= 3.92 \times 10^{-4}\left[\frac{10(3)}{7}(60)^{7/3} + 13.2(1/2)(60)^2\right] = 23.8,$$
$$E(t_2) = \int_0^{10} t_2 f(t_2)dt_2$$
$$= 3.92 \times 10^{-4}[176.2(1/2)(10)^2 + 60(5/11)(10)^{11/5}] = 5.1,$$
$$E(t_1,t_2) = \int_0^{t_2=10}\int_0^{t_1=60} 3.92 \times 10^{-4} t_1 t_2 [t_1^{1/3} + t_2^{1/5}]dt_1 dt_2 = 169,$$

thus

$$Cov(t_1,t_2) = E(t_1,t_2) - E(t_1)E(t_2)$$
$$= 169 - (23.8)(5.1) = 46.6.$$

Similarly,

$$E(t_1^2) = \int_0^{60} 3.92 \times 10^{-4}[10t_1^{7/3} + 13.2t_1^2]dt_1 = 1365.5,$$
$$E(t_2^2) = \int_0^{10} 3.92 \times 10^{-4}[176.2(t_2^2) + 60t_2^{11/5}]dt_2 = 34.6,$$
$$v(t_1) = E(t_1^2) - [E(t_1)]^2 = 1365.5 - (23.8)^2 = 799,$$
$$v(t_2) = E(t_2^2) - [E(t_2)]^2 = 34.6 - (5.1)^2 = 8.2,$$
$$\rho(t_1,t_2) = \frac{Cov(t_1,t_2)}{[v(t_1)v(t_2)]^{1/2}} = \frac{46.6}{[799 \times 8.2]^{1/2}} = 0.58.$$

This indicates that there is a reasonably high positive correlation between the time an operator spends to mitigate a loss-of-coolant accident and the length of time he or she needs to spend on the related procedures.

---

*Example 2.24*

Two identical pumps are needed in a process plant to provide a sufficient flow to cool a reactor. The flow out of each pump is known to be normally distributed with a mean of 540 gpm and a standard deviation of 65 gpm. Calculate the

a) the distribution of the resulting flow from both pumps,
b) probability that the total flow is less than 1,000 gpm.

*Solution:*

a) If $M_1$ and $M_2$ are flows from each pump, then the total flow is $M = M_1 + M_2$. Since each of the r.v.s $M_1$ and $M_2$ are normally distributed, then M is also normally distributed. From (2.66), the mean of r.v. $M$ is obtained

$$M_m = E(M) = E(M_1) + E(M_2)$$
$$= 540 + 540 = 1,080.$$

Assuming $M_1$ and $M_2$ are independent, then using (2.72) the variance r.v. $M$ is obtained.

$$v_M = v(M) = v(M_1) + v(M_2) = (65)^2 + (65)^2 = 8,450, \text{or}$$
$$\sigma_M = 91.9.$$

b) Using standard normal distribution transformation (2.36)

$$z = \frac{1,000 - 1,080}{91.9} = -0.87.$$

This corresponds to

$$\Pr(M \leq 1,000) = \Pr(Z \leq -0.87) = 0.19.$$

---

## 2.8 Estimation

Our observation from experiments, if recorded, would provide a basis for performing a statistical analysis known as **estimation**. The collection of all our observations in an experiment is called the **population** (which does not necessarily refer to human population).

Each observed value in a population is a value that a r.v., say $X$, can take. For example, the number of pump failures following a demand in a large plant provides a data population. To check the applicability of a given distribution, e.g., binomial distribution in the pump failure case, and to estimate the parameters of such a distribution, one needs to perform a statistical analysis known as **estimation**. In certain cases where the observable population is large or difficult to obtain, statistical estimation can be performed on a sample taken from the total population. Certainly the sample should be representative of the population selected at random. That is, the observations must be made independently. The number of observations in a random sample is called the **sample size** and is denoted by $n$.

Let's consider a sample of size $n$, represented by r.v. $X_i$, $i = 1, 2, \ldots, n$, where $X_i$ represents the value of the $i^{th}$ observation in the sample. If the elements of the sample are obtained independently and under the same conditions, then the possible values of any $X_i$ will be exactly the same as for the original population represented by a r.v. $X$. The question arises here as to how one can estimate the distribution of r.v. $X$ and the distribution parameters. This leads us to the next section, in which we elaborate on this question.

### *2.8.1 Point Estimation*

First, let us discuss two important measurements related to observed samples. These are the **sample mean** and **sample variance**. In statistics, sample mean and sample variance are called **statistics**. There are, however, many other statistics used in estimation theory besides the sample mean and sample variance. One major use of statistics is to estimate the parameters of the distribution of the population from which the sample is taken. For example, sample mean can be used to estimate the distribution mean $\mu$ and the sample variance to estimate the distribution variance $\sigma^2$. There are two methods of inferring through statistical analysis: (1) the estimation of population distributions and their characteristics; and (2) the test of statistical hypotheses. We address the first method in Chapter 3 and the second method in this chapter.

The sample mean is simply the arithmetic average of sample values, i.e., the sum of the sample values, $x_i$, divided by the sample size. The sample mean in statistics is denoted by $\bar{x}$, and is mathematically shown by

$$\bar{x} = \frac{\sum_{i=1}^{n} x_i}{n}. \tag{2.73}$$

The sample variance is the most common way of indicating dispersion in the sample data. It is denoted by $S^2$ and is defined as

$$S^2 = \frac{\sum_{i=1}^{n}(x_i - \bar{x})^2}{n-1}. \tag{2.74}$$

In (2.74) the value of $n-1$ in the denominator is used to make the $S^2$ estimation unbiased. However, in a biased version of (2.74), $n$ replaces $n-1$ in the denominator. (We discuss the bias concept further later in this chapter.) If we treat $\bar{x}$ as a r.v., then the sampling distribution of means is obtained. This r.v. is used to show variability between different sets of samples taken from the same population. It is possible to show that the distribution of r.v. $\bar{x}$ is normal. Accordingly, the expected value of the sampling distribution of means is

$$\mu_{\bar{x}} = E(\bar{x}) = \mu, \tag{2.75}$$

where $\mu$ is the mean of the population. If the population is very large, or if samples are obtained with replacements (see Chapter 3 for further discussion on replacement), then the variance of the sampling distribution of means is

$$\sigma_{\bar{x}}^2 = E[(\bar{x} - \mu)^2] = \sigma^2/n, \tag{2.76}$$

where $\sigma^2$ is the variance of the population. If the population is of size $N$ or if sampling is without replacement, and given that $n \leq N$, then (2.76) is replaced by

$$\sigma_{\bar{x}}^2 = \frac{\sigma^2}{n}\left(\frac{N-n}{N-1}\right). \tag{2.77}$$

---

*Example 2.25*

A sample of eight manufactured shafts is taken from a plant. The diameters of the shafts are 1.01, 1.08, 1.05, 1.01, 1.00, 1.02, 0.99, and 1.02 inches. Find the sample mean and variance.

*Solution:*

From (2.73) $\bar{x} = 1.0225$.
From (2.74) $S^2 = 0.00085$.

---

Reliability analysis often involves distribution models that have one or more parameters. The values of these parameters often are not known a priori. However, statistical data may be available from which one can infer an estimate of the parameters of interest. In the remainder of this section, we discuss the techniques for statistical inference, i.e., ways used to infer, from a specific sample, certain characteristics of the general population from which the sample is obtained. In Chapter 3, we examine specific application of these techniques as applied in reliability analysis.

### *2.8.2 Statistical Point Estimation*

Suppose we are interested in inferring parameters of a probability distribution [e.g., $f(t_i;\theta)$] that represents a general population. For this purpose, we wish to determine a point estimate value of $\hat{\theta}$ that is reasonably close to the unknown parameter $\theta$. Since we are inferring characteristics of the overall population from a sample the point estimate $\hat{\theta}$ may differ from the true value $\theta$, especially when the sample size is small. Because of such variations in all possible estimates, $\hat{\theta}$ can also be thought of as a r.v. also called an **estimator** of $\theta$.

Estimators have characteristics such as lack of bias. For example, an estimator is unbiased for estimating $\theta$ if $E(\hat{\theta}) = \theta$. When sampling from a normal distribution, the sample mean $\bar{x}$ from (2.73) can be used as an unbiased estimator of $\mu$. It can also be shown that an unbiased estimator of the sample variance shown in (2.74) is an estimator of $\sigma^2$. Below, discuss two general methods of determining estimators: the maximum likelihood method and the method of moments.

### Maximum Likelihood Method

The likelihood estimate of a sample can be thought of as the joint pdf of the sample. Let $t_1, t_2, \ldots, t_n$ represent a random sample taken from the population represented by the pdf $f(t;\theta)$. Accordingly, the likelihood function for the random sample is the joint pdf of r.v.'s $T_1, T_2, \ldots, T_n$, such that

$$L(\theta; t_1, t_2, \ldots, t_n) = \prod_{i=1}^{n} f(t_i;\theta). \tag{2.78}$$

The maximum likelihood method consists of finding $\hat{\theta}$ as an estimate of $\theta$ such that, for every value of $\theta$,

$$L(t, \hat{\theta}) \geq L(t, \theta). \tag{2.79}$$

Therefore, the only value of $\theta$ in the function $L(t,\theta)$, that satisfies $\theta$ is the maximum value of $L(t,\theta)$, denoted as $\hat{\theta}$. Accordingly, $\hat{\theta}$ is called the **maximum likelihood estimator** of $\theta$. Therefore,

$$\left.\frac{\partial L(t,\theta)}{\partial \theta}\right|_{\theta=\hat{\theta}} = 0, \tag{2.80}$$

from which $\hat{\theta}$ can be obtained. Notice that (2.80) is also equivalent to

$$\left.\frac{\partial \ln L(t,\theta)}{\partial \theta}\right|_{\theta=\hat{\theta}} = 0. \tag{2.81}$$

For certain cases, (2.79) can be solved analytically; in other cases it should be solved numerically.

---

*Example 2.26*

Consider a sample of $n$ times to failure of a component whose time to failure is known to be exponentially distributed with parameter $\lambda$ (the failure rate). Find the maximum likelihood estimator for $\lambda$.

*Solution:*

Using (2.69) and (2.72)

$$L(t,\lambda) = \prod_{i=1}^{n} \lambda \exp(-\lambda t_i) = \lambda^n \exp\left[-\lambda\left(\sum_{i=1}^{n} t_i\right)\right],$$

$$\ln L = n(\ln \lambda) - \lambda \sum_{i=1}^{n} t_i,$$

$$\left.\frac{\partial \ln L}{\partial \lambda}\right|_{\lambda=\hat{\lambda}} = \frac{n}{\hat{\lambda}} - \sum_{i=1}^{n} t_i = 0,$$

$$\hat{\lambda} = \frac{n}{\sum_{i=1}^{n} t_i}.$$

Since $(\partial^2 \ln L/\partial \lambda^2) = -n/\lambda^2 < 0$, then $\hat{\lambda} = \frac{n}{\sum_{i=1}^{n} t_i}$ is indeed the maximum.

---

Example 2.26 also illustrates the concept of point estimation. That is, for a small sample of $n$ times-to-failure, the estimator $\hat{\lambda}$ can be thought of as a **point estimate** of the failure rate. In other words, the point estimate of the failure rate is the number of failures, $n$, divided by the observed total accumulated test time $\sum_{i=1}^{n} t_i$.

### Method of Moments

This method is based on equating the moments of the population distribution to the moments of the sample. For a distribution with $m$ parameters, the moments are equated to generate a system of $m$ equations and unknowns. This method, in general, is less accurate than the maximum likelihood method. Hence, it should be used when the maximum likelihood method is difficult to apply. The following example describes the process used.

---

*Example 2.27*

The population of the diameters of a manufactured shaft is known to have a normal distribution. Find the estimators for $\mu$ and $\sigma^2$ using a small sample of these shafts. Apply the method of moments.

*Solution:*

Assume $X$ is a r.v. representing the diameter. Since the normal distribution is a two-parameter distribution, we need the first two moments about the origin of the normal pdf.

$$E(X) = \mu, \quad E(X^2) = \sigma^2 + \mu^2.$$

Thus,

$$\mu = \frac{\sum_{i=1}^{n} X_i}{n} = \bar{X} \mid_{\mu=\hat{\mu}} \quad (n = \text{sample size}).$$

Similarly,

$$\sigma^2 + \mu^2 = \frac{\sum_{i=1}^{n} X_i^2}{n} \Big|_{\sigma=\hat{\sigma}}.$$

Therefore,

$$\hat{\mu} = \bar{X}$$

$$\hat{\sigma}^2 = \frac{\sum_{i=1}^{n} X_i^2 - \bar{X}^2}{n} = \frac{\sum_{i=1}^{n}(X_i - \bar{X})^2}{n}.$$

---

*2.8.3 Estimation of Confidence Interval for Normal Distribution*

As discussed earlier, the point estimate $\hat{\theta}$ may differ from the true value $\theta$. In this case, it is always advisable to determine a confidence level $1-\alpha$. In other words, if $\theta_u$ and $\theta_l$ are associated with the upper and lower limits of the interval, then

$$\Pr[\theta_l \leq \theta \leq \theta_u] = 1 - \alpha, \quad 0 < \alpha < 1. \tag{2.82}$$

For example, the percentiles for estimating the $\mu$ parameter of normal distribution with unknown $\mu$ and $\sigma$ when sample size $n$ is small can be obtained from

$$\Pr[\bar{X} - t_{\alpha/2}S/\sqrt{n} < \mu < \bar{X} + t_{\alpha/2}S/\sqrt{n}] = 1 - \alpha, \tag{2.83}$$

where $\bar{X}$ and $S$ are obtained from (2.73) and (2.74), and $t_{\alpha/2}$ is the $t$-student distribution value with $n-1$ degrees of freedom. Values of $t$ for different degrees of freedom are shown in Table A.2. The concept of confidence interval estimation is discussed further in Chapter 3. If the samples are taken from a normal distribution with unknown $\mu$, but known $\sigma$, the intervals of $\mu$ can be estimated from

$$\Pr[\bar{X} - z_{\alpha/2}\sigma/\sqrt{n} < \mu < \bar{X} + z_{\alpha/2}\sigma/\sqrt{n}] = 1 - \alpha \tag{2.84}$$

where $z_{\alpha/2}$ is the transformation (2.36) taken from Table A.1 corresponding to a probability of $\alpha/2$. Similarly, confidence intervals of $\sigma^2$ for a normal distribution can be obtained from

$$\frac{(n-1)S^2}{\chi^2_{1-\alpha/2}(n-1)} < \sigma^2 < \frac{(n-1)S^2}{\chi^2_{\alpha/2}(n-1)}. \tag{2.85}$$

Where $\chi^2_f(\gamma)$ is obtained from Table A.3 for f-degrees of freedom at a level of $\gamma$. Estimation of parameters of other types of distribution is more involved. Some of these intervals will be further discussed in Chapter 3. For more discussion in this topic, the readers are referred to Mann et al. (1974).

It is important not to confuse confidence interval with other types of intervals in statistics. (e.g., tolerance interval or prediction interval). The confidence interval displays limits with some confidence (e.g., 95%) within which a population parameter (e.g., mean or standard deviation) exists. On the other hand, tolerance interval displays limits (with some confidence) within which some prespecified percentage of the population is contained. Finally, prediction interval displays limits within which some future observations are expected to occur. For further discussion in different types of intervals refer to Hill and Prane (1984).

## 2.9 Goodness-of-Fit Tests

Consider the problem of determining whether a sample belongs to a hypothesized theoretical distribution. For this purpose, we need to perform a test that indicates how good a fit we have between the frequency of occurrence of a r.v. characterized by an observed sample and the expected frequencies obtained from the hypothesized distribution. For this purpose, we need to perform a goodness-of-fit test.

Below describe two procedures often used as goodness-of-fit tests: the Chi-square and Kolmogorov goodness-of-fit tests.

### *2.9.1 Chi-Square Method*

As the name implies, this test is based on a statistic that has an approximate chi-square distribution. To perform this test, the observed sample taken from the population representing a r.v. $X$ must be split into $k$ non-overlapping cells to form an empirical distribution. The assumed (hypothesized) distribution model is then used to determine the probability $p_i$ that the r.v. $X$ would fall into each cell $i$. This process was described to some extent in Section 2.4. By multiplying $p_i$ by the sample size $n$, we arrive at the expected frequency for each cell. If we denote the expected frequency by $e_i$, then mathematically $e_i = np_i$. If the observed frequency for each cell $i$ of the sample is denoted by $o_i$, then the magnitude of differences between $e_i$ and $o_i$ can characterize the adequacy of the fit.

The chi-square test uses the statistic $X^2$ which approximately follows a chi-square distribution. The value of statistic $X^2$ is obtained from

$$W = X^2 = \sum_{i=1}^{n} \left[ \frac{(o_i - e_i)^2}{e_i} \right]. \tag{2.86}$$

If the observed frequencies $o_i$ differ considerably from the expected frequencies $e_i$, then $W$ will be large and the fit is considered poor. A good fit would obviously lead to not rejecting the hypothesized distribution, whereas a poor fit leads to its rejection. It is important to note that one can only fail to support the hypothesis so the person rejects it rather than positively affirm its truth. Therefore, the hypothesis is either rejected or not rejected rather than accepted or not accepted. The test can be summarized as follows:

STEP 1. The r.v. $X$ is hypothesized to have a known distribution $f(x)$.

STEP 2. Determine the significance of the test denoted by $\alpha$.

STEP 3. Establish the rejection region $R \geq \chi^2_{1-\alpha}(k-1-m)$, where $\alpha$ is the level of significance of the test, and $m$ is the number of times a parameter is used to describe the hypothesized distribution, i.e., the number of times that data are used to estimate parameters in $f(x)$. $\chi^2_{1-\alpha}(k-1-m)$ is obtained from Table A.3.

STEP 4. If $W > R$, where $W$ is obtained from (2.86), reject the hypothesized distribution; otherwise do not reject the distribution.

It is important at this point to specify the role of $\alpha$ in the chi-square test. Suppose the calculated value of $W$ in (2.86) exceeds the $1-\alpha = 0.95$ tabulated value of $\chi^2_{1-\alpha}(\cdot)$ in Table A.3. This indicates that chances are lower than 1 in 20 that the observed data emanate from the hypothesized distribution. In this case, the model is frequently rejected. On the other hand, if the calculated value of $W$ is smaller than $\chi^2_{0.95}(\cdot)$, chances are greater than 1 in 20 that the observed data match the hypothesized distribution model. In other words, there is less than 1 in 20 chances that a distribution that should have been rejected will not be rejected.

One instructive step in the chi-square test is to compare the observed data with the expected frequency to note which classes (cells) contributed most to the value of $W$. This would help indicate the nature of deviations. If the expected frequencies of several adjacent cells are individually small in comparison with other cells, it is necessary to combine these cells into one that is not comparatively small. By closely examining (2.86), this becomes evident; for small cells, the denominator of (2.86) would be small, resulting in a high contribution to $W$. Combining several adjacent cells would increase the expected frequency and eliminate an unduly high contribution to the value of $W$.

---

*Example 2.28*

The number of parts ordered per week by a maintenance department in a manufacturing plant is believed to follow a Poisson distribution. Use a chi-square goodness-of-fit test to determine the adequacy of the Poisson distribution. (Use the following sample data.)
*Solution:*

| No. parts per week ($x$) | No. observed frequency ($o_i$) | Expected frequency ($e_i$) | $(o_i - e_i)^2/e_i$ |
|---|---|---|---|
| 0 | 18 | 15.7 | 0.337 |
| 1 | 18 | 18.8 | 0.034 |
| 2 | 8 | 11.3 | 0.964 |
| 3 | 5 | 4.5 | |
| 4 | 2 | 1.4 | 0.523 |
| 5 | 1 | 0.3 | |
| Total | 52 | 52 | $W = 1.86$ |

Since in Poisson distribution events occur at a constant rate, then an estimate of $\rho$ is

$$\hat{\rho} = \frac{\text{No. of parts used}}{\text{No. of weeks}} = \frac{62}{52} = 1.2 \text{ parts/week.}$$

From the Poisson distribution,

$$\Pr(X = x_i) = \frac{\rho^x \exp(-\rho)}{x!}.$$

Using $\rho = \hat{\rho} = 1.2$, $\Pr(X = 0) = 0.301$. Therefore, $e_i = 0.301 \cdot 52 = 15.7$. Other $e_i$s are calculated similarly. The last three cells are small; therefore, we combine them into one cell. Since we obtained one parameter ($\rho$) from the sample, $m = 1$. Due to comparatively small $e_i$s in the last three cells, we combine them to generate one cell. Therefore, $\chi^2_{0.95}(4 - 1 - 1) = 5.99$, from Table A.3. Thus, $R = 5.99$. Since $W < R$, we will not reject the hypothesis that the data are originated from a Poisson distribution.

---

### *2.9.2 Kolmogorov Method*

When dealing with small samples, it is difficult to order data into sufficient cells with large frequencies. For such cases, the chi-square method is not effective, and the Kolmogorov method is more appropriate. In this method, the individual data points are treated independently without clustering them into cells. Similar to the chi-square method, the empirical distribution of the sample (in the form of a cumulative pdf) is compared with a hypothesized cumulative pdf.

A sample cumulative pdf is defined for an ordered sample of data $t_1 < t_2 < t_3 < \ldots < t_n$ by

$$\begin{aligned} S_n(t_i) &= \frac{i}{n}; \quad i = 1, 2, \ldots, n \\ S_n(t) &= 0 \quad \text{for } t < t_1, \quad \text{and} \\ S_n(t) &= 1 \quad \text{for } t > t_n. \end{aligned} \tag{2.87}$$

Statistic $W$ is used in the Kolmogorov test to measure the maximum difference between $S_n(t_i)$ and a hypothesized cdf, $F(t)$. For this case,

$$W = \max_i \left[ |F(t_i) - S_n(t_i)|, |F(t_i) - S_n(t_{i-1})| \right]. \tag{2.88}$$

Similar to the chi-square test, the following steps compose the test:

STEP 1. The r.v. $T$ is hypothesized to have a known cumulative distribution $F(t)$.
STEP 2. Determine the significance of the test denoted by $\alpha$.
STEP 3. Establish the rejection region $R > D_n(\alpha)$, where $D_n(\alpha)$ can be obtained from Table A.4.
STEP 4. If $W > D_n(\alpha)$, then reject the hypothesized distribution and conclude that $F(t)$ does not describe the data; otherwise, do not reject the hypothesis.

---

*Example 2.29*

Time to failure of an electronic device is measured in a life test. The failure times are 254, 586, 809, 862, 1381, 1923, 2542, and 4211 hours. Is an exponential distribution with $\lambda = 5 \times 10^{-4}\text{hr}^{-1}$ an adequate representation of this sample?

*Solution:*

For an exponential distribution with $\lambda = 5 \times 10^{-4}$, we get $F(t) = 1 - \exp(-5 \times 10^{-4}t)$. For $\alpha = 0.05$, $D_8(0.05) = 0.457$. Thus, the rejection area is $R > 0.457$.

| $i$ | $t_i$ | $S_n(t_i)$ | $S_n(t_{i-1})$ | $F_n(t_i)$ | $\lvert F_n(t_i) - S_n(t_i)\rvert$ | $\lvert F_n(t_i) - S_n(t_{i-1})\rvert$ |
|---|---|---|---|---|---|---|
| 1 | 254 | 0.125 | 0.000 | 0.119 | 0.006 | 0.119 |
| 2 | 586 | 0.250 | 0.125 | 0.254 | 0.004 | 0.129 |
| 3 | 809 | 0.375 | 0.250 | 0.333 | 0.042 | 0.083 |
| 4 | 862 | 0.500 | 0.375 | 0.350 | 0.150 | 0.025 |
| 5 | 1381 | 0.625 | 0.500 | 0.499 | 0.126 | 0.001 |
| 6 | 1923 | 0.750 | 0.625 | 0.618 | 0.132 | 0.007 |
| 7 | 2542 | 0.875 | 0.750 | 0.719 | **0.156** | 0.031 |
| 8 | 4211 | 1.000 | 0.875 | 0.878 | 0.122 | 0.003 |

Since $W = 0.156 < 0.457$, we should not reject the hypothesized exponential distribution model.

---

## BIBLIOGRAPHY

1. Cox, R.T. (1946). Probability, frequency and reasonable expectation, American Journal of Physics., 14: 1.
2. Hahn, G.J. and S.S. Shapiro (1967). *Statistical Models in Engineering,* John and Wiley and Son, New York.

3. Hill, H.E. and J.W. Prane (1984). *Applied Techniques in statistics for selected industries: coatings, paints and pigments*, John Wiley and Son, New York.
4. Johnson, N.L. and S. Kotz (1970). *Distribution in statistics*, (4 vols.), John Wiley and Son, New York.
5. Lindley, D.V. (1965). *Introduction to probability and statistics from a Bayesian viewpoint*, (2 vols.), Cambridge Press, Cambridge.
6. Mann, N., R.E. Schafer and N.D. Singpurwalla (1974). *Methods for statistical analysis of reliability and life data*, John Wiley and Son, New York.

## Exercises

2.1 Simplify the following Boolean functions:

a) $(\overline{\overline{A \cdot B + C}\bar{B}}$

b) $(A + B)\bar{A} + \bar{B}\bar{A}$

c) $\overline{ABBC}\bar{B}$

2.2 Reduce the following Boolean function:

$$A \cap B \cap (\overline{C \cup \bar{C} \cap A \cup \bar{B}})$$

2.3 Simplify the following Boolean expressions:

a) $\overline{(\overline{(A \cap B) \cup C}) \cap \bar{B}}$

b) $((A \cup B) \cap \bar{A}) \cup \bar{B} \cap \bar{A}$

2.4 Reduce Boolean function $G = (A \cup B \cup C) \cap (\overline{A \cap \bar{B} \cap \bar{C}}) \cap \bar{C}$. If $\Pr(A) = \Pr(B) = \Pr(C) = 0.9$, what is $\Pr(G)$?

2.5 Simplify the following Boolean equations:

a) $(A \cup B \cup C) \cap (\overline{A \cap \bar{B} \cap \bar{C}}) \cap \bar{C}$

b) $(A \cup B) \cap \bar{B}$

2.6 Reduce the following Boolean equation:

$$(\overline{A \cup (B \cap C)}) \cap (\overline{B \cup (D \cap A)})$$

2.7 Use both equations (2.17) and (2.21) to find the reliability $\Pr(s)$. Which equation is preferred for numerical solution?

$$\Pr(s) = \Pr(E_1 \cup E_2 \cup E_3), \quad \Pr(E_1) = 0.8, \quad \Pr(E_2) = 0.9, \quad \Pr(E_3) = 0.95$$

2.8 A stockpile of 40 relays contain 8 defective relays. If five relays are selected at random and the number of defective relays is known to be greater than two, what is the probability that exactly four relays are defective?

2.9 Given that $P = .006$ = probability of an engine failure on a flight between two cities, find the probability of:

a) No engine failure in 1000 flights

b) At least one failure in 1000 flights

c) At least two failures in 1000 flights

2.10 A random sample of 10 resistors is to be tested. From past experience, it is known that the probability of a given resistor being defective is 0.08. Let X be the number of defective resistors.

a) What kind of distribution function would be recommended for modeling the random variable X?

b) According to the distribution function in (a), what is the probability that in the sample of 10 resistors, there are more than 1 defective resistors in the sample?

2.11 How many different license plates can be made if each consists of three numbers and three letters, and no number or letter can appear more than once on a single plate?

2.12 The consumption of maneuvering jet fuel in a satellite is known to be normally distributed with a mean of 10,000 hours and a standard deviation of 1,000 hours. What is the probability of being able to maneuver the satellite for the duration of a 1-year mission?

2.13 Suppose a process produces electronic components, 20% of which are defective. Find the distribution of x, the number of defective components in a sample size of five. Given that the sample contains at least three defective components, find the probability that four components are defective.

2.14 If the heights of 300 students are normally distributed, with a mean of 68 inches and standard deviation of 3 inches, how many students have:

a) Heights of more than 70 inches?

b) Heights between 67 and 68 inches?

2.15 Assume that 1% of a certain type of resistor are bad when purchased. What is the probability that a circuit with 10 resistors has exactly 1 bad resistor?

2.16 Between the hours of 2 and 4 p.m. the average number of phone calls per minute coming into an office is two and one-half. Find the probability that during a particular minute, there will be more than five phone calls.

2.17 A guard works between 5 p.m. and 12 midnight; he sleeps an average of 1 hour before 9 p.m., and 1.5 hours between 9 and 12. An inspector finds him asleep, what is the probability that this happens before 9 p.m.?

2.18 The number of system breakdowns occurring with a constant rate in a given length of time has a mean value of two breakdowns. What is the probability that in the same length of time, two breakdowns will occur?

2.19 An electronic assembly consists of two subsystems, A and B. Each assembly is given one preliminary checkout test. Records on 100 preliminary checkout tests show that subsystem A failed 10 times. Subsystem B alone failed 15 times. Both subsystems A and B failed together five times.

a) What is the probability of A failing, given that B has failed.
b) What is the probability that A alone fails.

2.20 A presidential election poll shows one candidate leading with 60% of the vote. If the poll is taken from 200 random voters throughout the U.S., what is the probability that the candidate will get less than 50% of the votes in the election? (Assume the 200 voters sampled are true representatives of the voting profile.)

2.21 A newspaper article reports that a New York medical team has introduced a new male contraceptive method. The effectiveness of this method was tested using a number of couples over a period of 5 years. The following statistics are obtained:

| Year | Total number of times the method was employed | Number of unwanted pregnancies |
|---|---|---|
| 1 | 8,200 | 19 |
| 2 | 10,100 | 18 |
| 3 | 2,120 | 1 |
| 4 | 6,120 | 9 |
| 5 | 18,130 | 30 |

a) Estimate the mean probability of an unwanted pregnancy per use. What is the standard deviation of the estimate?
b) What are the 95% upper and lower confidence limits of the mean and standard deviation?

2.22 Suppose the lengths of the individual links of a chain distribute themselves with a uniform distribution, shown below.

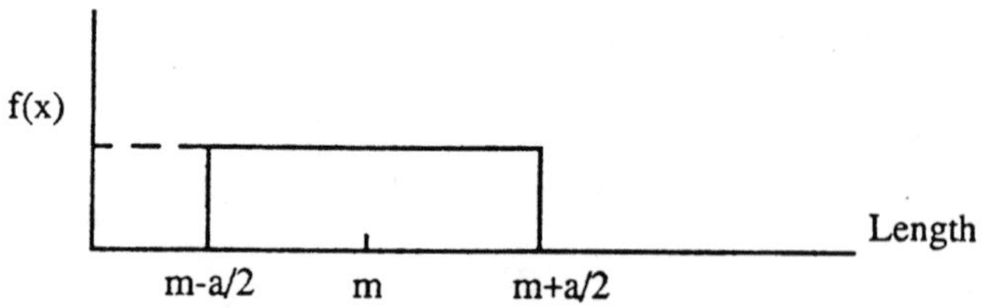


a) What is the height of the rectangle?

b) Find the cumulative pdf for the above distribution. Make a sketch of the distribution and label the axes.
c) If numerous chains are made from two such links hooked together, what is the pdf of two-link chains ?
d) Consider a 100-link chain. What is the probability that the length of the chain will be less than 100.5 m if a = 0.1m?

2.23 If $f(x,y) = \frac{1}{2}xy^2 + \frac{1}{2}yx^2, \qquad 0 < x < 1, \quad 0 < y < 2:$

a) Show that $f(x, y)$ is a joint probability density function.
b) Find $\Pr(x > y), \Pr(y > x), \Pr(x = y)$.

2.24 A company is studying the feasibility of buying an elevator for a building under construction. One proposal is a 10-passenger elevator that, on average, would arrive in the lobby once per minute. The company rejects this proposal because it expects an average of five passengers per minute to use the elevator.

a) Support the proposal by calculating the probability that in any given minute, the elevator does not show up, and 10 or more passengers arrive.
b) Determine the probability that the elevator arrives only once in a 5-minute period.

2.25 The frequency distribution of time to establish the root causes of a failure by a group of experts is observed and given below.

| Time (hrs.) | Frequency |
|---|---|
| 45-55 | 7 |
| 55-65 | 18 |
| 65-75 | 35 |
| 75-85 | 28 |
| 85-95 | 12 |

Test whether a normal distribution with known $\sigma = 10$ is an appropriate model for these data.

2.26 A random number generator yields the following sample of 50 digits:

| Digit | 0 | 1 | 2 | 3 | 4 | 5 | 6 | 7 | 8 | 9 |
|---|---|---|---|---|---|---|---|---|---|---|
| Frequency | 4 | 8 | 8 | 4 | 10 | 3 | 2 | 2 | 4 | 5 |

Is there any reason to doubt the digits are uniformly distributed? (Use the Chi-square goodness-of-fit test.)

2.27 A set of 40 high-efficiency pumps is tested, all of the pumps fail (F = 40) after 400 pump-hours (T = 400). It is believed that the time to failure of the pumps follows an exponential distribution. Using the following table and the goodness-of-fit method, and determine if the exponential distribution is a good choice.

| Time Interval (hrs.) | # of Observed Failures |
|---|---|
| 0 – 2 | 6 |
| 2 – 6 | 12 |
| 6–10 | 7 |
| 10–15 | 6 |
| 15–25 | 7 |
| 25–100 | 2 |
| | Total = 40 |

2.28 Use (2.65) and calculate mean and variance of a Weibull distribution.

2.29 Consider the following repair times

| Repair Time ($y$) | 0-4 | 4-24 | 24-72 | 72-300 | 300-5400 |
|---|---|---|---|---|---|
| No. Observed Frequency | 17 | 41 | 12 | 7 | 9 |

Use a chi-square goodness-of-fit test to determine the adequacy of a lognormal distribution:

$$f(y) = \frac{1}{y\sigma_t\sqrt{2\pi}} \exp\left[-(\ln y - \nu_t)^2/2\sigma_t^2\right], \quad \sigma_t = 1.837, \quad \nu_t = 2.986.$$

a) For 5% level of significance
b) For 1% level of significance

2.30 Consider the following time to failure data with the ranked value of $t_i$. Test the hypothesis that the data fit a normal distribution. (Use the Kolmogorov test for this purpose.)

| Event | 1 | 2 | 3 | 4 | 5 | 6 | 7 | 8 | 9 | 10 |
|---|---|---|---|---|---|---|---|---|---|---|
| Time to Failure (hr) | 10.3 | 12.4 | 13.7 | 13.9 | 14.1 | 14.2 | 14.4 | 15.0 | 15.9 | 16.1 |

# 3
# Elements of Component Reliability

In this chapter we discuss the basic elements of component reliability evaluation. The discussion centers primarily around the classical frequency approach to component reliability. However, we also present the fundamental aspects of component reliability analysis from the Bayesian point of view.

We start with a formal definition of reliability and define commonly used terms and metrics. These formal definitions are not necessarily limited to component reliability; rather, they encompass a broad group of items (i.e., components, subsystems, systems, etc.). We then focus on important aspects of component reliability analysis in the rest of this chapter. In chapter 4, we shift our attention to simple and complex system reliability considerations.

## 3.1 Concept of Reliability

Reliability has many connotations. In general, it refers to an item's capability to perform an intended function. The better the item performs its intended function, the more reliable it is. Formally, reliability is viewed as both an **engineering** and **probabilistic** notion. Indeed, both of these views form the fundamental basis for reliability studies. The reliability engineering notion deals with those design and analysis activities that extend an item's life by controlling its potential means of failure. Examples include designing stronger elements, recognizing harmful environments and reducing the harm, minimizing loads and stresses applied to the item during use, and providing a preventive maintenance program to minimize the occurence of failures.

To quantitatively measure the reliability of an item, we use the probabilistic notion, which treats reliability as a conditional probability of the successful achievement of an item's intended function. The probabilistic definition of reliability is given in Chapter 1 and (1.1) is its mathematical representation. The right-hand side of (1.1) denotes the probability that a specified failure time $T$ exceeds a specified mission time $t$ given that conditions $c_1, c_2, \ldots$ exist (or are met).

Practically, r.v. $T$ represents **time-to-failure** of the item, and conditions $c_1, c_2, \ldots$ represent conditions (e.g., design-related conditions) that are required, a priori, for successful performance of the item. Other representations of r.v. $T$ include **cycle-to-failure, stress-to-failure** and so on. In the remainder of this book, we consider only time-to-failure representation. Conditions $c_1, c_2, \ldots$ are often implicitly considered; therefore, (1.1) is written in a nonconditional form of (1.2). We use (1.2) in the remainder of this book.

*3.1.1 Reliability Function*

Let's start with the formal definition that we just gave in expression (1.1). Furthermore, let $f(t)$ denote a pdf representing the r.v. $T$. $f(t)$ is obviously characterized by the design (e.g., strength), operational, and environmental (e.g., stress) effects of the item. According to (2.33), the probability of failure of the item as a function of time can be defined by

$$\Pr(T \leq t) = \int_0^t f(\theta)\, d\theta = F(t), \quad \text{for } t \geq 0, \tag{3.1}$$

where $F(t)$ denotes the probability that the item will fail sometime up to time $t$. According to our formalism presented in (1.2), (3.1) is the **unreliability** of the item. Formally, we call $F(t)$ the **unreliability function**. Conversely, we can find the **reliability function** by writing

$$R(t) = 1 - F(t) = \int_t^\infty f(\theta)\, d\theta. \tag{3.2}$$

Provided we can obtain the pdf $f(\theta)$, we can obtain $R(t)$. Basic characteristics of the pdf and $R(t)$ can be helpful in defining some useful definitions. The **mean time to failure** ($MTTF$), for example, is defined as the expected value of $f(t)$. This illustrates the expected time during which the item will perform its function successfully (sometimes called expected life). According to (2.58),

$$MTTF = E(t) = \int_0^\infty t f(t)\, dt. \tag{3.3}$$

If $\lim_{t \to \infty} R(t) = 0$, then its easy to show that a more compact form of (3.3) is given by

$$E(t) = \int_0^\infty R(t)\, dt. \tag{3.4}$$

It is important to make a distinction, at this point, between $MTTF$ and **mean time between failure** ($MTBF$). In this case, the integrands of (3.4) will be 0 and $t_m$ instead of 0 and $\infty$, where $t_m$ is the time between two surveillances. If we have surveillance and the item is renewed through replacement, maintenance, or repair, the value of $E(t)$ is referred to as *MTBF*. Theoretically, the renewal process is assumed to be perfect. That is, the item that goes through repair or maintenance is assumed to exhibit characteristics of a new item. In practice this may not be true, in which case one needs to determine a new MTBF for the item for each renewal cycle ($i^{th}$ failure). However, the **averaged** approach represented by assuming **as-good-as-new** items is quite adequate for many reliability considerations.

---

*Example 3.1*

A device has a time to failure represented by an exponential distribution. If the device has survived up to time $t$, determine the *MTTF* of the device.

*Solution:*

According to (2.12) and (3.3),

$$MTTF = \frac{\int_0^\infty \tau f(t+\tau)\, d(\tau)}{\int_t^\infty f(\tau)\, d(\tau)}.$$

---

*3.1.2 Failure Rate and Hazard Rate*

Failures can occur due to many physical processes or mechanisms. In probabilistic reliability, all these mechanisms of failure are accounted for through a function called **instantaneous failure rate** or **hazard rate** $h(t)$. The hazard rate can be interpreted as the probability of the first and only failure of an item in the next instant of time, given that the item is presently operating. Clearly, this only applies to nonrepairable items in which only one failure can occur. For repairable items the most appropriate term is failure rate or more correctly the **rate of occurrence of failure**. This topic will be further discussed in Chapter 5. Hazard rate function is obtained via **life test data**. We define the rate of failure for a component c as $\lambda = 1/\Delta t$ (probability that c will fail between $t$ and $t + \Delta t$ given that no failure is observed before $t$). **The instantaneous failure rate** (hazard rate) for component c is defined as the limit of the rate of failure as the interval $\Delta t$ approaches zero. According to the conditional probability definition (2.12),

$$\begin{aligned} h(t) &= \lim_{\Delta t \to 0} \frac{1}{\Delta t} \frac{\Pr(t \le T \le t + \Delta t)}{\Pr(T \ge t)} = \lim_{\Delta t \to 0} \frac{\Delta F(t)/\Delta t}{R(t)} \\ &= \frac{dF(t)/dt}{R(t)} = \frac{f(t)}{R(t)}. \end{aligned} \tag{3.5}$$

Here, $f(t)$ represents the time to failure pdf of component c. Accordingly, $h(t)dt$ represents the conditional probability that given c has survived up to time $t$, it will fail sometime between $t$ and $t + dt$. Hazard rate is an important function in reliability analysis since it shows changes in the probability of failure over the lifetime of a component.

In practice, $h(t)$ often exhibits a bathtub shape and is referred to as a **bathtub** curve. A typical bathtub curve is shown in Fig. 3.1.

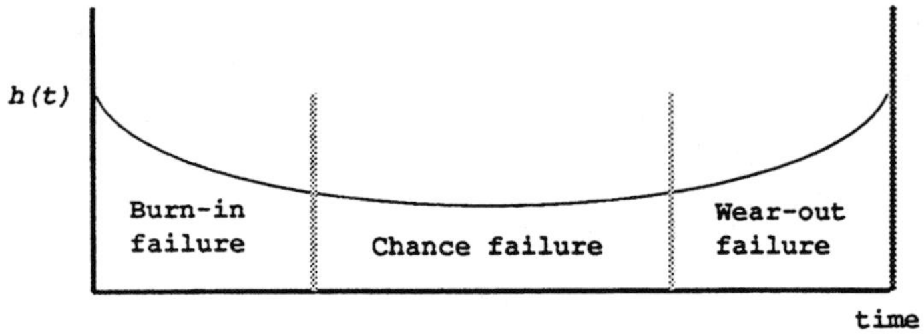


*Fig. 3.1 Typical Bathtub Curve*

Generally, a bathtub curve can be divided into three regions. The **burn-in** early failure region exhibits a **decreasing rate of failure**, characterized by early failures attributable to defects in design, manufacturing, or construction. The **chance failure** region exhibits a reasonably **constant rate of failure**, characterized by random failures of the component. In this period, many mechanisms of failure due to complex underlying physical, chemical, or nuclear phenomena give rise to this approximately constant rate of failure. The third region, called **wear-out failure**, exhibits an **increasing rate of failure**, characterized mainly by complex aging phenomena. Here the component deteriorates (e.g., due to accumulated fatigue) and is more vulnerable to outside shocks. One interesting observation is that these three regions are different for different types of components. Fig. 3.2 and Fig. 3.3 show typical bathtub curves for electrical and mechanical devices, respectively. It is evident that electrical devices exhibit a relatively much larger chance failure period. Fig. 3.4 shows the effect of various levels of stress on a device. It is clear that as stress level increases, the chance failure region decreases and, premature wear-out occurs. Therefore, it is important to minimize stress factors such as harsh operating environment to maximize reliability.

We now explain the mathematical relationship between $h(t)$ and

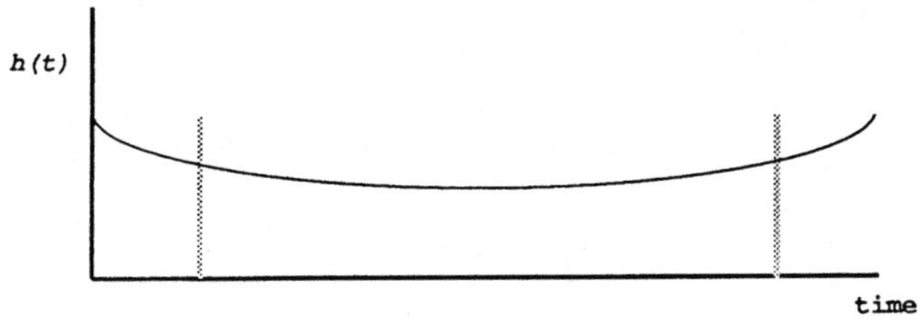


*Fig. 3.2 Bathtub Curve for Typical Electrical Devices*

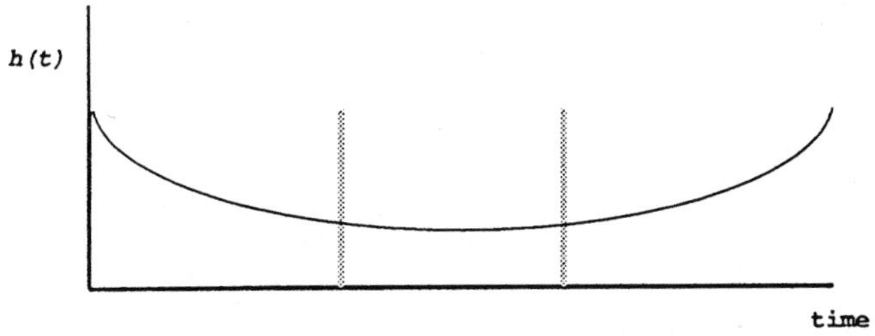


*Fig. 3.3 Bathtub Curve for Typical Mechanical Devices*

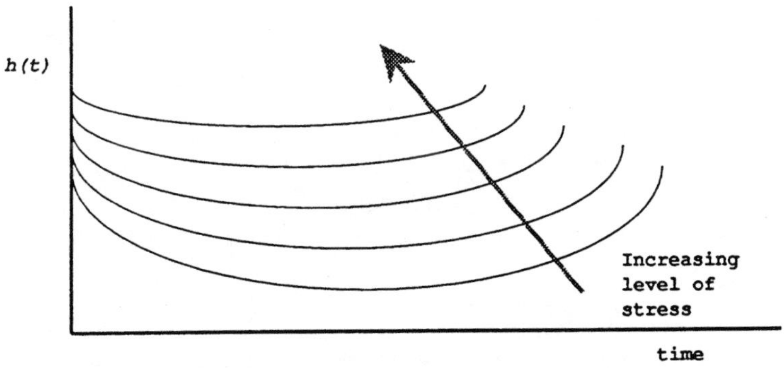


*Fig. 3.4 Effect of Stress on a Typical Bathtub Curve*

$R(t)$. Recall that from (3.2), $R(t) = 1 - F(t)$ or $dR(t) = -dF(t)$. By integrating both sides of (3.5) from 0 to t,

$$\int_0^t h(x)\,dx = \int_0^t \frac{dF(\theta)/d(\theta)}{R(\theta)}\,d\theta$$
$$= \int_0^t \frac{-dR(\theta)/d\theta}{R(\theta)}\,d\theta = -\ln R(\theta)|_0^t$$
$$= -\ln R(t) + \ln R(0).$$

By assuming that a component is totally reliable at the beginning of life (i.e., $R(0) = 1$) then

$$\int_0^t h(t)\,dt = -\ln R(t), \quad \text{and} \tag{3.6}$$

$$R(t) = \exp\left(-\int_0^t h(x)\,dx\right). \tag{3.7}$$

If the beginning of the life is not at $t = 0$, but rather at $t = t_1$, then the integrands of (3.6) will be between $t_1$ and $t$. If one of $f(t)$, $h(t)$, or $R(t)$ is known, we can compute the other two by using (3.5), (3.6) and (3.7). Table 3.1 lists the unreliability and hazard functions for important pdfs.

---

*Example 3.2*

A device is known to have an instantaneous failure rate $h(t)$ as shown below. Determine the pdf and the reliability function.

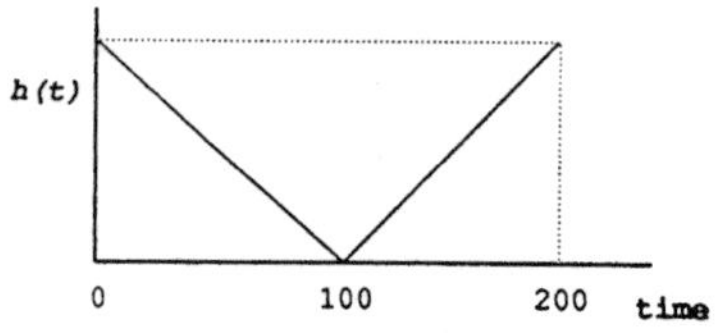


*Solution:*

For $0 \le t \le 100$, $\quad h(t) = 0.1 - 0.0009t$,

$$\int_0^t h(t)dt = \int_0^t (0.1 - 0.0009\theta)\,d\theta = [0.1\theta - (0.0009/2)\theta^2]|_0^t = 0.1t - 4.5\times 10^{-4}t^2,$$

thus

$$R(t) = \exp(-0.1t + 4.5\times 10^{-4}t^2).$$

## *Table 3.1 Important pdfs and Their Reliability Characteristics*

| Distribution Parameters | Exponential | Normal | Lognormal |
|---|---|---|---|
| pdf. $f(t)$ | $\lambda \exp(-\lambda t)$ | $\frac{1}{\sigma\sqrt{2\pi}} \exp\left[-\frac{1}{2}\left(\frac{t-\mu}{\sigma}\right)^2\right]$ | $\frac{1}{\sigma_t t\sqrt{2\pi}} \exp\left[-\frac{1}{2\sigma_t^2}(\ln t - \mu_t)^2\right]$ |
| Unreliability $F(t)$ | $1 - \exp(-\lambda t)$ | $\frac{1}{\sigma\sqrt{2\pi}} \int_0^t \exp\left[-\frac{(t-\mu)^2}{2\sigma^2}\right] dt$ | $\frac{1}{\sigma_t\sqrt{2\pi}} \int_0^t \frac{1}{\theta} \exp\left[-\frac{1}{2}\frac{(\ln\theta - \mu_t)^2}{\sigma_t^2}\right] d\theta$ |
| Instantaneous failure rate $h(t)$ | $\lambda$ | $\frac{f(t)}{1 - F(t)}$ | $\frac{f(t)}{1 - F(t)}$ |
| Mean time to failure (MTTF) | $\frac{1}{\lambda}$ | $\mu$ | $\exp\left[\left(\mu_t + \frac{1}{2}\sigma_t\right)\right]$ |
| $f(t)$ | 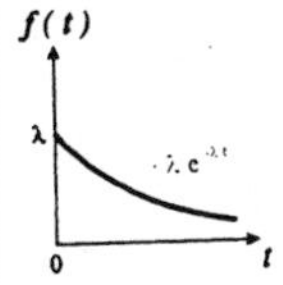  | 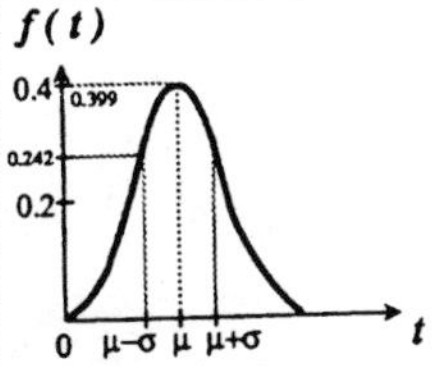  | 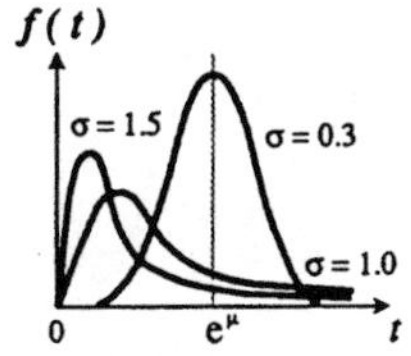  |
| $F(t)$ | 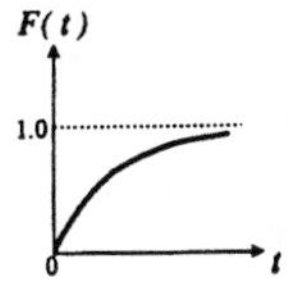  | 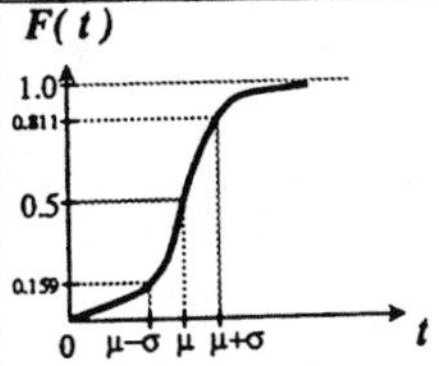  | 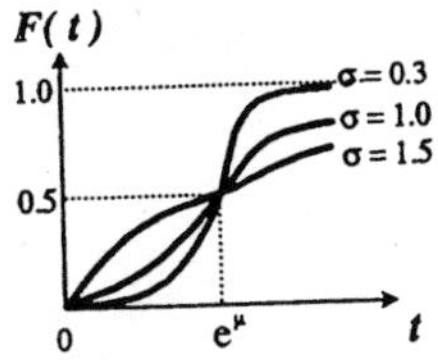  |
| $h(t)$ | 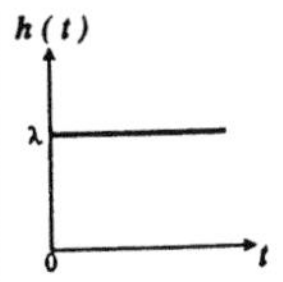  | 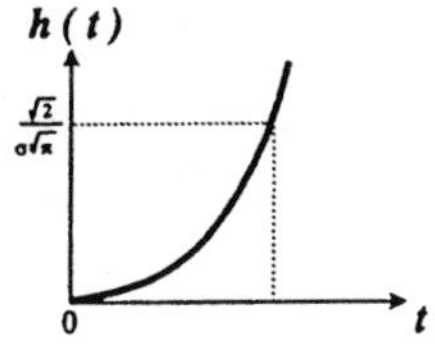  | 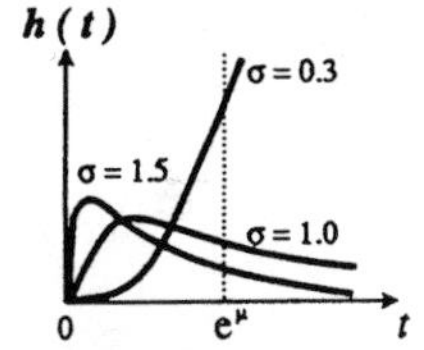  |
| Major applications in component reliability | - Life distribution of complex nonrepairable systems<br>- Life distribution "burn-in" of some components | - Life distribution of high stress components | - Size distribution of breaks (in pipes, etc.)<br>- Life distribution of some transistors<br>- Distribution of the failure rates |

*Table 3.1 Important pdfs and Their Reliability Characteristics*

| Distribution Parameters | Weibull | Gamma | Smallest Extreme Value |
|---|---|---|---|
| pdf. $f(t)$ | $\frac{\beta(t)^{\beta-1}}{\alpha^{\beta}}\exp\left[-\left(\frac{t}{\alpha}\right)^{\beta}\right]$ | $\frac{1}{\beta\alpha\Gamma(\alpha)}t^{\alpha-1}\exp(-t/\beta)$ | $\frac{1}{\delta}\exp\left[\frac{1}{\delta}(t-\lambda)-\exp\left(\frac{t-\lambda}{\delta}\right)\right]$ |
| Unreliability $F(t)$ | $1-\exp\left[-\left(\frac{t}{\alpha}\right)^{\beta}\right]$ | $\frac{\int_0^{t/\beta} y^{\alpha-1}\exp(-y)\,dy}{\Gamma(\alpha)}$ | $1-\exp\left[-\exp\left(\frac{t-\lambda}{\delta}\right)\right]$ |
| Instantaneous failure rate $h(t)$ | $\beta/\alpha\left(\frac{t}{\alpha}\right)^{\beta-1}$ | $\frac{t^{\alpha-1}\exp(-t/\beta)}{\beta^{\alpha}[\Gamma(\alpha)-\int_0^{t/\beta} y^{\alpha-1}\exp(-y)dy]}$ | $\frac{1}{\delta}\exp\left[\frac{t-\lambda}{\delta}\right]$ |
| Mean time to failure (MTTF) | $\alpha\Gamma\left(\frac{1+\beta}{\beta}\right)$ | $\alpha\beta$ | $\lambda+0.5776$ |
| $f(t)$ | 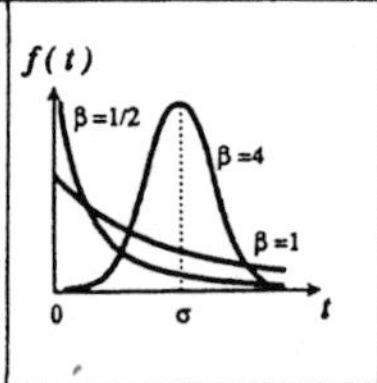 | 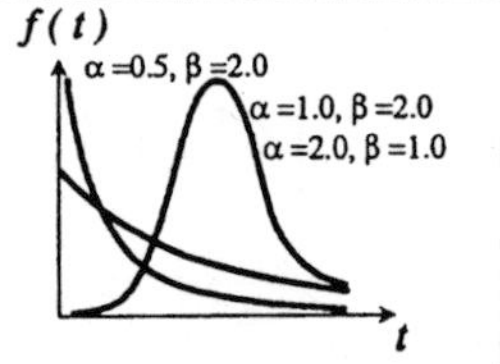 | 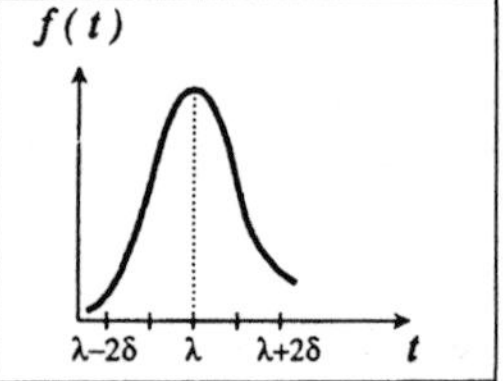 |
| $F(t)$ | 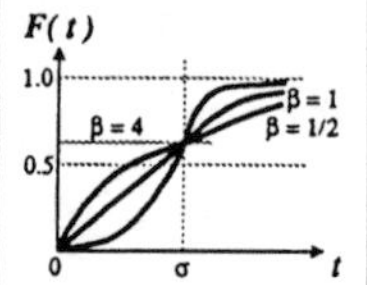 | 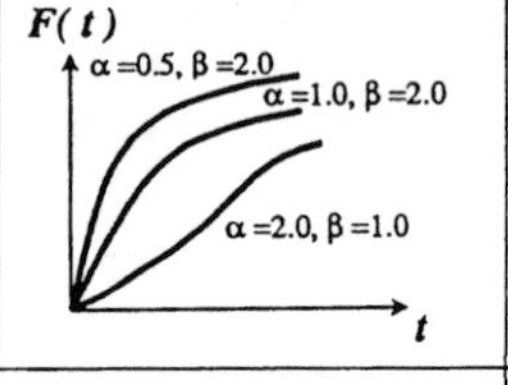 | 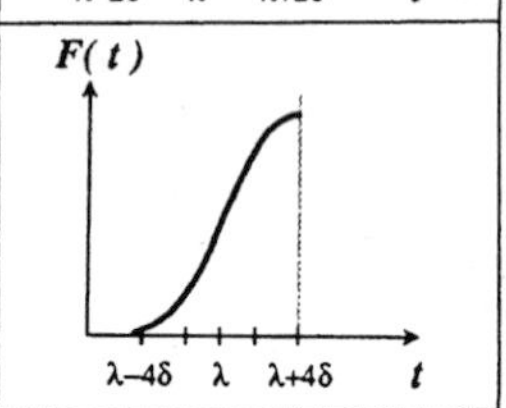 |
| $h(t)$ | 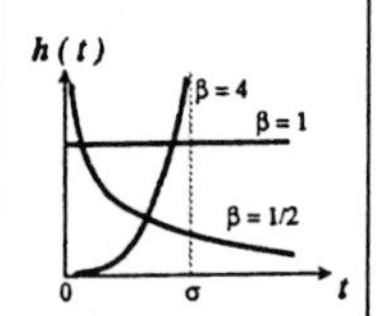 | 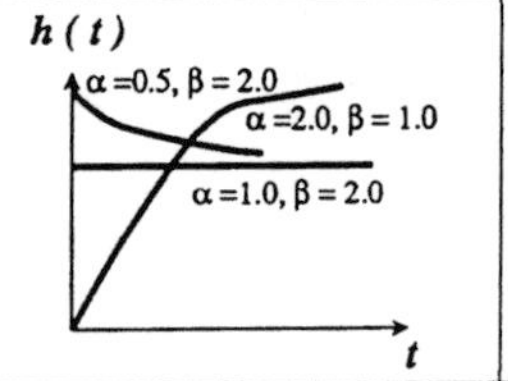 | 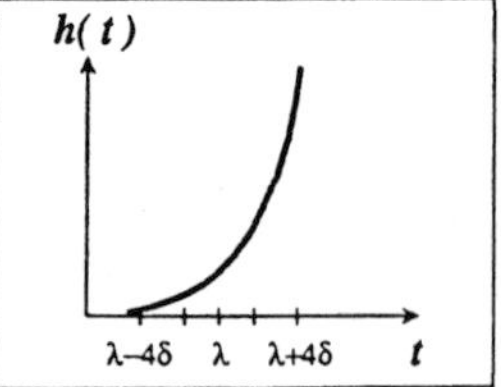 |
| Major applications in component reliability | - Corrosion resistance<br>- Life distribution of many basic components, such as capacitors, relays ballbearings, and certain motors | - Distribution of time between recalibration or maintenance of components<br>- Time to failure of systems with standby components | - Distribution of breaking strength of some components<br>- Breakdown voltage of capacitors |

Using (3.5),

$$f(t) = (0.1 - 0.0009t)\exp(-0.1t + 4.5 \times 10^{-4}t^2).$$

For $100 \leq t \leq 200$,

$$h(t) = -0.1 + 0.0009t.$$

Accordingly,

$$R(t) = \exp(0.1t - 4.5 \times 10^{-4}t^2).$$

$$f(t) = (-0.1 - 0.0009t)\exp(0.1t - 4.5 \times 10^{-4}t^2).$$

---

## 3.2 Common Distributions in Component Reliability

In general, Table 3.1 displays reliability characteristics of components whose times to failure are exponential, normal, lognormal, Weibull, or gamma. Some of the more subtle characteristics of each of these distributions are further discussed in this section.

### *3.2.1 Exponential Distribution*

The exponential distribution is the most commonly used distribution in reliability analysis. This can be attributed primarily to its simplicity and the fact that it exhibits the simple, constant, hazard rate model, a situation that is often realistic. In the context of the bathtub curve, this distribution can simply represent the chance failure region. It is evident that for components whose chance failure (or random failure) characteristic is long, in comparison to the other two regions, this distribution is adequate. This is often the case for electronic components and mechanical components. Especially in certain applications, new components are screened and only those that are determined to have passed (burn-in period) are used. For such components, exponential distribution is a reasonable choice. In general, exponential distribution is a good model for representing systems and complex, nonredundant components consisting of many interacting parts.

The exponential distribution can be described by a Poisson process using the random shocks concept for the occurrence of failures. That is, by assuming that the failure is caused by a random shock, and the shock occurs in a time interval of length t described by a Poisson process, then

$$\Pr[X = n] = \frac{\exp(-\lambda t)(\lambda t)^n}{n!}, \qquad n = 0, 1, 2, \ldots; \qquad \lambda, t > 0,$$

where $n$ is the random number of shocks occurring in the interval $[0,t]$ and $\lambda$ is the rate at which the shocks occur. Accordingly, since in nonrepairable items the first shock causes component failure, then the component works only when no shocks occur. Thus,

$$R(t;\lambda) = \Pr[X = 0] = \exp(-\lambda t). \tag{3.8}$$

Using (3.2), the exponential pdf will be

$$f(t) = \lambda \exp(-\lambda t). \tag{3.9}$$

Let us now look at one of the interesting characteristics of an exponential distribution: **failure process represented by an exponential distribution has no memory.**

Consider the law of conditional probability and assume that an item has survived after operating for a time $t$. The probability that the item would fail sometime between $t$ and $t + \Delta t$ is

$$\Pr(t \le T \le t + \Delta t | T > t) = \frac{\exp(-\lambda t) - \exp(-\lambda(t + \Delta t))}{\exp(-\lambda t)} = 1 - \exp(-\lambda \Delta t), \tag{3.10}$$

which is independent of $t$. In other words, the component that has worked up to time $t$ has no memory of its past. This phenomenon can also be easily described by the shock model. That is, at any point along time $t$, the rate at which fatal shocks occur is the same regardless of whether any shock has occurred up to time $t$.

### *3.2.2 Weibull Distribution*

The Weibull distribution has a wide range of applications in reliability analysis. This distribution covers a variety of shapes. Due to its (life-period) flexibility for showing hazard rates, all three regions of the bathtub curve can be represented by the Weibull distribution. It is possible to show that the Weibull distribution is appropriate for a system or complex component composed of a number of components or parts whose failure is governed by the most severe defect of its components or parts. Or in a sample of size N, the Weibull distribution is a good candidate to represent the pdf of the weakest element of the sample. The distribution is given by

$$f(t) = \frac{\beta t^{(\beta-1)}}{\alpha^{\beta}} \exp[-(t/\alpha)^{\beta}], \quad t, \alpha, \beta > 0. \tag{3.11}$$

Using (3.5), $h(t)$ is derived.

$$h(t) = (\beta/\alpha)(t/\alpha)^{\beta-1}, \quad \beta, \alpha > 0, \quad 0 < t < \infty. \tag{3.12}$$

Parameters $\alpha$ and $\beta$ are referred to as the **scale** and **shape** parameters. In (3.12), if $0 < \beta < 1$, then the Weibull distribution characterizes burn-in (early) type failures (decreasing hazard rate). For $\beta = 1$, the Weibull distribution reduces to an exponential distribution. If $1 < \beta < \infty$, the Weibull distribution characterizes the wear-out characteristics of a component (increasing hazard rate). Applications of the Weibull pdf include:

- Corrosion resistance.
- Number of component downtimes per shift.
- Time to failure of many types of equipment, including capacitors, relays, electron tubes, germanium transistors, photoconductive cells, ball bearings, and certain motors.
- Time to failure of basic elements of a system (components, parts, etc.), although the time to failure of the system itself is better represented by an exponential pdf.

In some cases, a parameter $\theta$ called **location parameter** is used in the Weibull distribution to account for a period of guaranteed life. For example, if a component has a warranty period $\theta$, or a period with no initial failures, the hazard rate will be

$$h(t, \alpha, \beta, \theta) = (\beta/\alpha)\left(\frac{t-\theta}{\alpha}\right)^{\beta-1}, \quad \beta, \alpha > 0, \quad 0 < \theta \leq t < \infty. \tag{3.13}$$

Accordingly, the pdf and reliability functions become

$$f(t) = \frac{\beta}{\alpha}\left(\frac{t-\theta}{\alpha}\right)^{\beta-1} \exp\left[-\left(\frac{t-\theta}{\alpha}\right)^{\beta}\right], \quad t \geq \theta, \text{ and} \tag{3.14}$$

$$R(t) = \exp - \left(\frac{t-\theta}{\alpha}\right)^{\beta}. \tag{3.15}$$

Sometimes a transformation $\lambda = 1/\alpha^{\beta}$ is used in which case (3.12) changes to $h(t) = \lambda\beta t^{\beta-1}$. This form will be used later in Chapter 5.

### *3.2.3 Gamma Distribution*

Gamma distribution is sometimes used as a distribution of time required for exactly $k$ independent failures to occur, provided that failures take place at a constant rate. This is clearly a natural outgrowth of the Poisson distribution. Examples of its application include the distribution of times between recalibration of an instrument that needs recalibration after $k$ uses; time between maintenance of items that require maintenance after $k$ uses; and time to failure of a system with standby components.

The gamma distribution is obtained from the distribution of time to the occurrence of the $k^{th}$ event in a Poisson process. The gamma distribution is also suitable for representing the so-called shock models discussed earlier. In these models, an item fails after it receives a given number of shocks. If one assumes that the shocks occur according to Poisson process, a gamma distribution adequately models the distribution of time to failure of the item, given that it fails after receiving $k$ shocks. According to our discussion in Section 3.2.1, the exponential distribution could also reasonably represent cases for which an item fails after receiving only one shock. Therefore, it is reasonable to expect exponential distribution to be a special case of gamma distribution.

The pdf of a gamma distribution has two parameters, $\alpha$ and $\beta$, such that

$$f(t) = \frac{1}{\beta^{\alpha}\Gamma(\alpha)} t^{\alpha-1} \exp(-t/\beta), \quad \alpha, \beta, t > 0, \tag{3.16}$$

where $\Gamma(\alpha)$ denotes the gamma function evaluated at $\alpha$. For the special case where $\alpha = 1$, the gamma pdf reduces to exponential pdf. Also, for all integer values of $\alpha$, the gamma pdf is known as **Erlangian** pdf. For the special case where $\alpha = n/2$ (n being an integer) and $\beta = 2$, the gamma distribution changes to the **chi-square distribution** with n degrees of freedom.

For cases where $\alpha$ is an integer, the reliability and hazard rates become simpler.

$$R(t) = \sum_{k=0}^{\alpha-1} \frac{(t/\beta)^k \exp(-t/\beta)}{k!}, \tag{3.17}$$

$$h(t) = \frac{t^{\alpha-1}}{\beta^{\alpha}\Gamma(\alpha)\sum_{k=0}^{\alpha-1}\frac{(t/\beta)^k}{k!}}. \tag{3.18}$$

It is evident from (3.18) that $h(t)$ will be a decreasing function of $t$ for $\alpha < 1$, a constant for $\alpha = 1$, and an increasing function for $\alpha > 1$. Thus, the gamma distribution represents three regions of the bathtub curve.

---

*Example 3.3*

The mean time to adjustment of an engine in a fighter plane is M = 100 hours. (Assume time to adjustment follows an exponential distribution.) Suppose there is a rule to replace certain parts of the engine after three consecutive adjustments.

1. What is the distribution of the mean time to replace?
2. What is the probability that a given engine does not require part replacement for at least 200 hours?

3. What is the mean time to replace?

*Solution:*

1. Use gamma distribution for $T$ with $\alpha = 3$, $\beta = 100$.
2. $R = \sum_{k=0}^{2} \frac{(t/100)^k \exp(-t/100)}{k!}$
   $= \frac{(\frac{200}{100})^0 \exp(-2)}{0!} + \frac{(2)^1 \exp(-2)}{1!} + \frac{2^2 \exp(-2)}{2!}$
   $= 0.135 + 0.271 + 0.271 = 0.677$,
3. Mean time to replace= $E(f) = \alpha\beta = 3 \times 100 = 300$ hours.

---

### *3.2.4 Normal Distribution*

The normal distribution is a basic distribution of statistics. According to the **central limit theorem**, the average of the values for $n$ random observations approaches a normal distribution. This distribution is an appropriate model for many physical phenomena, e.g., the distribution of the diameters of manufactured shafts. Since the normal distribution supports values of the random variable in $(-\infty, \infty)$ range, it has limited applications in reliability-type problems that involve time-to-failure estimations. However, for cases where the mean $\mu$ is positive and larger than $\sigma$ by several folds, the probability that the r.v. $T$ takes negative values will be negligible. For cases where the probability that r.v. $T$ takes negative values is not negligible, the distribution can be truncated below zero and rescaled (Johnson and Kotz, 1970). The normal pdf is given by

$$f(t) = \frac{1}{\sigma_t\sqrt{2\pi}} \exp[-\frac{1}{2\sigma_t^2}(t - \mu_t)^2], \quad -\infty < t < \infty, \ -\infty < \mu < \infty, \ \sigma^2 > 0, \tag{3.19}$$

where $\mu$ represents the $MTTF$ and $\sigma$ represents the standard deviation of the failure time. The instantaneous failure rate will always be a monotonically increasing function of time $t$. Normal pdf thus represents the wear-out region of the bathtub curve. Normal distribution is a reasonable model to represent failure of components under high stress when **stress** rather than time to failure is used as the r.v.

### *3.2.5 Lognormal Distribution*

The lognormal distribution has many uses in reliability analysis. In essence, the lognormal distribution represents a r.v. whose logarithm follows a normal distribution. This model is particularly suitable for failure processes that are the result of many small multiplicative errors. Specific applications of this pdf include time to failure of components due to fatigue cracks (Mann et al., 1974).

Also, failures attributed to maintenance activities are often represented by a lognormal distribution. Other applications include the distribution of break sizes (e.g., pipe breaks in process industries) and life distribution of some transistor types. The lognormal distribution can also be used to represent the distribution of a constant failure rate (by considering the failure rate as a r.v.) estimated from raw data or by Bayesian reliability analysis. We discuss this topic further in Section 3.6.

The lognormal pdf is a two-parameter distribution. For a r.v. $T$, the lognormal pdf is

$$f(t) = \frac{1}{\sigma_t t\sqrt{2\pi}} \exp[-\frac{1}{2\sigma_t^2}(\ln t - \mu_t)^2]; \quad 0 \le t < \infty, \ -\infty < \mu_t < \infty, \ \sigma^2 > 0, \tag{3.20}$$

where $\mu_t = E(\ln t)$ and $\sigma_t^2 = v(\ln t)$. The hazard rate for a lognormal pdf initially increases over time and then decreases. The rate of increase and decrease depends on the values of the parameters $\mu_t$ and $\sigma_t$. In general, this distribution is appropriate for representing time to failure for a component whose early failure (or processes that lead to failure) dominates its failure behavior.

### *3.2.6 Extreme Value Distribution*

The extreme value distribution can be used to model component failures due to extreme phenomena, such as minimum temperatures, great strength of materials and equipment. This distribution is related primarily to the Weibull distribution since both are suitable for "weakest link" type failures, and Weibull data can conveniently be analyzed in terms of the simpler extreme value distribution. There are two types of extreme value distributions: smallest extreme value distribution and largest extreme value distribution. The smallest extreme value distribution is more appropriate for component time-to-failure analysis. The largest extreme value distribution is seldom used in life and failure data analysis. The largest extreme value distribution is appropriate for estimating extreme flood heights, extreme wind velocities, and the age of the oldest person dying each year in a community. The smallest value *pdf* is given by

$$f(t) = \frac{1}{\delta} \exp\left[\frac{1}{\delta}(t - \lambda) - \exp\left(\frac{t-\lambda}{\delta}\right)\right], \tag{3.21}$$

$$-\infty < \lambda < \infty, \quad \delta > 0, \quad -\infty < t < \infty.$$

The parameter $\lambda$ is called the **location** parameter and can take any value. The parameter $\delta$ is called the **scale** parameter and is always positive. The hazard rate for the smallest extreme value distribution is

$$h(t) = (1/\delta)\exp[(t - \lambda)/\delta], \tag{3.22}$$

which is an increasing failure rate (wear-out). This can be characterized as component failure due to aging. That is, in this model, the component's wear-out period is characterized by an exponentially increasing rate of failure. Similarly, the extreme value reliability function is

$$R(t) = \exp\{-\exp[(t - \lambda)/\delta]\} \quad -\infty < t < \infty. \tag{3.23}$$

Clearly, negative values of $t$ are not meaningful when it is representing time to failure. It can be shown that the smallest extreme value distribution reduces to a Weibull distribution by using

$$\lambda = \ln(\alpha), \tag{3.24}$$

$$\delta = 1/\beta. \tag{3.25}$$

The largest extreme value distribution, though not very useful for component failure, is useful for estimating natural extreme phenomena and is presented here for the purpose of comparison with the smallest value distribution. The largest extreme value distribution is given by

$$f(t) = \frac{1}{\delta}\exp\left\{-\frac{1}{\delta}(t - \lambda) - \exp\left[-\frac{(t-\mu)}{\delta}\right]\right\}, \tag{3.26}$$

$$-\infty < t < \infty, \quad \delta > 0, \quad -\infty < \lambda < \infty.$$

For further discussions regarding the extreme value distribution, see Gumble (1958), Hahn and Shapiro (1967), and Johnson and Kotz (1970). Also, Castillo (1988) provides several useful examples.

---

*Example 3.4*

The maximum demand for electric power at any given time during a year is directly related to extreme weather conditions. An electric utility has determined that the distribution of maximum power demands can be presented by the largest extreme value distribution with $\lambda = 1200(MW)$ and $\delta = 480(MW)$. Determine the probability (per year) that the demand will exceed the utility's maximum installed power of 3000(MW).

*Solution:*

Since this is the largest extreme value distribution, we should integrate (3.26) from 3000 to $\infty$.

$$\Pr(t > 3000) = \int_{3000}^{\infty} f(t)dt = 1 - \exp\{-\exp[t - \lambda]/\delta\}.$$

Since

$$\frac{t - \lambda}{\delta} = \frac{3000 - 1200}{480} = 3.75,$$

then

$$\Pr(t > 3000) = 0.023.$$

---

## 3.3 Component Reliability Model Selection

In the previous section, we discussed several useful pdf models for reliability analysis of components. A probability model is referred to a mathematical expression that describes in terms of probabilities how a r.v. is spread over its range. It is necessary at this point to discuss how field data can support the selection of a probability model for reliability analysis. In this section, we elaborate on several methods for selecting and evaluating the models using observed failure data. These methods can be divided into four groups: nonparametric methods, parametric methods (probability plotting), total-time-on-test plots, and goodness-of-fit tests. We discuss each of these methods in more detail.

### *3.3.1. Nonparametric Methods*

The nonparametric method, in principle, attempts to directly estimate the reliability characteristic of an item (e.g., the pdf, reliability, and hazard rates) from a sample. The shape of these functions, however, is often used as an indication of the most appropriate pdf. Two cases exist: small samples and large samples. For a small sample of failure data, it is most convenient to use a set of unbiased estimators to estimate $h(t)$, $R(t)$, and the pdf. For a larger sample of failure data, the data can be grouped into intervals, and $h(t)$, $R(t)$, and the pdf can be estimated from another set of nonparametric estimators. It is important to mention that failure data from a maintained item can be used as the sample only if a maintained component is assumed to be **as good as new** following maintenance. Then each failure time can be considered as a sample point independent of the previously observed failure times. Therefore, $n$ observed times to failure of such a maintained component is equivalent to putting $n$ independent new components under test.

## Small Sample

Suppose $n$ times to failure make a small sample (e.g., $n < 15$). Let the data be ordered such that $t_1 \leq t_2 \leq \ldots \leq t_n$. Blom (1958) recommends the following nonparametric estimators for the reliability functions of interest:

$$\hat{\lambda}(t_i) = \frac{1}{(n - i + 0.625)(t_{i+1} - t_i)}, \quad i = 1, 2, \ldots, n - 1, \tag{3.27}$$

$$\hat{R}(t_i) = \frac{n - i + 0.625}{n + 0.25}, \quad i = 1, 2, \ldots, n, \text{and} \tag{3.28}$$

$$\hat{f}(t_i) = \frac{1}{(n + 0.25)(t_{i+1} - t_i)}, \quad i = 1, 2, \ldots, n - 1. \tag{3.29}$$

Although there are other estimators besides those above, Kimball (1960) concludes that estimators in (3.27)–(3.29) are optimal and recommends their use. All of the characteristics necessary for a pdf should exist in (3.29). Namely, it is evident from (3.29) that $\hat{f}(t_i)(t_{i+1} - t_i)$ adds up to approximately 1 for all values of i, as is expected.

---

*Example 3.5*

A high-pressure pump in a process plant has the failure times $t_i$ (shown in the following table). Plot $\hat{h}(t)$, $\hat{R}(t)$, and $\hat{f}(t)$ and discuss the results.

*Solution:*

| $i$ | $t_i$ | $t_{i+1} - t_i$ | $\hat{h}(t_i)$ | $R(\hat{t}_i)$ | $\hat{f}(t_i)$ |
|---|---|---|---|---|---|
| 1 | 0.20 | 0.60 | $\frac{1}{(6.625)(0.6)} = 0.25$ | $\frac{6.625}{7.25} = 0.91$ | $\frac{1}{(7.25)(0.6)} = .23$ |
| 2 | 0.80 | 0.30 | 0.59 | 0.78 | 0.46 |
| 3 | 1.10 | 0.41 | 0.53 | 0.64 | 0.34 |
| 4 | 1.51 | 0.32 | 0.86 | 0.50 | 0.43 |
| 5 | 1.83 | 0.69 | 0.55 | 0.36 | 0.20 |
| 6 | 2.52 | 0.46 | 1.34 | 0.22 | 0.30 |
| 7 | 2.98 | | | 0.09 | |

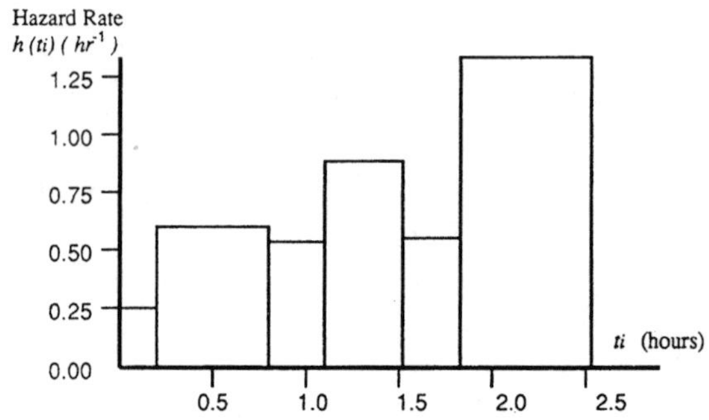


From the histogram above, one can conclude that the instantaneous failure rate is reasonably constant over the operating period of the component, with an increase toward the end. One point of caution: although a constant hazard rate can be concluded from the result, several other tests and additional observations may be needed to support the conclusion. Additionally, the histogram is only a representative of the case under study. An extension of the result to future times or other cases (e.g., other high-pressure pumps) may not be accurate.

---

## Large Sample

Suppose $n$ times to failure comprise a large sample. Suppose the sample is grouped into a number of similar time-to-failure increments $\Delta t$. According to the definition of reliability, a nonparametric estimate of the reliability function becomes

$$\hat{R}(t_i) = \frac{N_s(t_i)}{N}, \tag{3.30}$$

where $N_s(t_i)$ represents the number of surviving components or the number of times that a maintained component remains operational beyond time $t_i$. Time $t_i$ is usually taken to be the upper endpoint of each interval. Similarly, the pdf is estimated by

$$\hat{f}(t_i) = \frac{N_f(t_i)}{N\Delta t}, \tag{3.31}$$

where $N_f(t_i)$ is the number of failures observed in the interval $t_i$ to $t_i + \Delta t$. Finally, according to (3.5),

$$\hat{h}(t_i) = \frac{N_f(t_i)}{N_s(t_i)\Delta t}. \tag{3.32}$$

It is clear that for $i = 1$, $N_s(t_i) = N$; and for $i > 1$, $N_s(t_i) = N_s(t_{i-1}) - N_f(t_i)$. Equation (3.32) gives an average failure rate during the interval $t_i$ to $t_i + \Delta t$. As expected, when $N_s(t_i) \to \infty$ and $\Delta T \to 0$, then $\hat{h}(t_i)$ approaches the hazard rate $h(t)$. This was illustrated earlier in this chapter.

In (3.32), $N_f(t_i)/N_s(t_i)$ is the probability that the component will fail in the interval $t_i$ to $t_i + \Delta t$, since $N_s(t_i)$ represents the number of components working at $t_i$. By dividing this probability by $\Delta t$, the probability of failure per time period $\Delta t$ is obtained. Thus, (3.32) is an estimated average rate of failure in each period. It is important to notice that the accuracy of this estimate depends on $\Delta t$. Therefore, if smaller $\Delta t$'s are used, we would, theoretically, expect to obtain a better answer. However, drawback of using smaller $\Delta t$'s is the lack of reasonable data for the estimation of $\hat{R}(t)$, $\hat{f}(t)$, and $\hat{h}(t)$. Therefore, selecting $\Delta t$ requires consideration of both of these opposing factors.

---

*Example 3.6*

Times to failure for an electronic component are obtained during three stages of failure. The first stage is believed to be influenced by infant mortality of the component; the second stage represents random failure; and the third stage represents the wear-out period. Plot the instantaneous failure rate for this component, using the data provided.

| $i$ | Data obtained during early stage of the component operation | | Data obtained during useful life of the component operation | | Data obtained during the wear-out period | |
|---|---|---|---|---|---|---|
| | Interval (in hrs) | $N_f(t_i)$ | Interval (in 100 hrs) starting at 150 hrs | $N_f(t_i)$ | Interval (in 1,000 hrs) starting at 11,000 hrs | $N_f(t_i)$ |
| 1 | 0–20 | 79 | 0–20 | 211 | 0–100 | 34 |
| 2 | 20–40 | 37 | 20–40 | 142 | 100-200 | 74 |
| 3 | 40–60 | 15 | 40–60 | 67 | 200–300 | 110 |
| 4 | 60–80 | 6 | 60–80 | 28 | over 300 | 82 |
| 5 | 80–100 | 2 | 80–100 | 21 | | |
| 6 | 100–120 | 1 | over 100 | 31 | | |
| 7 | over 120 | 10 | | | | |
| | Total | 150 | Total | 500 | Total | 300 |

*Solution:*

For each of the three stages, the results along with a plot of $\hat{h}(t_i)$ are presented below.

Early Stage Results

| $i$ | Interval (in hrs) | $N_f(t_i)$ | $N_s(t_i)$ | $\hat{h}(t_i)$ see (3.26) | $\hat{f}(t_i)$ see (3.25) | $\hat{R}(t_i)$ see (3.24) |
|---|---|---|---|---|---|---|
| 1 | 0–20 | 79 | $150 - 79 = 71$ | $\frac{79}{150\times20} = 0.026$ | $\frac{79}{150\times20} = 0.0260$ | $\frac{150}{150} = 1.000$ |
| 2 | 20–40 | 37 | $71 - 37 = 34$ | $\frac{37}{71\times20} = 0.026$ | $\frac{37}{150\times20} = 0.0120$ | $\frac{71}{150} = 0.470$ |
| 3 | 40–60 | 15 | $34 - 15 = 19$ | $\frac{15}{34\times20} = 0.022$ | $\frac{15}{150\times20} = 0.0050$ | $\frac{34}{150} = 0.230$ |
| 4 | 60–80 | 6 | $19 - 6 = 13$ | $\frac{6}{19\times20} = 0.016$ | $\frac{6}{150\times20} = 0.0020$ | $\frac{19}{150} = 0.130$ |
| 5 | 80–100 | 2 | $13 - 2 = 11$ | $\frac{2}{13\times20} = 0.008$ | $\frac{2}{150\times20} = 0.0007$ | $\frac{13}{150} = 0.087$ |
| 6 | 100–120 | 1 | $11 - 1 = 10$ | $\frac{1}{11\times20} = 0.005$ | $\frac{1}{150\times20} = 0.0003$ | $\frac{11}{150} = 0.073$ |
| 7 | over 120 | 10 | $10 - 10 = 0$ | – | – | $\frac{10}{150} = 0.070$ |

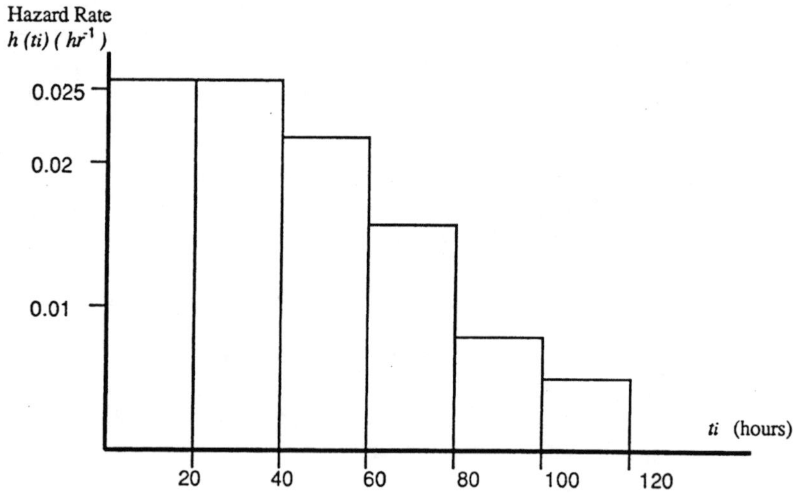

## Useful Life Stage Results

| $i$ | Interval (in hrs) | $N_f(t_i)$ | $N_s(t_i)$ | $\hat{h}(t_i)$ see (3.26) | $\hat{f}(t_i)$ see (3.25) | $\hat{R}(t_i)$ see (3.24) |
|---|---|---|---|---|---|---|
| 1 | 0–20 | 211 | $500 - 211 = 289$ | $\frac{211}{500\times20\times100} = .000211$ | $\frac{211}{500\times20\times100} = .000211$ | $\frac{500}{500} = 1.00$ |
| 2 | 20–40 | 142 | $289 - 142 = 147$ | $\frac{142}{289\times20\times100} = .000246$ | $\frac{142}{500\times20\times100} = .000142$ | $\frac{289}{500} = 0.58$ |
| 3 | 40–60 | 67 | $147 - 67 = 80$ | $\frac{67}{147\times20\times100} = .000228$ | $\frac{67}{500\times20\times100} = .000067$ | $\frac{147}{500} = 0.29$ |
| 4 | 60–80 | 28 | $80 - 28 = 52$ | $\frac{28}{80\times20\times100} = .000175$ | $\frac{28}{500\times20\times100} = .000028$ | $\frac{80}{500} = 0.16$ |
| 5 | 80–100 | 21 | $52 - 21 = 31$ | $\frac{21}{52\times20\times100} = .000202$ | $\frac{21}{500\times20\times100} = .000021$ | $\frac{52}{500} = 0.10$ |
| 6 | over 100 | 31 | $31 - 31 = 0$ | – | – | $\frac{31}{500} = 0.06$ |

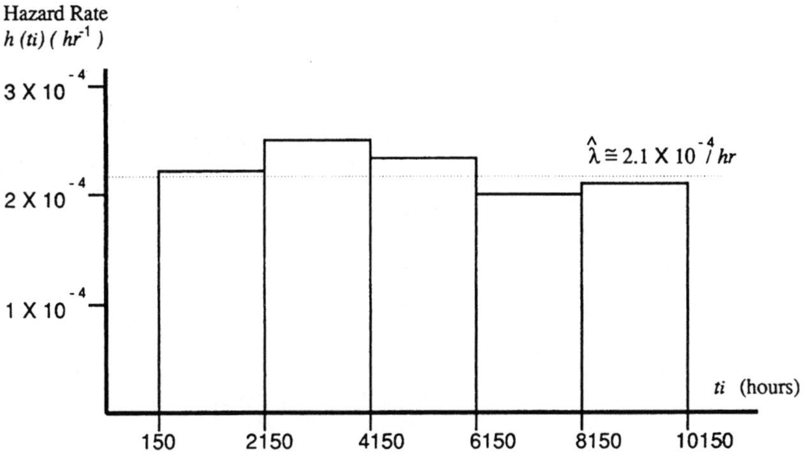

Wear-Out Stage Results

| $i$ | Interval (in hrs) | $N_f(t_i)$ | $N_s(t_i)$ | $\hat{h}(t_i)$ see (3.26) | $\hat{f}(t_i)$ see (3.25) | $\hat{R}(t_i)$ see (3.24) |
|---|---|---|---|---|---|---|
| 1 | 0–100 | 34 | $300 - 34 = 266$ | $\frac{34}{300 \times 100} = .0011$ | $\frac{34}{300 \times 100} = .00111$ | $\frac{300}{300} = 1.00$ |
| 2 | 100–200 | 74 | $266 - 74 = 192$ | $\frac{74}{266 \times 100} = .0028$ | $\frac{74}{300 \times 100} = .00250$ | $\frac{266}{300} = 0.89$ |
| 3 | 200–300 | 140 | $192 - 140 = 52$ | $\frac{140}{192 \times 100} = .0073$ | $\frac{140}{300 \times 100} = .00460$ | $\frac{192}{300} = 0.64$ |
| 4 | 300–400 | 52 | $52 - 52 = 0$ | $\frac{52}{52 \times 100} = .0100$ | $\frac{52}{300 \times 100} = .00173$ | $\frac{52}{300} = 0.17$ |

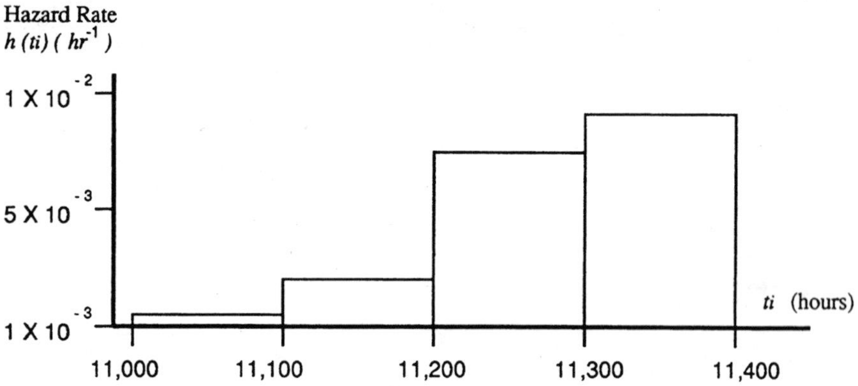


### *3.3.2. Probability Plotting*

Probability plotting is a simple graphical method of displaying and analyzing observed data. The data are plotted on special probability papers such that the resulting curve failure is a straight line. Each type of distribution has its own graph paper. If a set of data is hypothesized to originate from a known distribution, the graph can be used to conclude whether or not the hypothesis should be rejected. From the plotted line, one can also determine the parameters of the hypothesized distribution. Probability plotting is often used in reliability analysis to test the appropriateness of using known distributions to present a set of observed data. This method is used because it provides simple and visual representation of the data. However, the conclusion based on probability plotting is not as strong as the statistical goodness-of-fit tests discussed in Chapter 2. Therefore, the method should be used with care and preferably as a supplemental method to other data analysis techniques. When appropriate, we will discuss various factors that should be considered when the probability plotting method is used.

There are probability papers for a number of the distributions discussed in this book. The reader is referred to Nelson(1979), Nelson(1982), Martz and Waller(1982) and Kececioglu (1991) for further discussion regarding other distributions and various plotting techniques.

## Exponential Plot

If the logarithm of both sides of (3.8) is taken, it follows that

$$\ln R(t) = -\lambda t. \tag{3.33}$$

If $\ln R(t)$ is plotted against time $t$ on regular plotting paper, or if $R(t)$ is plotted against $t$ on semilogarithmic plotting paper, according to (3.33) the resulting plot will be a straight line with slope of $(-\lambda)$. To better understand the mechanics of the process, consider the following $n$ times to failure observed from a life test:

$$t_1 \leq t_2 \ldots \leq t_n.$$

According to (3.33), an estimate of the reliability $R(t_i)$ can be made for each $t_i$. A crude estimation of $R(t_i)$ is clearly $(n-i)/n$ (sample statistic). However, (3.28) provides better estimation statistic for $R(t_i)$. Graphically, the y-axis shows $R(t_i)$ and the $x$-axis shows $t_i$. The resulting points should reasonably fall on a straight line if these data can be described by an exponential distribution. Since the slope of $R(t)$ vs. $t$ is negative, it is also possible to plot $1/R(t)$ vs. $t$ in which the slope is positive. Other, appropriate estimators of $R(t_i)$ include $(n-i+1)/(n+1)$ mean rank and $(n-i+0.7)/(n+0.4)$ median rank. Limited studies of estimators (e.g., Kimball,1960) have shown that (3.28) is close to optimal and thus is used in our context.

It is also possible to get an estimation of the $MTTF$ from the plotted graph. For this purpose, at the level of $R = 0.368$ (or $1/R = 2.717$), a line parallel to the $x$-axis is drawn. At the junction of this line and the fitted line, another line vertical to the $x$-axis is drawn. The value of $t$ read along the the $x$-axis is an estimate of $MTTF$, and its inverse is $\hat{\lambda}$.

---

*Example 3.7*

Nine times to failure of a diesel generator are recorded as 31.3, 45.9, 78.3, 22.1, 2.3, 4.8, 8.1, 11.3, and 17.3 days. If the diesel is restored to "as good as new" after each failure, determine whether the data represent an exponential distribution. Determine $\hat{\lambda}$ and $\hat{R}$ (193 hours).

*Solution:*

First arrange the data in increasing order and then calculate the corresponding $\hat{R}(t_i)$.

| $i$ | $t$ | $\frac{n-i+.625}{n+.25}$ | $\frac{n+.25}{n-i+.625}$ |
|---|---|---|---|
| 1 | 2.3 | 0.93 | 1.07 |
| 2 | 4.8 | 0.82 | 1.21 |
| 3 | 8.1 | 0.72 | 1.40 |
| 4 | 11.3 | 0.61 | 1.64 |
| 5 | 17.3 | 0.50 | 2.00 |
| 6 | 22.1 | 0.39 | 2.55 |
| 7 | 31.3 | 0.28 | 3.53 |
| 8 | 45.9 | 0.18 | 5.69 |
| 9 | 78.3 | 0.07 | 14.80 |

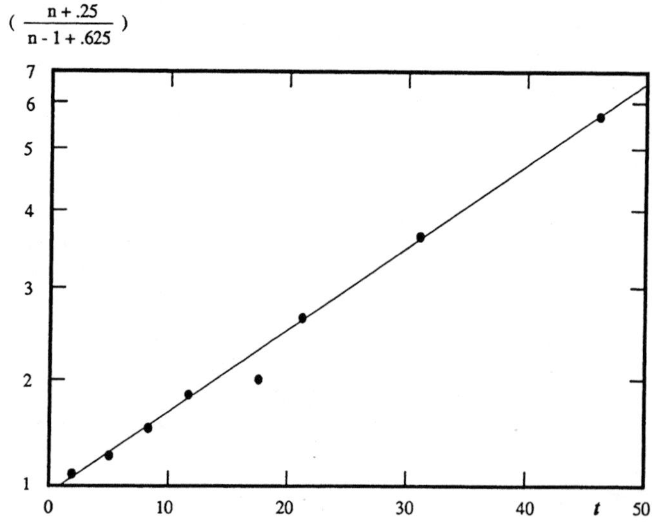


*Exponential Plot for Example 3.7*

A plot of the data on semilogarithm paper is shown above.

$$\hat{\lambda} = \frac{\ln 7 - \ln 3}{48.6 - 28} = 0.041 \text{ failures/days} \quad (\text{or } 1.71 \times 10^{-3} \text{ failures/hour}),$$

$$\hat{R}(193) = \exp[-(1.71 \times 10^{-3} \times 193)] = 0.72.$$

---

## Weibull Plot

Similar to plots of an exponential distribution, plots of a Weibull distribution require special probability papers. If the observed data form a reasonably straight line, the Weibull distribution can be assumed. From Table 3.1,

$$1/R(t) = \exp[(t/\alpha)^\beta], \quad \text{and}$$

$$\log \ln[1/R(t)] = \beta \log t - \beta \log \alpha. \tag{3.34}$$

This provides the basis for the Weibull plots. It is evident that $\log \ln[1/R(t)]$ plots as a straight line against $\log t$ with slope $\beta$ and $y$-intercept of $-\beta \log \alpha$. Accordingly, the values of the Weibull parameters $\alpha$ and $\beta$ can be obtained from the $y$-intercept and the slope of the graph, respectively.

Several estimators of $1/R(t)$ exist when converting $1/R(t)$ to $\frac{1}{1-F(t)}$. The most appropriately used estimator of $F(t)$ is $\frac{i-.375}{n+.25}$ (similar to the discussions presented for exponential plots). The mechanics of the process are simple. On the special Weibull paper (Fig. 3.5), $t_i$ is plotted in the logarithmic $x$-axis and $[(i - .375)/(n + .25)] \times 100$ is plotted on the $y$-axis (often labeled % failure). The third scale shown in Fig. 3.5 is for $\log \ln \frac{1}{R(t)}$, but it is more convenient to use the $F(t)$ shown earlier. (The two have a one-to-one correlation.) It is important to note that a variety of Weibull papers are available and they are somewhat different. They include Ford, General Motors, and commercially available TEAM[1] probability papers. These papers are fundamentally the same. The user should determine ahead of time the appropriate number of log cycles for $t$ and the percent failure needed so as to select the correct paper. The Ford Weibull paper has two cycles along the $X$-axis and a percent failure range of 0.01 to 100.

The degree to which the plotted data fall on a straight line determines the conformance of the data to a Weibull distribution. If the data reasonably plot as a straight line, the Weibull distribution is a reasonable fit, and the shape parameter $\beta$ and the scale parameter $\alpha$ can be obtained. If a line is drawn parallel to the plotted straight

---

[1] Technical and Engineering Aids for Management (TEAM), P.O. Box 25, Tamworth, New Hampshire 03886.

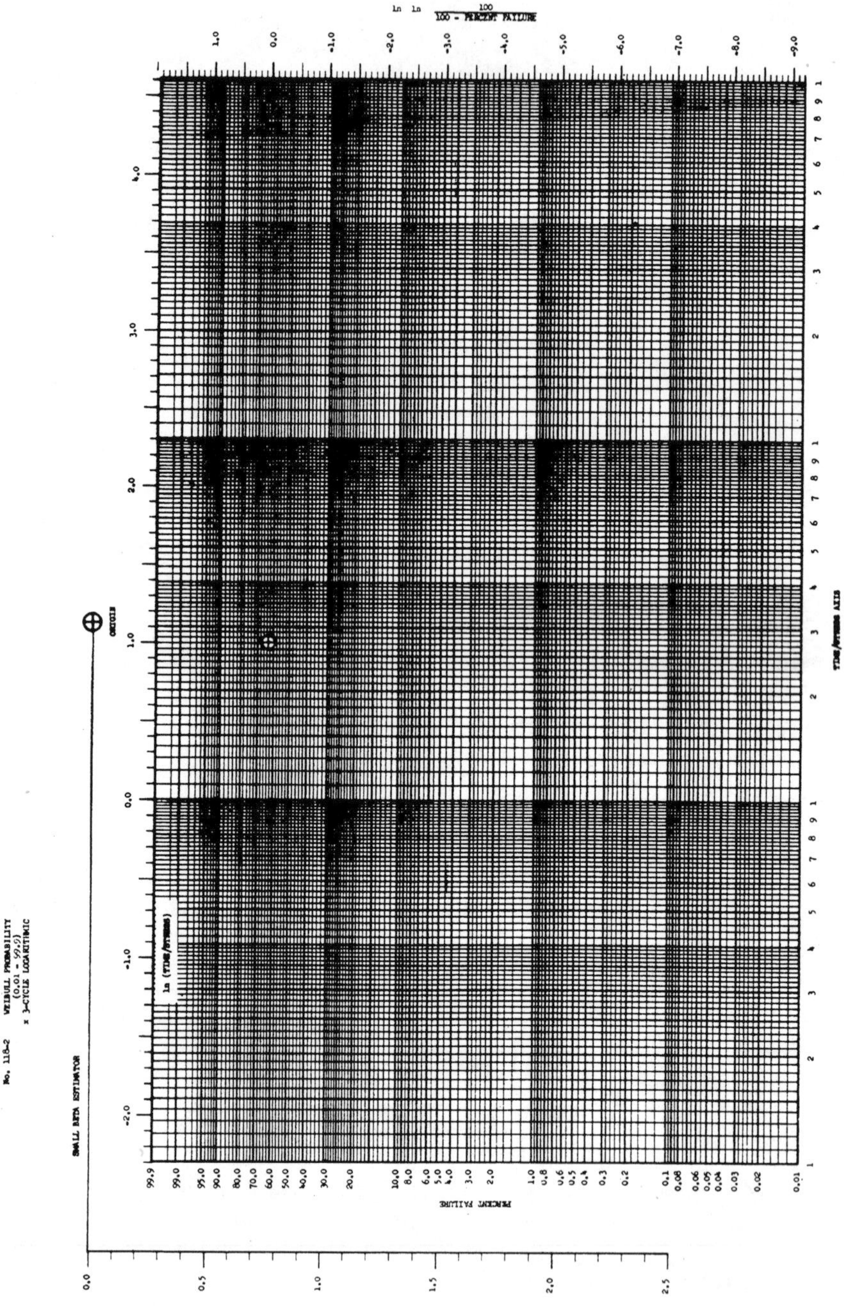


*Fig. 3.5 TEAM Weibull Probability Paper*

line from the center of a small circle $\oplus$ (in the TEAM paper called ORIGIN) until it crosses the WEIBULL SLOPE scale on the Ford and General Motors papers, and the SMALL BETA ESTIMATOR on the TEAM paper, the value of $\beta$ can be obtained. To find $\alpha$, draw a line from the 63.2(0.0 level at the right vertical line of the TEAM paper) until it intersects with the fitted straight line. From this intersection, draw a vertical line until it intersects with the $x$-axis, and read off the $x$- axis value at this intersection. This value represents the scale parameter $\alpha$.

---

*Example 3.8*

Time to failure of an electronic device is considered to follow a Weibull distribution. Ten of these devices are put to life testing. The times to failure (in hours) are: 89, 132, 202, 263, 321, 362, 421, 473, 575, and 663. If the Weibull distribution is the correct choice to model the data, what are the parameters of this pdf? What is the reliability of a typical device at 1,000 hours.

*Solution:*

| $i$ | 1 | 2 | 3 | 4 | 5 | 6 | 7 | 8 | 9 | 10 |
|---|---|---|---|---|---|---|---|---|---|---|
| $t_i$ | 89 | 132 | 202 | 263 | 321 | 362 | 421 | 473 | 575 | 663 |
| $\frac{i-.375}{n+.25} \times 100$ | 6.10 | 15.85 | 25.61 | 35.37 | 45.12 | 54.88 | 64.46 | 74.39 | 84.15 | 93.90 |

Fig. 3.6 shows a straight line on the Weibull probability paper. Clearly the conjecture is reasonable. The graphical estimate of $\beta$ is approximately 1.8, and the estimate of $\alpha$ is approximately 420 hours. Percent failure at 1,000 hours is about 99.1%; the reliability is about 0.9% $[R(t = 1000) = 0.009]$.

---

In cases where the data do not statistically fall on a straight line but are concave or convex in shape, it is possible to find a **location parameter** $\theta$ that might straighten out these points. For this procedure, see Kececioglu (1991) and Nelson (1979). The location parameter represents the $\theta$ parameter of a two-parameter Weibull distribution described by (3.14). Often the $\theta$ parameter represents the period of guaranteed life.

If the failure data are grouped, the class midpoints $t'_i$ (rather than $t_i$) should be used for plotting, where $t'_i = \frac{t_{i-1}+t_i}{2}$. One can also use class endpoints instead of midpoints. Recent studies suggest that the Weibull parameters obtained by using class endpoints in

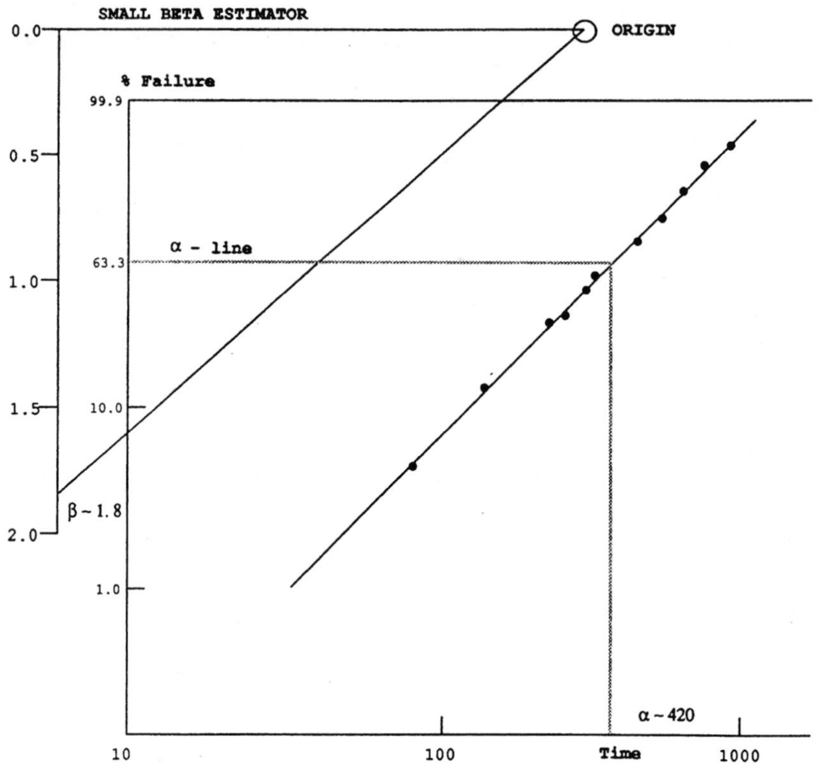


*Fig. 3.6 Weibull plot*

the plots are better matched with those of the maximum likelihood estimators method.

## Normal and Lognormal Plots

The same special probability papers can be used for both normal and lognormal plots. On the $x$-axis, $t_i$ (for normal) and $\ln t_i$ (for lognormal) is expressed, on the $y$-axis, the cumulative probability estimated by $\frac{i-.375}{n+.25} \times 100$ (for both normal and lognormal distribution) is expressed. Some lognormal papers are logarithmic on the $x$-axis, in which case $t_i$ can be directly expressed. Kimball (1960) shows that $\frac{i-.375}{n+.25} \times 100$ is better than the two more commonly used estimators

$\frac{i-.5}{n} \times 100$ and $[i/(n+1)] \times 100$. If the plotted data fall on a straight line, a normal or lognormal distribution can be conformed. To estimate the mean parameter $\mu$, the value of 50% is selected from the $x$-axis and a line parallel to the $y$-axis is drawn to intersect with the plotted straight line. From the intersection, a horizontal line to the $x$-axis is drawn. Its intersection with the $y$-axis is the parameter $\mu$ (mean or median of the normal distribution and $m$ (median) of the lognormal distribution). Similarly, if the corresponding $y$-axis intersection for the 84% value is selected from the $x$-axis, the parameter $\sigma$ can be obtained for the normal distribution from $\sigma \cong t_{84\%} - t_{50\%}$ or $\sigma \cong t_{84\%} - \mu$. For the lognormal distribution, $\hat{\sigma} \cong \ln t_{84\%} - \hat{\mu}$.

---

*Example 3.9*

The time it takes for a thermocouple to drift upward or downward to an unacceptable level is measured and recorded in a process plant. (See the following table.) Determine whether the drifting process can be modeled by a normal distribution.

| $i$ | $t_i$ (months) | $\frac{i-.375}{n+.25} \times 100$ | $i$ | $t_i$ (months) | $\frac{i-.375}{n+.25} \times 100$ |
|---|---|---|---|---|---|
| 1 | 11.2 | 4.39 | 8 | 17.2 | 53.50 |
| 2 | 12.8 | 11.40 | 9 | 18.1 | 60.53 |
| 3 | 14.4 | 18.42 | 10 | 18.9 | 67.54 |
| 4 | 15.1 | 25.44 | 11 | 19.3 | 74.56 |
| 5 | 16.2 | 32.46 | 12 | 20.0 | 81.58 |
| 6 | 16.3 | 39.47 | 13 | 21.8 | 88.60 |
| 7 | 17.0 | 46.49 | 14 | 22.7 | 95.61 |

*Solution:*

As shown in Fig. 3.7, the data conform to a normal distribution, with $\mu = t_{50\%} = 17.25$ months and $t_{84\%} = 20.75$. Therefore $\sigma \cong 20.75 - 17.25 = 3.5$ months.

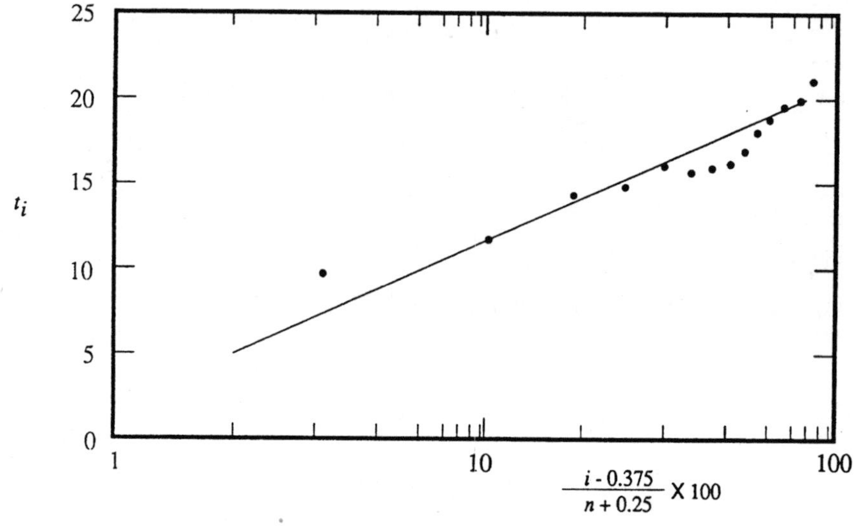


*Fig. 3.7 Normal Distribution Plot for Example 3.9*

---

*Example 3.10*

Five components are placed under fatigue crack tests. The following failure times are observed. Determine the conformance of the data with a lognormal distribution. Determine the parameters of the lognormal distribution.

*Solution:*

| $i$ | $t_i$ (hours) | $\frac{i-.375}{n+.25} \times 100$ |
|---|---|---|
| 1 | 363 | 11.90 |
| 2 | 1,115 | 30.95 |
| 3 | 1,982 | 50.00 |
| 4 | 4,241 | 69.05 |
| 5 | 9,738 | 88.10 |

The parameters $\hat{\mu}$ and $\hat{\sigma}$ are estimated as follows: $\hat{\mu}_y = 2000$, and similarly,

$$\hat{\mu}_t = \ln(2000) = 7.6 \quad \text{and} \quad \hat{\sigma}_t \cong \ln(6700) - 7.6 \cong 1.21.$$

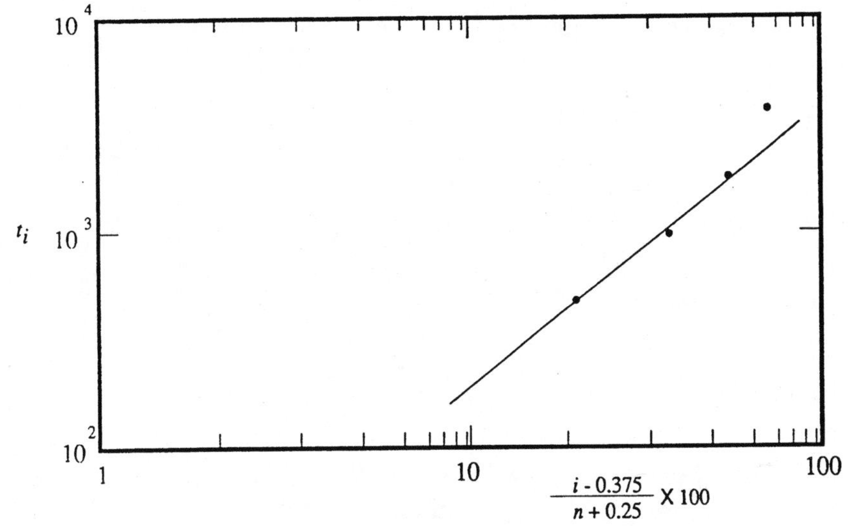


*Fig. 3.8 Lognormal Distribution Plot for Example 3.10*

### *3.3.3 Total-Time-On-Test Plots*

The total-time-on-test plot is a graphical technique and a good tool for model identification. This method has been discussed in detail by Barlow and Campo (1975) and Barlow (1978). Additionally, Davis (1977) discussed the use of this method in determining an optimal replacement policy in maintainability. In essence, the method is used to determine whether the data exhibit an increasing failure rate, a constant failure rate, or a decreasing failure rate. Although it is possible to treat grouped failure data using this method, we discuss its use for ungrouped failure data only. Consider the observed failure times of $n$ components under test such that

$$t_1 \leq t_2 \leq \ldots \leq t_n.$$

If the number of survivors to time $t_i$ is denoted by $N_s(t_i)$ then the survival probability (reliability) is also obtained from

$$R(t_i) = \frac{N_s(t_i)}{n}.$$

The **total time on test** to age $t_i$ denoted by $H(t_i)$ is obtained from

$$H(t_i) = \int_0^{t_i} N_s(t)dt. \tag{3.35}$$

For convenience, (3.29) can be expressed as follows:

$$\int_0^{t_i} N_s(t)dt = nt_1 + (n-1)(t_2 - t_1) + \ldots + (n-i+1)(t_i - t_{i-1}). \quad (3.36)$$

Equation (3.36) is valid since $n$ components have survived up to time $t_1$ (time to the first failure), $(n-1)$ of the components survived during the period between $t_1$ and $t_2$, and so on. A scale transform is commonly obtained by dividing (3.36) by the total time to failure. Thus, the scaled total time on test at time $t_i$ is defined as

$$\tilde{H}(t_i) \cong \frac{\int_0^{t_i} N_s(t)dt}{\int_0^{t_n} N_s(t)dt}. \quad (3.37)$$

It is easy to show that for the exponential distribution, $\tilde{H}(t_i) = i/n$. A graphical representation of total time on test is formed by plotting $i/n$ on the $x$-axis and $\tilde{H}(t_i)$ on the $y$-axis. Its deviation from the reference line $\tilde{H}(t_i) = i/n$ is then determined. If the data fall on the $\tilde{H}(t_i) = i/n$ line, an exponential distribution can be assumed. If the plot is concave over most of the graph, there is the possibility of an increasing failure rate. If the plot is convex over most of the graph, then it raises the possibility of a decreasing failure rate. If the plot does reasonably fall on a straight line and does not demonstrate pronounced concavity or convexity, one can reasonably assume an exponential distribution.

The total-time-on-test plot, similar to other plotting and analysis techniques, should not be used as the only test for determining model accuracy. This is even more important when only a small sample is available. The total-time-on-test plots are simple to carry out and provide a good alternative to more elaborate hazard and probability plots. These plots are scale invariant and, unlike probability plots, no special plotting papers are needed.

---

*Example 3.11*

In a nuclear plant, the times to failure (in hours) of the feedwater pumps are recorded. Use the total-time-on-test plot to evaluate the failure data.

*Solution:*

| $i$ | $t_i$ | $(n-i+1)(t_i - t_{i-1})$ | $H(t_i)$ | $\tilde{H}(t_i)$ |
|---|---|---|---|---|
| 1 | $0.14 \times 10^4$ | $1.40 \times 10^4$ | $1.40 \times 10^4$ | 0.14 |
| 2 | $0.35 \times 10^4$ | $1.89 \times 10^4$ | $3.29 \times 10^4$ | 0.33 |
| 3 | $0.59 \times 10^4$ | $1.92 \times 10^4$ | $5.21 \times 10^4$ | 0.53 |
| 4 | $0.76 \times 10^4$ | $1.19 \times 10^4$ | $6.40 \times 10^4$ | 0.65 |
| 5 | $0.86 \times 10^4$ | $0.60 \times 10^4$ | $7.00 \times 10^4$ | 0.71 |
| 6 | $0.90 \times 10^4$ | $0.20 \times 10^4$ | $7.20 \times 10^4$ | 0.73 |
| 7 | $1.16 \times 10^4$ | $1.04 \times 10^4$ | $8.24 \times 10^4$ | 0.83 |
| 8 | $1.20 \times 10^4$ | $0.12 \times 10^4$ | $8.36 \times 10^4$ | 0.84 |
| 9 | $1.91 \times 10^4$ | $1.42 \times 10^4$ | $9.78 \times 10^4$ | 0.99 |
| 10 | $2.04 \times 10^4$ | $0.13 \times 10^4$ | $9.91 \times 10^4$ | 1.00 |

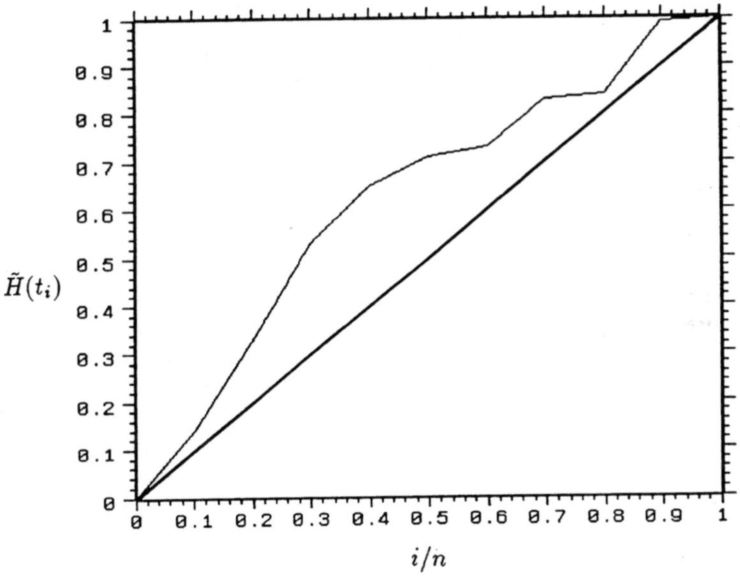


This graph indicates a mild tendency toward an increasing failure rate over time.

---

## *3.3.4 Goodness-Of-Fit Tests*

The goodness-of-fit tests provide a parametric and more formal structure for model identification and selection. They do not, however, provide the advantages of a visual representation of the data as do the graphs. The goodness-of-fit tests are used to judge the agreement between observed data and a postulated probability distribution model. The two most commonly used tests are the **chi-square** and **Kolmogorov** tests. In Chapter 2, we elaborated on the mechanics of these two tests. In this section, we emphasize some of the key aspects of these two tests as they apply to model identification and selection.

In a goodness-of-fit test, a hypothesis is postulated and, through a statistical procedure, the hypothesis is tested to determine whether the evidence supports the truth of the hypothesis. In the model selection process, the hypothesis takes the form of a statement in which a well-defined or known pdf is believed to characterize the time-to-failure distribution of a component. To support this hypothesis, a goodness-of-fit test is used along with a set of observed failure data.

A **significance level** is associated with each test. This level determines the stringency of the criteria for rejecting or not rejecting the hypothesis. In essence, the significance level controls the probability of error in deciding to reject a hypothesis. For example, when a significance level of a test is 0.05, there is 1 chance in 20 that rejection of the hypothesis (the failure model) will be incorrect. Accordingly, at a 0.01 significance level, there is 1 chance in 100 that the rejection will be incorrect. Thus, a significance level of 0.01 puts a more stringent requirement on the hypothesis test.

## 3.4 Life Testing

Life testing is used to obtain failure data for reliability estimation methods. In life testing, a sample of components from a hypothesized population of such components is placed under test using the environment in which the components are expected to function, and their times to failure are recorded. In general, two major types of tests are conducted; the first **with replacement** of the failed components, and the second **without replacement** of the failed components. The test with replacement is sometimes called **monitored**; as soon as a component fails, it is replaced by a new identical one. In the tests without replacement, a component is not replaced when it fails, and the test continues with the remaining components.

When $n$ components are placed under life test, whether with replacement or not, it is sometimes necessary, due to the long life of certain components, to terminate the test and perform the reliability analysis based on the observed data up to the point of termination. There are two types of possible life tests with termination. The first is **time terminated** (Type I), and the second is **failure termi-**

**nated** (Type II). In the time-terminated life test, $n$ units are placed under test and the test is terminated after a predetermined number of clock hours have elapsed. The number of components failed during the test time and the corresponding time to failure of each component are recorded. In the failure-terminated life tests, $n$ units are placed under test and the test is terminated when a predetermined number of component failures have occurred. The time to failure of each failed component and the time that the last failure occurred are recorded.

For each Type I and Type II life test, there are two possibilities: with replacement or without replacement. Therefore, four types of life test experiments are possible. Each of these are discussed in more detail in the remainder of this section.

### *3.4.1 Type I Life Test with Replacement*

Suppose $n$ components are placed under test with replacement (i.e., monitored), and the test is terminated after a specified time $t_0$. The accumulated component hours or other time units of test $T$, by both failed and nonfailed components, is given by:

$$T = nt_0. \tag{3.38}$$

Equation (3.38) indicates that at each time, from the beginning of the test up to time $t_0$, exactly $n$ components have been under test. Accordingly, if $R$ failures are observed up to time $t_0$, then assuming an exponential distribution, the point estimate of the failure rate of the component, according to Example 2.26, is

$$\hat{\lambda} = \frac{R}{T}. \tag{3.39}$$

Or, the mean-time-to-failure estimate is

$$\widehat{MTTF} = \frac{T}{R}. \tag{3.40}$$

Also, the number of units in the test $n'$ is

$$n' = n + R. \tag{3.41}$$

### *3.4.2 Type I Life Test without Replacement*

Suppose $n$ components are placed under test without replacement, and the test is terminated after a prespecified time $t_0$ during which $R$ failures have occurred. The accumulated component hours of the failed and survived components, $T$, can be obtained by

$$T = \sum_{i=1}^{R} T_i + (n - R)t_0, \tag{3.42}$$

where $\sum_{i=1}^{R} T_i$ represents the accumulated component hours of the $R$ failed components ($R$ is random here), and $(n - R)t_0$ represents the accumulated component hours of the surviving components. The failure rate and $MTTF$ estimates can be obtained from (3.39) and (3.40), respectively. Since no replacement has taken place, the total number of components in the test is $n' = n$.

*3.4.3 Type II Life Test with Replacement*

Consider a situation where $n$ components are being tested with replacement (i.e., monitored), and a component is replaced with an identical component as soon as it fails (except for the last failure). If the test is terminated after a time $T_r$ when the $r^{th}$ failure has occurred (i.e., $r$ is specified but $T_r$ is random), then the total accumulated component hours of operation, $T$, due to both failed and nonfailed components, is given by

$$T = nT_r. \tag{3.43}$$

Note that $T_r$, unlike $t_0$, is a variable in this case. If the time to failure follows an exponential distribution from an estimate of $\hat{\lambda}$, similarly $\lambda$ is estimated from

$$\hat{\lambda} = \frac{r}{T}, \tag{3.44}$$

and accordingly,

$$\widehat{MTTF} = \frac{T}{r}. \tag{3.45}$$

The total number of units in this test, $n'$, is determined by

$$n' = n + r - 1, \tag{3.46}$$

where $(r - 1)$ is the total number of failed and replaced components. All failed components are replaced except the last one, because the test is terminated when the last component(i.e., the $r^{th}$) fails.

*3.4.4 Type II Life Test without Replacement*

Consider a situation where $n$ components are being tested without replacement, i.e., when a failure occurs, the failed component is not replaced by a new one. The test is terminated at time $T_r$ when the $r^{th}$ failure has occurred (i.e., $r$ is specified but $T_r$ is random). The accumulated component hours of operation of both failed and nonfailed components is obtained from

$$T = \sum_{i=1}^{r} T_i + (n-r)T_r, \tag{3.47}$$

where $\sum_{i=1}^{r} T_i$ is the accumulated time contribution from the failed components, and $(n-r)T_r$ is the accumulated time contribution from the survived components. Accordingly, the failure rate and $MTTF$ estimates for an exponentially distributed time to failure can be obtained from (3.44) and (3.45). It should also be noted that the total number of units in this test is

$$n' = n, \tag{3.48}$$

since no components are being replaced.

---

*Example 3.12*

Ten light bulbs are placed under life test. The test is terminated at $t_0 = 850$ hours. Eight components fail before 850 hours have elapsed. Determine the accumulated component hours and an estimate of the failure rate and $MTTF$ for the following situation:

a) The components are replaced when they fail.
b) The components are not replaced when they fail.
c) Repeat a) and b), assuming the test is terminated when the eighth component fails.

The failure times obtained are: 183, 318, 412, 432, 553, 680, 689, and 748.

*Solution:*

a) Type I test: using (3.38), $T = 10 \times 850 = 8500$ component hours.
From (3.39), $\hat{\lambda} = 8/8500 = 9.4 \times 10^{-4}\text{hr}^{-1}$, or
From (3.40), $M\hat{T}TF = 1062.5$ hrs.

b) Type I test: using (3.42), $\sum_{i=1}^{8} T_i = 4015$, $(n-r)t_0 = (10-8)850 = 1700$. Thus, $T = 4015 + 1700 = 5715$ component hours.
From (3.39), $\hat{\lambda} = 8/5715 = 1.4 \times 10^{-3}\text{hr}^{-1}$.
From (3.40), $M\hat{T}TF = 714.4$ hrs.

c) Type II test:
1. Here, $T_r$ is the time of the eighth failure, which is 748.

Using (3.43), $T = 10 \times 748 = 7480$ component hours.
From (3.44), $\hat{\lambda} = 8/7480 = 1.1 \times 10^{-3}\text{hr}^{-1}$.
From (3.45), $\hat{MTTF} = 935$ hrs.

2. Using (3.47), $\sum_{i=1}^{8} T_i = 4015$,
$(n - r)T_r = (10 - 8)748 = 1496$. Thus,
$T = 4015 + 1496 = 5511$component hours.
From (3.44), $\hat{\lambda} = 8/5511 = 1.5 \times 10^{-3}$ hr$^{-1}$.
From (3.45), $\hat{MTTF} = 688.8$ hrs.

A simple comparison of results indicates that although the same set of data are used, the effect of the type of the test and the effect of the replacement of the failed units can have a reasonable effect on the estimated parameters.

---

## 3.5 Parameter Estimation

This section deals with statistical methods for estimating model parameters, such as $\lambda$ of the exponential pdf, $\mu$ and $\sigma$ of the normal and lognormal pdfs, p of the binomial distribution, and $\alpha$ and $\beta$ of the Weibull pdf. The objective is to find a **point estimate** and a **confidence interval** for the parameters of interest. We briefly discussed this topic in Chapter 2 for estimating parameters and confidence intervals associated with a normal distribution. In this section we expand the discussion to include estimation of parameters of other distributions useful to reliability analysis.

It is important to realize why we need to consider confidence intervals in the estimation process. In essence, this need stems from the fact that we have a limited amount of information (e.g., times to failure), and thus we cannot state our estimation with certainty. Therefore, the confidence interval is highly influenced by the amount of data available. Of course other factors, such as diversity in the sources of data and accuracy of the selected model and the data sources, also influence the state of our uncertainty regarding the estimated parameters. In discussing the goodness-of-fit tests, we dealt with uncertainty due to the accuracy of the model by using the concept of **levels of significance**. However, uncertainty due to diversity and accuracy of the data sources is a much more difficult issue to deal with this context.

The methods of parameter estimation discussed in this section are more formal and accurate methods of determining distribution parameters than the estimation methods described previously (such as the plotting methods).

### 3.5.1 Parameter Estimation When Failures Occur in Time

When failures occur in time for a component, the most common distributions used are exponential and lognormal pdfs. The exponential distribution is used to model component time to failure when failures have a constant "arrival" rate. This concept, known as the shock model, was discussed in Section 3.2.1. When the time to failure of a component is believed to occur at a constant rate, the exponential model is reasonable and the parameter estimation should proceed. The lognormal distribution is used to model the occurrences of events in time that may vary significantly (sometimes even by orders of magnitude). For example, time to failure of certain components as well as repair time needed for a component can be modeled by a lognormal distribution.

**Exponential Distribution**

In Example 2.26, we discussed the maximum likelihood estimator for the parameter $\lambda$ (failure rate, or constant hazard rate) of the exponential distribution. This point estimator is $\hat{\lambda} = r/T$, where $r$ is the number of failures observed and $T$ is the total accumulated test time. Epstin (1960) has shown that if the time to failure is exponentially distributed with parameter $\lambda$, the quantity $2r\lambda/\hat{\lambda} = 2\lambda T$ has a chi-square distribution with $2r$ degrees of freedom for the Type II (failure-terminated test)and $(2r+2)$ degrees of freedom for the Type I (time -terminated) test. Thus, one can determine confidence intervals, e.g., $1-\alpha$, associated with $\hat{\lambda}$ from the chi-square distribution from

$$1-\alpha = \sum_{\substack{k=r(\text{for Type I})\\ k=r+1(\text{for Type II})}}^{\infty} \frac{(\lambda t)^k \exp(-\lambda t)}{k!} = \int_0^{2\lambda t} f(\chi^2)d\chi^2. \qquad (3.49)$$

Consequently, for the **two-sided confidence interval** for the $1-\alpha/2$ and $\alpha/2$ confidence levels of the chi-square distribution,

$$\Pr\left[\chi^2_{\alpha/2}(2r) \le \frac{2r\lambda}{\hat{\lambda}} \le \chi^2_{1-\alpha/2}(2r)\right] = 1-\alpha. \qquad (3.50)$$

By rearranging and using $\hat{\lambda} = r/T$,

$$\Pr\left[\frac{\chi^2_{\alpha/2}(2r)}{2T} \le \lambda \le \frac{\chi^2_{1-\alpha/2}(2r)}{2T}\right] = 1-\alpha. \qquad (3.51)$$

Equation (3.51) shows the **two-sided confidence interval** for the true value of $\lambda$. Another alternative to (3.50) is the **one-sided confidence interval**,

$$\Pr\left[0 \le \lambda \le \frac{\chi^2_{1-\alpha}(2r)}{2T}\right] = 1 - \alpha. \tag{3.52}$$

Accordingly, confidence intervals of $MTTF$ and $R(t)$ at $t = t_0$ can also be obtained for both one-sided and two-sided confidence intervals from (3.51) and (3.52). The results are summarized in Table 3.2.

*Table 3.2* $100(1-\alpha)\%$ *Confidence Limits on* $\lambda$, $MTTF$, *and* $R(t_0)$

| Parameter | Type I (Time Terminated Test) | | | |
|---|---|---|---|---|
| | One-sided confidence limits | | Two-sided confidence limits | |
| | Lower limit | Upper limit | Lower limit | Upper limit |
| $\lambda$ | 0 | $\frac{\chi^2_{1-\alpha}(2r+2)}{2T}$ | $\frac{\chi^2_{\alpha/2}(2r)}{2T}$ | $\frac{\chi^2_{1-\alpha/2}(2r+2)}{2T}$ |
| $MTTF$ | $\frac{2T}{\chi^2_{1-\alpha}(2r+2)}$ | $\infty$ | $\frac{2T}{\chi^2_{1-\alpha/2}(2r+2)}$ | $\frac{2T}{\chi^2_{\alpha/2}(2r)}$ |
| $R(T_0)$ | $\exp\left[-\frac{\chi^2_{1-\alpha}(2r+2)}{2T}t_0\right]$ | 1 | $\exp\left[-\frac{\chi^2_{1-\alpha/2}(2r+2)}{2T}t_0\right]$ | $\exp\left[-\frac{\chi^2_{\alpha/2}(2r)}{2T}t_0\right]$ |

| Parameter | Type II (Failure Terminated Test) | | | |
|---|---|---|---|---|
| | One-sided confidence limits | | Two-sided confidence limits | |
| | Lower limit | Upper limit | Lower limit | Upper limit |
| $\lambda$ | 0 | $\frac{\chi^2_{1-\alpha}(2r)}{2T}$ | $\frac{\chi^2_{\alpha/2}(2r)}{2T}$ | $\frac{\chi^2_{1-\alpha/2}(2r)}{2T}$ |
| $MTTF$ | $\frac{2T}{\chi^2_{1-\alpha}(2r)}$ | $\infty$ | $\frac{2T}{\chi^2_{1-\alpha/2}(2r)}$ | $\frac{2T}{\chi^2_{\alpha/2}(2r)}$ |
| $R(T_0)$ | $\exp\left[-\frac{\chi^2_{1-\alpha}(2r)}{2T}t_0\right]$ | 1 | $\exp\left[-\frac{\chi^2_{1-\alpha/2}(2r)}{2T}t_0\right]$ | $\exp\left[-\frac{\chi^2_{\alpha/2}(2r)}{2T}t_0\right]$ |

It is possible to show that the upper confidence limit of $\lambda$ for the Type I (time-terminated) test is obtained from a chi-square distribution with $(2r+2)$ degrees of freedom. The lower confidence limit is obtained from the chi-square distribution with $2r$ degrees of freedom. Thus for Type I tests, the two-sided confidence interval is

$$\Pr\left[\frac{\chi^2_{\alpha/2}(2r)}{2T} \leq \lambda \leq \frac{\chi^2_{1-\alpha/2}(2r+2)}{2T}\right] = 1-\alpha. \tag{3.53}$$

The one-sided confidence interval is

$$\Pr\left[0 \leq \lambda \leq \frac{\chi^2_{1-\alpha}(2r+2)}{2T}\right] = 1-\alpha. \tag{3.54}$$

It should be emphasized here that (3.50) through (3.54) apply only when the failure rate is constant. Otherwise, a Weibull model or other appropriate distribution should be used. This subject is discussed in more detail in Chapter 5.

If no failure is observed during a test, $\hat{\lambda} = 0$ or $MTTF = \infty$. This cannot realistically be true; since we may have had a small or restricted test. Had the test been continued, it is highly likely that eventually a failure would be observed. An upper level estimate for both one-sided and two-sided confidence limits can be obtained with $r = 0$. However, the lower limit for the two-sided confidence limit cannot be obtained with $r = 0$. It is possible to relax this limitation by conservatively assuming that a failure occurs in the very next instant. Then $r = 1$ can be used to evaluate the lower two-sided confidence limit. This conservative modification, although sometimes used to allow a complete statistical analysis, lacks firm statistical basis. Welker and Lipow (1974) have shown methods to determine approximate point estimates in these cases.

---

*Example 3.13*

Twenty-five units are subjected to a reliability test that lasts 500 hours. In this test, eight failures occur at 75, 115, 192, 258, 312, 389, 410, and 496 hours. The failed units are replaced. Find $\hat{\lambda}$, one-sided and two- sided confidence limits on $\lambda$ and $MTTF$ at the 90% confidence level; one-sided and two-sided 90% confidence limits on reliability at $t_0 = 1000$ hours.

*Solution:*

This is a Type I test. The accumulated time $T$ is obtained from (3.38)

$$T = 25 \times 500 = 12,500 \text{ component hours.}$$

From (3.39), $\hat{\lambda} = 8/12,500 = 6.4 \times 10^{-4}\ \text{hr}^{-1}$.

One-sided confidence limits on $\lambda$ are

$$0 \leq \lambda \leq \frac{\chi^2(2 \times 8 + 2)}{2 \times 12,500}.$$

From Table A.3,

$$\chi^2_{.9}(18) = 25.99, \quad 0 \leq \lambda \leq 1.04 \times 10^{-3}.$$

Two-sided confidence limits on $\lambda$ are

$$\frac{\chi^2_{.05}(2 \times 8)}{2 \times 12,500} \leq \lambda \leq \frac{\chi^2_{.95}(2 \times 8 + 2)}{2 \times 12,500}.$$

From Table (A.3),

$$\chi^2_{.05}(16) = 7.96, \quad \text{and} \quad \chi^2_{.95}(18) = 28.87.$$

Thus,

$$3.18 \times 10^{-4} \leq \lambda \leq 1.15 \times 10^{-3}.$$

One-sided 90% confidence limits on $R(1000)$ are

$$\exp(-1.04 \times 10^{-3} \times 1000) \leq R(1000) \leq 1, \quad \text{or} \quad 0.35 \leq R(1000) \leq 1.$$

Two-sided 90% confidence limits on $R(t)$ are

$$\exp(-1.15 \times 10^{-3} \times 1000) \leq R(1000) \leq \exp(-3.18 \times 10^{-4} \times 1000),$$

or

$$0.32 \leq R(1000) \leq 0.73.$$

---

## Lognormal Distribution

A lognormal distribution is usually used to represent the occurrence of events in time. For example, a r.v. representing the length of time required for repair of hardware follows a lognormal distribution. Because the lognormal distribution has two parameters, parameter estimation poses a more challenging problem than for the exponential distribution. It is easy to prove through the maximum likelihood method that point estimators for the two parameters of the lognormal distribution can be obtained from

$$\hat{\mu}_t = \sum_{i=1}^{n} \frac{\ln(t_i)}{n}, \quad \text{and} \tag{3.55}$$

$$\hat{\sigma}_t^2 = \frac{\sum_{i=1}^{n}[\ln(t_i) - \hat{\mu}]^2}{n-1}. \tag{3.56}$$

The confidence interval for $\mu_t$ is obtained by using (2.83)

$$\Pr\left[\hat{\mu}_t - \frac{(\hat{\sigma}_t)t_{\alpha/2}}{\sqrt{n}} \leq \mu_t \leq \hat{\mu}_t + \frac{(\hat{\sigma}_t)t_{\alpha/2},}{\sqrt{n}}\right] = 1 - \alpha, \tag{3.57}$$

Similarly, the confidence interval on $\sigma_t$ is obtained by using (2.84).

$$\Pr\left[\frac{(n-1)\hat{\sigma}_t^2}{\chi^2_{(1-\alpha/2)}(n-1)} \leq \sigma_t^2 \leq \frac{(n-1)\hat{\sigma}_t^2}{\chi^2_{(\alpha/2)}(n-1)}\right] = 1 - \alpha. \tag{3.58}$$

## Weibull Distribution

A Weibull distribution can be used for data believed to have an increasing, decreasing, or constant rate of failure. This distribution is a two-parameter distribution and estimation of the parameters is rather involved. It can be shown that, under the situation that all $n$ units under test or observation have failed, the maximum likelihood estimates of $\beta$ and $\alpha$ parameters of the Weibull distribution can be obtained from

$$\frac{\sum_{i=1}^{n}(t_i)^{\hat{\beta}}\ln t_i}{\sum_{i=1}^{n}(t_i)^{\hat{\beta}}} - \frac{1}{\hat{\beta}} = \frac{1}{n}\sum_{i=1}^{n}\ln t_i, \quad \text{and} \tag{3.59}$$

$$\hat{\alpha} = \left(\frac{\sum_{i=1}^{n}(t_i)^{\hat{\beta}}}{n}\right)^{1/\hat{\beta}}. \tag{3.60}$$

The solution of (3.59) is not trivial and may require an iterative process to obtain $\hat{\beta}$. Estimation of the confidence intervals of $\beta$ and $\alpha$ is very involved. Readers are referred to Bain (1978) and Mann et.al (1974). In Chapter 5 we will discuss another form of estimating the Weibull parameters.

---

*Example 3.14*

Using the data in Example 3.8, obtain the maximum likelihood estimators for the parameters of a Weibull distribution.

*Solution:*

By trial and error, (3.59) is solved for $\hat{\beta}$ and the value of $\hat{\beta} = 2.54$. From (3.60), $\hat{\alpha} = 356.9$. Comparison of these results with the plot is reasonable, but it points out the approximate nature of these data analysis methods, and emphasizes the need for using more than one of the models discussed.

---

If a lot of data are available, the analyst should first fit the data into a distribution that best represents the data. The analyst should then proceed with the estimation of the distribution's parameters. Therefore, the exponential, lognormal, or other distribution should only be selected when the adequacy of a distribution fit can be justified. For example, when we have only five sample points for the times required for maintenance, one can theoretically fit these points to several types of distributions. However, we can assume that since the data occur in multiples of time, they can be adequately represented by a lognormal distribution. An accurate representation can only be determined if there are enough data to test the adequacy of the selected model.

### *3.5.2 Parameter Estimation When Failures Occur on Demand*

When the data are in the form of $X$ failures in $n$ trials, there is no time relationship, and the binomial distribution best represents the data. This is often the situation for standby components (or systems). (e.g., a redundant pump that is demanded for operation $n$ times in a given period of test or observation.) In a binomial distribution, the only parameter of interest is $p$. The best estimator for $p$ can be expressed as

$$\hat{p} = \frac{X}{n}. \tag{3.61}$$

The confidence limits on $p$ can be calculated from

$$p_l = \{1 + (n - X + 1)X^{-1}F_{(1-\alpha/2)}[2n - 2X + 2; 2X]\}^{-1}, \tag{3.62}$$

$$p_u = \{1 + (n - X)\{(X + 1)F_{(1-\alpha/2)}[2X + 2; 2n - 2X]\}^{-1}\}^{-1}, \tag{3.63}$$

where $F_{(1-\alpha/2)}(f_1; f_2)$ is the $F$ statistic of the $F$ distribution with $f_1$ degrees of freedom for the numerator, and $f_2$ degrees of freedom for the denominator. Table A.5 shows the values for $F$. When $n$ is large and $X$ is small (as a rule of thumb, when $X \leq n/10$), (3.53) can be used as an approximation for (3.62) and (3.63). In this case, $X$ is represented by $r$ and $n$ is represented by $T$ in (3.53).

---

*Example 3.15*

An emergency pump in a nuclear power plant is in standby mode. There have been 563 start tests for the pump and only 3 failures have been observed. No degradation in the pump's physical characteristics or operating environment is observed. Determine the estimation of the probability of failure per demand associated with the 95% confidence interval.

*Solution:*

$n = 563, \quad X = 3.$
Using (3.61),
$\hat{p} = \frac{3}{563} = 0.0053.$
$p_l = \{1 + (561)\frac{1}{3}F_{0.975}[1122; 6]\}^{-1} = 0.0011.$
From Table A.5, $F_{0.975}[1122; 6] = 4.85.$
Similarly, $p_u = 4/(4 + 560/F_{0.5}[8; 1120])$,
$p_u = \{1 + (560)\{4F_{0.975}[8; 1120]\}^{-1}\}^{-1} = 0.0255.$
$F_{0.975}[8; 1120] = 3.67,$
Therefore, $p_r[0.0011 \leq p \leq 0.0255] = 95\%.$

---

*Example 3.16*

In a commercial nuclear plant, the performance of the emergency diesel generators has been observed for about 5 years. During this time, there have been 35 real demands with 4 observed failures. Determine the 90% confidence limits and point estimate for the probability of failure per demand. What would the error be if we used (3.53) instead of (3.62) and (3.63)?

*Solution:*

Here,

$$X = 4 \text{ and } n = 35.$$

Using (3.61),

$$\hat{p} = \frac{4}{35} = 0.114.$$

To find lower and upper limits, we use (3.62) and (3.63). Thus,

$$p_l = \{1 + 8F_{0.95}[64; 8]\}^{-1} = \{1 + 8 \times 3\}^{-1} = 0.040,$$

$$p_u = \{1 + 31\{5F_{0.95}[10; 62]\}^{-1}\}^{-1} = \{1 + 3.116\}^{-1} = 0.243.$$

If we used (3.53),

$$p_l = \frac{\chi^2_{0.05}(8)}{2 \times 35} = \frac{2.733}{70} = 0.039,$$

$$p_u = \frac{\chi^2_{0.95}(10)}{2 \times 35} = \frac{18.31}{70} = 0.262.$$

The error due to this approximation is

$$\text{lower limit error} = \frac{|0.04 - 0.039|}{0.04} \times 100 = 2.5\%,$$

$$\text{upper limit error} = \frac{|0.243 - 0.262|}{0.243} \times 100 = 7.8\%.$$

Notice that this is not a negligible error, and (3.53) should not be used. Since $X > n/10$, equation (3.53) is clearly not a good approximation.

## 3.6 Bayesian Parameter Estimation

In Section 3.5, we discussed the importance of quantifying uncertainties associated with the estimates of various distribution parameters. In the preceding sections, we discussed formal statistical methods for quantifying the uncertainties. Namely, we discussed the concept of confidence intervals, which in essence is a statistical treatment of available information. It is evident that the more data available, the more confident and accurate the statistical treatments For this reason, the statistical approach is sometimes called the **frequentist** method of treatment. In Bayesian term, the parameters of interest are treated as random variables, the exact values of which are unknown. Thus a pdf can be assigned to represent the parameter; the mean (or for some cases the median) of the pdf can be used as an estimate of the parameter. Accordingly, the uncertainty associated with the data can be represented by probability bounds obtained from the parameter pdf (i.e., the probability that the r.v. falls between two bounds). Clearly this is a **probabilistic** treatment, and the role of failure data is not as profound as in the statistical treatment. The pdf of a parameter in Bayesian terms can be obtained from a **prior** or **posterior** pdf. In practice, however, the prior pdf is used to represent the relevant and prior knowledge, including subjective expert judgment regarding the characteristics of the parameter and its distribution. When the prior knowledge is combined with other relevant information (often statistics obtained from tests and observations), a posterior distribution is obtained, which better represents the parameter of interest. Since the selection of the prior and the determination of the posterior often involve subjective judgments, the Bayesian estimation is sometimes called the **subjectivist** approach to parameter estimation.

The main concept of the Bayesian estimation was discussed in Chapter 2. In essence, the Bayes theorem can be written in one of three forms: discrete, continuous, or mixed. Martz and Waller (1982) have significantly elaborated on the concept of Bayesian techniques in parameter estimation as well as for other applications. The discrete form was discussed in Section 2.2.3. The continuous and mixed forms are the common forms was used for parameter estimation.

### Continuous Form

Let $g(\theta)$ be a pdf representing our prior state of knowledge regarding the distribution of parameter $\theta$. If the available data gained through life test, field data, or other sources are represented by a r.v. $X$, then $X$ can be represented by a pdf $f(X|\theta)$, called the **likelihood function**. The posterior $g(\theta|X)$ is then obtained from

$$g(\theta|X) = \frac{f(X|\theta)g(\theta)}{\int_{-\infty}^{\infty} f(X|\tau)g(\tau)d\tau}. \tag{3.64}$$

### Mixed Form

If the available data are such that they can best be described by a discrete pdf $\Pr(X|\theta)$ to form the likelihood function, then a mixed form of the Bayes' theorem can be used. In this form, the prior $g(\theta)$ is a continuous distribution that describes our prior state of knowledge. Thus, the posterior $g(\theta|X)$ is obtained from

$$g(\theta|X) = \frac{\Pr(X|\theta)g(\theta)}{\int_{-\infty}^{\infty} \Pr(X|\tau)g(\tau)d\tau}. \tag{3.65}$$

Often the integration of the denominators of (3.64) and (3.65) require numerical integration. Occasionally, close form calculations are possible.

### Parameter Mean and Median Estimates

Two important estimates for $\theta$, mean $\mu_\theta$ and median $m_\theta$, can be obtained from the posterior distribution

$$\mu_\theta = \int_{-\infty}^{\infty} \theta g(\theta|X)d\theta, \quad \text{and} \tag{3.66}$$

$$\int_{-\infty}^{m_\theta} g(\theta|X)d\theta = 0.5. \tag{3.67}$$

The values of $\mu_\theta$ and $m_\theta$ represent probabilistic point estimates of the parameter of interest.

### Parameter Interval Estimation

Similar to the mean estimation, the interval estimation is straightforward. Since the parameter of interest $\theta$ is a r.v., we can determine the probability that the parameter of interest falls between two bounds. If this probability is high enough (e.g., 90% and above), one can estimate, with this high certainty, the range of the parameter of interest. For a probability of $1 - \alpha$, the **symmetric posterior bounds** $\theta_L$ (for the lower bound) and $\theta_U$(for the upper bound) can be obtained from

$$\int_{\theta_L}^{\theta_U} g(\theta|X)d\theta = 1 - \alpha, \tag{3.68}$$

and the one sided bounds are

$$\int_{\theta_U}^{\infty} g(\theta|X)d\theta = \alpha/2, \tag{3.69}$$

and

$$\int_{-\infty}^{\theta_L} g(\theta|X)d\theta = \alpha/2.$$

Again, $\theta_L$ and $\theta_U$ are **probability bounds**, not **confidence intervals** used in the statistical estimation. A confidence interval in the statistical estimation can be interpreted as a range of values that includes a fixed but unknown quantity of the parameter of interest. The failure data observed in the statistical case show the variability in the sampling procedure and not the variability in the underlying parameter itself (as is the case in the Bayesian estimation technique). Thus, a procedure that allows more samples to be generated gives a tight confidence interval.

### Determining the Prior Distribution

A critical task in a Bayesian parameter estimation is the determination of the prior distribution. For this purpose, a source of data is needed for defining the prior pdf. Generally, two kinds of information may be available regarding a parameter of interest: generic information, including **expert knowledge** about the component for which the parameter is being estimated; and data obtained from tests or observations of similar components. For example, failure data obtained for a pump used in a nonnuclear application can be used as prior knowledge for a nuclear application. If one wishes to update the two kinds of prior with specific information regarding the parameter of interest (e.g., data obtained from a life test), then such data would form the likelihood function, and according to the Bayes' equation, a posterior pdf can be obtained. Fitting a distribution to generic information is a difficult task and often involves subjective judgments.

Three types of prior pdf exist: **noninformative**, **natural conjugate**, and **empirical**. Noninformative priors are those that make minimal or no assumptions (and do not provide prior information) about the parameter. Therefore, certain values of parameters are not favored over other values. Uniform distribution is clearly the most appropriate noninformative prior. In this case, any value of the parameter is equally likely. Since no information is provided, the posterior would be highly influenced by the likelihood function. Therefore, it yields results very similar to those of the statistical estimation. The statistical approach can thus be considered as a special case of the Bayesian approach. As a rule of thumb, noninformative priors should not be used when any prior information exists.

Natural conjugate priors make small assumptions about prior information and are more practical than noninformative priors. A natural conjugate can be a prior for a given likelihood, and the posterior would be the same kind of distribution. For example, assume a likelihood is binomially distributed with a parameter $p$. Then a prior represented by a beta distribution with the same parameter $p$ would yield a posterior distribution that is also a beta distribution. The advantage is that the posterior would be known in closed form. Another useful natural conjugate distribution is the exponential distribution. These natural conjugate priors are often flexible enough to model a wide range of prior information. The major disadvantage of natural conjugate priors is that they cannot be used if the form of the prior distribution is fixed and known.

Empirical priors, as the name indicates, involve cases where there is no theoretical justification for using a noninformative or natural conjugate prior. In these cases, the prior information (e.g., generic data sources or old data) influences the use of a prior distribution. The priors in these cases are often obtained by probability plotting, moment matching, goodness-of-fit tests, or a combination of these methods.

---

*Example 3.17*

Suppose a diesel generator failure to start on demand has a probability $p$. The prior data on the performance of the similar diesel are obtained from field data, and $p$ is assumed to follow a lognormal distribution with known parameters $\mu_t$ and $\sigma_t$ such that $\mu_t = -3.22$ and $\sigma_t^2 = 0.51$. A limited test of the diesel generators of interest shows that 5 failures are observed in 582 demands. Calculate a Bayesian point estimation of $p$ (mean and median) and the 90th percentiles of $p$. Compare these results with corresponding values for the prior distribution.

*Solution:*

Since we are dealing with a demand failure, a binomial distribution best represents the observed data. This is a mixed continuous and discrete situation, and (3.65) applies. The likelihood function is

$$\Pr(X|p) = \binom{582}{8} p^8(1-p)^{574},$$

and the prior is

$$f(p) = \frac{1}{\sigma_t p\sqrt{2\pi}} \exp\left[-\frac{1}{2}\left(\frac{\ln(p) - \mu_t^2}{\sigma_t}\right)\right], \qquad p > 0.$$

Since $0 \le p \le 1$, the posterior is

$$f(p|X) = \frac{p^7(1-p)^{574} \exp\left[-1/2\left(\frac{\ln(p)-\mu_t}{\sigma_t}\right)^2\right]}{\int_0^1 \tau^7(1-\tau)^{574} \exp\left[-1/2\left[\frac{\ln(\tau)-\mu_t}{\sigma_t}\right]^2 d\tau\right]}$$

By using a numerical integration, the denominator can be calculated. Thus,

$$f(p|X) = 7.62 \times 10^{18} p^7(1-p)^{574} \exp -\frac{1}{2}\left[\frac{\ln(p) - \mu_t}{\sigma_t}\right]^2.$$

From (3.66) and (3.67), $\mu_p = 0.0512$ and $m_p = 0.0398$. From (3.68) and (3.69), $p_l = 0.0069$ and $p_u = 0.0244$. The point estimate of the actual data using the classical inference method is

$$\hat{p} = \frac{8}{582} = 0.0137.$$

The comparison of the prior and posterior is given below.

| | Prior | Posterior |
|---|---|---|
| Mean | 0.0512 | 0.0168 |
| Median | 0.0398 | 0.0165 |
| 5th Percentile | 0.0124 | 0.0094 |
| 95th Percentile | 0.1280 | 0.0253 |

Example 3.17 shows that the smaller posterior mean ($\mu$) for the observed data result in a smaller value for $\hat{p}$. This example also demonstrates that the Bayesian estimation technique can rely on both prior and observed information. When there is not a lot observed information, the Bayesian method relies more on prior information, which incidently is obtained through expert judgment. This reliance have always been a source of controversy associated with the Bayesian method.

## 3.7 Methods of Generic Failure Rate Determination

Due to the lack of observed data, component reliability determination may require use of generic failure data adjusted for the various factors that influence the failure rate for the component under analysis. Generally, these factors are:

1. **Environmental Factors** – These factors affect the failure rate due to extreme mechanical, electrical, nuclear, and chemical environments. For example, a high-vibration environment, would lead to high stresses that promote failure of components.
2. **Design Factors** – These factors affect the failure rate due to the quality of material used and workmanship, material composition, functional requirements, geometry, and complexity.
3. **Operating Factors** – These factors affect the failure rate due to the applied stresses resulting from operation, testing, repair and maintenance practices, etc.

To a lesser extent the **Age Factor** is used to correct for early and the wear-out periods, and **Original Factor** is used to correct for the accuracy of the data source (generic data). For example, obtaining data from observed failure records as opposed to expert judgment may affect the failure rate dependability.

Accordingly,

$$\lambda_a = \lambda_g K_E K_D K_O \ldots, \tag{3.70}$$

where $\lambda_a$ is the actual failure rate and $\lambda_g$ is the generic base failure rate, and $K_E, K_D,$ and $K_O$ are correction factors for the environment, design, and operation, respectively. It is possible to subdivide each of the correction factors to their contributing subfunctions accordingly. For example, $K_E = f(k_a, k_b, ...)$, when $k_a$ and $k_b$ are factors such as vibration level, moisture, and ph level. These factors may be different for different types of components. For example, in Fig. 3.9, $k_E$ represents a high-temperature environment for a capacitor.

This concept is used in the procedure specified in government-contract for determining the actual failure rate of electronic components. The procedure is summarized in MIL-HDBK-217 (1991). In this procedure, a base failure rate of the component is obtained from a table, and then they are multiplied by the applicable adjusting factors for each type of component. For example, the actual failure rate of a tantalum electrolytic capacitor is given by

$$\lambda_P = \lambda_b(\Pi_E.\Pi_{SR}.\Pi_Q.\Pi_{CV}), \tag{3.71}$$

where $\lambda_P$ is the actual part failure rate and $\lambda_b$ is the base (or generic) failure rate, and the $\Pi$ factors are adjusting factors for the environment, series resistance, quality and capacitance factors. Values of $\lambda_b$ and the factors are given in MIL-HDBK-217 for many types of electronic components. Generally, $\lambda_b$ is obtained from an empirical model called the Arrhenius model

$$\lambda_b = K \exp(-E/kT), \tag{3.72}$$

where,

$E =$ activation energy for the process,
$k =$ $1.38 \times 10^{-23} J.K^{-1}$,
$T =$ absolute temperature ($^o$K),
$K =$ constant.

The Arhenius model forms the basis for a large portion of electronic components described in MIL-HDBK-217. However, care must be applied in using this data base, especially because the data in this handbook are derived from repairable systems (and hence, apply to such systems). Also, application of the various adjusting factors can drastically affect the actual failure rates. Therefore, proper care must be applied to ensure correct use of the factors and to verify the adequacy of the factors suggested. Also the appropriateness of the Arhenius model has been debated many times in the literature.

For other types of components, many different generic sources of data are available. Among them are IEEE-500 (1984), Nuclear Power Plant, and Probability Risk Assessment (PRA) data sources.

For example, Table B.1 (in Appendix B) shows a set of data obtained from NUREG/CR-4550 (1990).

## BIBLIOGRAPHY


1. Bain, L.J. (1978). *Statistical Analysis of Reliability and Life-Testing Models: Theory and Methods*, Marcel Dekker, New York.
2. Barlow, R.E. (1978). Analysis of Retrospective Failure Data using Computer Graphics, *Proceedings of the 1978 Annual Reliability and Maintainability Symposium*, pp. 113-116.
3. Barlow,R.E. and R.A. Campo (1975). *Total Time on Test Processes and Applications to Failure Data Analysis, Reliability and Fault Tree Analysis*, Eds. Barlow, Fussell and Singpurwalla, SIAM, Philadelphia, pp. 451–481.
4 Blom, (1958). *Statistical Estimates and Transformed Beta Variables*, John Wiley and Son, New York.
5. Castillo, E. (1988). *Extreme Value Theory in Engineering*, Academy Press, San Diego.
6. Davis, (1952). An Analysis of Some Failure Data, *J. Am. Stat. Assoc.*, 47, pp. 113–150.
7. Epstein, B. (1960). "Estimation from Life Test Data," Technometrics, 2, 447.
8. Gumble, E.J. (1958). *Statistics of Extremes*, Columbia University Press, New York.
9. Hahn,G.J. and S.S. Shapiro (1967). *Statistical Models in Engineering*, Wiley, New York.
10. IEEE Std. 500 (1984). Guide to the Collection and Presentation of Electrical, Electronic, Sensing Component and Mechanical Equipment Reliability Data for Nuclear Power Generating Stations, IEEE Standards, New York.
11. Johnson, N.L. and S. Kotz (1970). *Distribution in Statistics*, 4 Vols., John Wiley and Son, New York.
12. Kececioglu, D. (1991). *Reliability Engineering Handbook*, Prentice Hall, New Jersey.
13. Kimbal, (1960). On the Choice of Plotting Position on Reliability Paper, *J. Amer. Stat. Assoc.* **55**, pp. 546–560.
14. Mann, N. R.E. Schafer and N.D. Singpurwalla (1974). *Methods for Statistical Analysis of Reliability and Life Data*, John Wiley and Son, New York.
15. Martz, H.F. and R.A. Waller (1982). *Bayesian Reliability Analysis*, John Wiley and Son, New York.

16. Welker, E.L. and M. Lipow (1974). "Estimating The Exponential Failure Rate Dormant Data with No Failure Events," Proc. Rel. Maint. Symp.,Vol. 1 (2), p. 1194.
17. Nelson, W. (1979). How to Analyze Data with Simple Plots, ASQC Basic Reference in Quality Control: Statistical Techniques, *Am. Soc. Quality Control*, Milwaukee, WI.
18. Nelson, W. (1982). *Applied Life Data Analysis*, John Wiley and Son, New York.
19. NUREG/CR-4450 (1990). "Analysis of Core Damage Frequency From Internal Events," Vol. 1, U.S. Nuclear Regulatory Commission, Washington, DC.
20. O'Connor, P.D.T. (1991). *Practical Reliability Engineering*, 3rd ed., John Wiley and Son, New York.

## Exercises

3.1 For a gamma distribution with $\alpha = 400,\ \beta = 3.8$, determine $\Pr(x < 200)$?

3.2 Time to failure of a relay follows a Weibull distribution with $\alpha =$ 10years, $\beta = 0.5$.
Find the following:
a) Pr(failure after 1 year)
b) Pr (failure after 10 years)
c) The MTTF

3.3 The hazard rate of a device is $h(t) = 1/\sqrt{t}$. Find the following:
a) Probability density function
b) Reliability function
c) MTTF
c) Variance

3.4 Assume that 100 components are placed on test for 1,000 hours. From previous testing, we believe that the hazard rate is constant, and the MTTF = 500 hours. Estimate the number of components that will fail in the time interval of 100 to 200 hours. How many components will fail if it is known that 15 components failed in $T < 100$ hours?

3.5 Assume that t, the random variable that denotes life in hours of a specified component, has a cumulative density function (cdf) of

$$\begin{aligned} F(t) &= 1 - \frac{100}{t} \quad t \geq 100 \\ &= 0, \qquad t < 100. \end{aligned}$$

Determine the following:
a) pdf $f(t)$
b) Reliability function $R(t)$
c) MTTF

3.6 Show whether a uniform distribution represents an increasing failure rate, decreasing failure rate, or constant failure rate.

3.7 Consider the following Rayleigh distribution:

$$f(t) = \frac{2t}{\alpha^2} \exp\left[\frac{-t^2}{\alpha^2}\right] \qquad t \geq 0, \quad \alpha > 0.$$

a) Find the hazard rate h(t) corresponding to this distribution.
b) Find the Reliability function R(t).
c) Find the MTTF. (Notice: $\int_0^\infty \exp[-ax^2] = \frac{1}{2}\sqrt{\frac{\pi}{a}}$.)
d) For which part of the bathtub curve is this distribution adequate?

3.8 Due to the aging process, the instantaneous rate of failure of a nonrepairable item is increasing according to $\lambda(t) = \lambda\beta t^{\beta-1}$. Assume that the value of $\lambda$ and $\beta$ are estimated as $\hat{\beta} = 1.62$ and $\hat{\lambda} = 1.2 \times 10^{-5}$/hour. Determine the probability that the item will fail sometime between 100 and 200 hours. Assume an operation beginning immediately after the onset of aging.

3.9 Suppose r.v. X has the exponential pdf $f(x) = \lambda \exp[-\lambda x]$, for $t > 0$, and $f(x) = 0$, for $t \leq 0$. Find $\Pr(x > a + b | x > a)$ given $a, b > 0$.

3.10 The following time to failure data are found when 158 transformer units are put under test. Use a nonparametric method to find f(t), h(t), and R(t) of the transformers. No failures are observed prior to 1,750 hours.

| Age Range (hr.) | No. of Failures |
|---|---|
| 1750 - 2250 | 17 |
| 2250 - 2750 | 54 |
| 2750 - 3250 | 27 |
| 3250 - 3750 | 17 |
| 3750 - 4250 | 19 |
| 4250 - 4750 | 24 |

3.11 A test was run on 10 electric motors under high temperature. The test was run for 60 hours, during which six motors failed. The failures occurred at the following times: 37.5, 46.0, 48.0, 51.5, 53.0, and 54.5. We don't know whether an exponential distribution or a Weibull distribution model is better for representing these data. Use the plotting method as the main tool to discuss the appropriateness of these two models.

3.12 A test of 25 integrated circuits over 500 hours yields the following data:

| Time Interval | No. of Failures in Each Interval |
|---|---|
| 0-100 | 10 |
| 100-200 | 7 |
| 200-300 | 3 |
| 300-400 | 3 |
| 400-500 | 2 |

Find and plot the pdf, hazard rate, and reliability function for each interval of these integrated circuits using a nonparametric method.

3.13 Total test time of a device is 50,000 hours, The test is terminated after the first failure. If the p.d.f. of the device is known to be exponentially distributed, what is the probability that the estimated failure rate is not greater than $4.6 \times 10^{-5}(\text{hours}^{-1})$.

3.14 A manufacturer uses exponential distribution to model "cycle to failure" of its products. In this case, r.v. T in the exponential pdf represents the number of cycles to failure. $\lambda = 0.003$ f/cycle.

a) What is the mean cycle to failure for this product?

b) If a component survives for 300 cycles, what is the probability that it will fail sometime after 500 cycles ? Accordingly, if 1,000 components have survived 300 cycles, how many would one expect to fail after 500 cycles?

3.15 The diameter of a sample of 25 shafts are measured. The mean diameter is 0.102 centimeter, with a standard deviation of 0.005 centimeter. What is the upper 95% confidence limit on the mean diameter of all shafts produced by this process, assuming the distribution of shaft diameters is normal?

3.16 The mean life of a sample of 10 car batteries is of 102.5 months, with a standard deviation of 9.45 months. What are the 80% confidence limits for the actual mean and standard deviation of a pdf that represents these batteries?

3.17 The breaking strengths of 5 specimens of a rope of 1/4 inch diameter are 660, 460, 540, 580, and 550 lbs. Estimate the following:

a) The mean breaking strength by a 95% confidence level assuming normally distributed strength.

b) The point estimate at which only 5% of such specimens would be expected to break if $\bar{x}$ is assumed to be an unbiased estimate of the true mean, and $s^2$ is assumed to be the true standard deviation. (Assume $x$ is normally distributed.)

c) The 90% confidence interval of the estimate of the standard deviation.

3.18 One hundred and twenty four devices are placed on an overstress test with failures occurring at the following times.

| Time (hours) | Total # of Failures | Time (hours) | Total # of Failures |
|---|---|---|---|
| 0.4 | 1 | 8.0 | 20 |
| 1.0 | 3 | 12.0 | 30 |
| 2.0 | 5 | 25.0 | 50 |
| 5.0 | 15 | | |

Based on previous work on this type of product, we believe the failure rate follows a Weibull distribution.

a) Plot the data on Weibull probability paper.
b) Find the shape parameter.
c) Find the scale parameter.
d) What other distributions may also represent these failure data?

3.19 Seven pumps have failure times (in months) of 15.1, 10.7, 8.8, 11.3, 12.6, 14.4, and 8.7. (Assume an exponential distribution.)

a) Estimate a point estimate of the MTTF.
b) Estimate the reliability of a pump for $t = 12$ months.
c) Calculate the 95% two-sided estimate of $\lambda$.

3.20 The average life of a certain type of small motor is 10 years, with a standard deviation of 2 years. The manufacturer replaces free of charge all motors that fail while under guarantee. If the manufacturer is willing to replace only 3% of the motors that fail, how long a guarantee should be offered? Assume the time to failure of the motors follows a normal distribution.

3.21 A manufacturer claims that certain machine parts will have a mean diameter of 4 cm, with a standard deviation of 0.01 mm. The diameters of five parts are measured and found to be (in mm): 39.98, 40.01, 39.96, 40.03, and 40.02. Do you accept this claim with a 90% confidence level?

3.22 You are to determine a life test experiment to estimate the failure rate of a new device. Your boss asks you to make sure that the 80% upper and lower limits of the estimate (two-sided estimate) do not differ by more than a factor of 2. Due to cost constraints, the components will be put under test until they fail. Design this experiment, i.e., determine how many components should be put under test. Find a point estimate of the failure rate if the last item will fail at $t = 1248$ hours after the beginning of the test.

3.23 For an experiment, 25 relays are allowed to run until the first failure occurred at t = 15 hours. At this point, the experimenters decide to continue the test for another 5 hours. No failures occur during this extended period, and the test is terminated. Using the 90% confidence level, determine the following:
   a) Point estimate of MTTF
   b) Two-sided confidence interval for MTTF
   c) Two-sided reliability of the component at t = 25 hours

3.24 A locomotive control system fails 15 times out of the 96 times it is activated to function. Determine the following:
   a) A point estimate for failure probability of the system
   b) 95% confidence intervals two-sided estimate of the true value of the probability (Assume that after each failure, the system is repaired and put back to an as-good-as-new state.)

3.25 A circuit board design has 60 units of an electronic part. Experience shows that each unit time to failure is exponentially distributed, with $\lambda = 1 \times 10^{-4}$/hour. The manufacturer acquires a new electronic part that fails five times slower than the old units (i.e., $\lambda = 2 \times 10^{-5}$/hr). The manufacturer decides to replace an old one with new ones only after one of them fails. Predict the total number of unit replacements (new and old unit replacements together) during a 1,000 hours of operation.

3.26 A sample of 10 measurements of the diameter of a sphere gives a mean of 4.38 inches, with a standard deviation of 0.06 inch. Find the 99% confidence limits of the actual mean and standard deviation.

3.27 The following sample is taken from an industrial process, which follows a normal distribution: 8.9, 9.8, 10.8, 10.7, 11.0, 8.0, and 10.8. For this sample, the 95% confidence error on estimating the mean ($\mu$) is 2.2. What sample size should be taken if we want the 99% confidence error to be 1.5?

3.28 Suppose the generic failure rate of a component corresponding to an exponential time to failure model is $\lambda_g = 10^{-3}(\text{hr}^{-1})$ with a standard deviation of $\lambda_g/2$. Assume that ten components are closely observed for 1500 hours and one failure is observed. Using the Bayesian method, calculate the mean and variance of $\lambda$ from the posterior distribution. Calculate the 90 percent lower confidence limit.

3.29 In the reactor safety study, the failure rate of a diesel generator has a lognormal distribution with the upper and lower 90% bounds of $3 \times 10^{-2}$

and $3 \times 10^{-4}$ respectively. If a given nuclear plant experiences 2 failures in 8,760 hours of operation, determine the upper and lower 90% bounds given this plant experience (Consider the reactor safety study valves as prior information.).

# 4
# System Reliability Analysis

Assessment of the reliability of a system from its basic elements is one of the most important aspects of reliability analysis. A system is a collection of items (subsystems, components, units, blocks, etc.) whose proper, coordinated function leads to the proper functioning of the system. In reliability analysis, it is therefore important to model the relationship between various items as well as the reliability of the individual items to determine the reliability of the system as a whole. In Chapter 3, we elaborated on the reliability analysis at a basic item level (one for which enough information is available to predict its reliability). In this chapter, we discuss methods to model the relationship between system components, which allows us to determine overall system reliability.

The physical configuration of an item that belongs to a system is often used to model system reliability. In some cases, the manner in which an item fails is important for system failure, and should be considered in the system reliability analysis. For example, in a system composed of two parallel electronic units, if a unit fails short, the system will fail, but for most other types of failures of the unit, the system will still be functional since the other unit functions properly.

There are several system modeling schemes for reliability analysis. In this chapter we describe the following modeling schemes: **reliability block diagram**, which includes parallel, series, standby, shared load, and complex systems; **fault tree and success tree methods**, which include the method of construction and evaluation of the tree; **event tree method**, which includes modeling of multi-system designs and complex systems whose individual units should work in a chronological or approximately chronological manner to achieve a mission; **failure mode and effect analysis**; and **Master Logic Diagram (MLD)**. We assume here that items composing a system are statistically independent according to the definition provided in Chapter 2. However, in Chapter 6 we elaborate on system reliability considerations when components are statistically dependent.

## 4.1 Reliability Block Diagram Method

Reliability block diagrams are frequently used to model the effect of item failures (or functioning) on system performance. It often corresponds to the physical arrangement of items in the system. However, in certain cases, it may be different. For instance, when two resistors are in parallel, the system fails if one fails short. Therefore, the reliability block diagram of this system for the "fail short" mode of failure would be composed of two series blocks. However, for other modes of failure of one unit, such as "open" failure mode, the reliability block diagram is composed of two parallel blocks. In the remainder of this section, we discuss the reliability of the system for several types of the system functional configurations.

### *4.1.1 Series System*

A reliability block diagram is in a series configuration when failure of any one item (according to the failure mode of each item based on which the reliability block diagram is developed) results in the failure of the system. Accordingly, for functional success of a series system, all of its items must successfully function during the intended mission time of the system. Fig. 4.1 shows the reliability block diagram of a series system consisting of $N$ units.

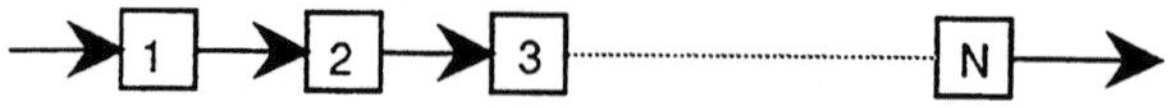


*Fig. 4.1 Series System Reliability Block Diagram*

The reliability of the system in Fig. 4.1 is the probability that all $N$ units succeed during its intended mission time $t$. Thus, probabilistically, the system reliability $R_s(t)$ for independent units is obtained from

$$R_s(t) = R_1(t) \cdot R_2(t) \cdot \ldots R_N(t) = \prod_{i=1}^{N} R_i(t), \tag{4.1}$$

where $R_i(t)$ represents the reliability of the $i^{th}$ unit. The hazard rate (instantaneous failure rate) for a series system is also a convenient expression. Since $H(t) = -d \ln R(t)/dt$, according to (4.1), the hazard rate of the system $h_s(t)$ is

$$h_s(t) = \frac{-d \ln \prod_{i=1}^{N} R_i(t)}{dt} = \sum_{i=1}^{N} \frac{-d \ln R_i(t)}{dt} = \sum_{i=1}^{N} h_i(t). \tag{4.2}$$

Let's assume a constant hazard rate model for each unit (e.g., assume an exponential time to failure for each unit). Thus, $h_i(t) = \lambda_i$. According to (4.2), the system constant rate of failure is

$$\lambda_s = \sum_{i=1}^{N} \lambda_i. \tag{4.3}$$

Expression (4.3) can also be easily obtained from (4.1) by using the constant failure rate reliability model for each unit, $R_i(t) = \exp(-\lambda_i t)$.

$$R_s(t) = \prod_{i=1}^{N} \exp(-\lambda_i t) = \exp\left(-\sum_{i=1}^{N} \lambda_i\right) t. \tag{4.4}$$

Since $R_s(t) = \exp(-\lambda_s t)$, $\lambda_s = \sum_{i=1}^{N} \lambda_i$. Accordingly, the $MTTF$ of the system ($MTTF_s$) can be obtained from

$$MTTF_s = \frac{1}{\lambda_s} = \frac{1}{\sum_{i=1}^{N} \lambda_i}. \tag{4.5}$$

---

*Example 4.1*

A system consists of three units whose reliability block diagram is in a series. The failure rate for each unit is constant as follows: $\lambda_1 = 4.0 \times 10^{-6}\text{hr}^{-1}$, $\lambda_2 = 3.2 \times 10^{-6}\text{hr}^{-1}$, and $\lambda_3 = 9.8 \times 10^{-6}\text{hr}^{-1}$. Determine the following parameters of the system:

a) $\lambda_s$
b) $R_s(1000 \text{ hours})$
c) $MTTF_s$.

*Solution:*

a) According to (4.3), $\lambda_s = 4.0 \times 10^{-6} + 3.2 \times 10^{-6} + 9.8 \times 10^{-6} = 1.7 \times 10^{-5}\text{hr}^{-1}$.
b) $R_s(t) = \exp(-\lambda_s t) = \exp(-1.7 \times 10^{-5} \times 1000) = 0.983$, or unreliability of $\bar{R}(1000) = 0.017$.
c) According to (4.5), $MTTFs = \frac{1}{\lambda_s} = \frac{1}{1.7 \times 10^{-5}} = 58{,}823.5$ hours.

---

*4.1.2 Parallel Systems*

A reliability block diagram is in a parallel configuration when the failure of all units in the system results in system failure. Accordingly, for a parallel system, success of only one unit would be

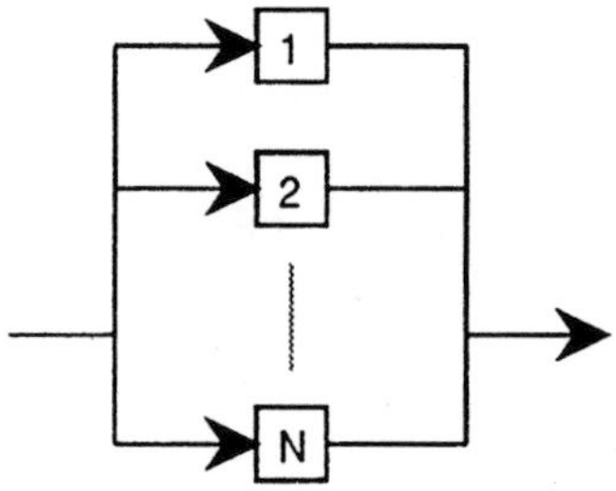


*Fig. 4.2 Parallel System Block Diagram*

sufficient to guarantee the success of the system. Fig. 4.2 shows a parallel system consisting of $N$ units.

According to the definition of a parallel system, failure of all units results in the failure of the system. Thus, for a set of $N$ independent units,

$$F_s(t) = F_1(t) \cdot F_2(t) \ldots F_N(t) = \prod_{i=1}^{N} F_N(t). \tag{4.6}$$

Since $R_i(t) = 1 - F_i(t)$ (i.e., $F_i(t)$ is the unreliability of the units), then

$$R_s(t) = 1 - F_s(t) = 1 - \prod_{i=1}^{N}[1 - R_i(t)]. \tag{4.7}$$

The hazard rate can also be determined by using $h(t) = -d \ln R(t)/dt$. The result of $h(t)$ is rather complex.

For consideration of various characteristics of system reliability, let's consider a special case where the instantaneous failure rate is constant for each unit (exponential time to failure model), and the system is composed of only two units. Since $R_i(t) = \exp(-\lambda_i t)$, then according to (4.7),

$$\begin{aligned} R_s(t) &= 1 - (1 - \exp(-\lambda_1 t))(1 - \exp(-\lambda_2 t)) \\ &= \exp(-\lambda_1 t) + \exp(-\lambda_2 t) - \exp(-(\lambda_1 + \lambda_2)t). \end{aligned} \tag{4.8}$$

Since $h_s(t) = \frac{f_s(t)}{R_s(t)}$ and $f_s(t) = -\frac{d[R_s(t)]}{dt}$, then using (4.8),

$$f_s(t) = \lambda_1 \exp(-\lambda_1 t) + \lambda_2 \exp(-\lambda_2 t) - (\lambda_1 + \lambda_2) \exp(-(\lambda_1 + \lambda_2)t).$$

Thus,

$$h_s(t) = \frac{\lambda_1 \exp(-\lambda_1 t) + \lambda_2 \exp(-\lambda_2 t) - (\lambda_1 + \lambda_2) \exp(-(\lambda_1 + \lambda_2)t)}{\exp(-\lambda_1 t) + \exp(-\lambda_2 t) - \exp(-(\lambda_1 + \lambda_2)t)}. \tag{4.9}$$

The $MTTF$ of the system can also be obtained from

$$\begin{aligned} MTTF_s &= \int_0^\infty R_s(t)dt \\ &= \int_0^\infty [\exp(-\lambda_1 t) + \exp(-\lambda_2 t) - \exp(-(\lambda_1 + \lambda_2)t)]dt \\ &= \frac{1}{\lambda_1} + \frac{1}{\lambda_2} - \frac{1}{\lambda_1 + \lambda_2}. \end{aligned} \tag{4.10}$$

Accordingly, one can use the binomial expansion to derive the $MTTF$ for $N$ parallel units:

$$\begin{aligned} MTTF_s = {} & (\frac{1}{\lambda_1} + \frac{1}{\lambda_2} + \ldots + \frac{1}{\lambda_N}) - (\frac{1}{\lambda_1 + \lambda_2} + \frac{1}{\lambda_1 + \lambda_3} + \ldots + \frac{1}{\lambda_{N-1} + \lambda_N}) \\ & + (\frac{1}{\lambda_1 + \lambda_2 + \lambda_3} + \ldots + \frac{1}{\lambda_{N-2} + \lambda_{N-1} + \lambda_N}) \ldots \\ & + (-1)^{N+1} \frac{1}{\lambda_1 + \lambda_2 + \ldots + \lambda_N}. \end{aligned} \tag{4.11}$$

In the special case where all units are identical with a constant failure rate $\lambda$ (e.g., in an active redundant system), (4.7) simplifies to the following form:

$$R_s(t) = 1 - (1 - \exp(-\lambda t))^N, \tag{4.12}$$

and from (4.11),

$$MTTF_s = MTTF(1 + \frac{1}{2} + \ldots + \frac{1}{N}). \tag{4.13}$$

It can be seen from (4.13) that in the design of active redundant systems, the $MTTF_s$ exceeds the $MTTF$ of an individual unit. However, the contribution to the $MTTF_s$ from the second unit, the third unit, and so on would have a diminishing return as $N$ increases. That is, there would be an optimum number of parallel units by which a designer can maximize the reliability and at the same time minimize the cost of the component in its life cycle.

Let's consider a more general form of a series and parallel system—the so-called K-out-of-N system. In this type of system, if any combination of $K$ units out of $N$ independent units work, it guarantees the success of the system. For simplicity, assume that all units are identical (which, by the way, is often the case). The binomial distribution can easily represent the probability that the system functions:

$$R_s(t) = \sum_{r=K}^{N} \binom{N}{r} [R(t)]^r [1 - R(t)]^{N-r} = 1 - \sum_{r=0}^{K-1} \binom{N}{r} [R(t)]^r [1 - R(t)]^{N-r}. \tag{4.14}$$

---

*Example 4.2*

A system is composed of the same units as in Example 4.1. However, these units are in parallel. Find the unreliability and $MTTFs$.

*Solution:*

According to (4.7),

$$R_s(t) = 1 - (1 - e^{-\lambda_1 t})(1 - e^{-\lambda_2 t})(1 - e^{-\lambda_3 t}).$$

$$R_s(1000) = (1 - e^{-4.0\times10^{-6}\times1000})(1 - e^{-3.2\times10^{-6}\times1000})(1 - e^{-9.8\times10^{-6}\times1000})$$
$$= 1.25 \times 10^{-7}.$$

$$MTTFs = (\frac{1}{\lambda_1} + \frac{1}{\lambda_2} + \frac{1}{\lambda_3}) - (\frac{1}{\lambda_1 + \lambda_2} + \frac{1}{\lambda_1 + \lambda_3} + \frac{1}{\lambda_2 + \lambda_3})$$
$$+ (\frac{1}{\lambda_1 + \lambda_2 + \lambda_3}) = 4.35 \times 10^5 \text{hours}.$$

---

*Example 4.3*

How many components should be used in an active redundancy design to achieve a reliability of 0.999 such that, for successful system operation, a minimum of two components is required? Assume a mission of $t = 720$ hours for a set of components that are identical and have a failure rate of 0.00015/hour.

*Solution:*

For each component $R(t) = \exp(-\lambda t) = \exp(-0.00015 \times 720) = 0.8976$. According to (4.14),

$$0.999 = 1 - \sum_{r=0}^{1} \binom{N}{r} [0.8976]^r [0.1024]^{N-r}$$
$$= 1 - [0.1024]^N - N[0.8976][0.1024]^{N-1}.$$

From the above equation, $N = 5$, which means that at least five components should be used to achieve the desired reliability over the specified mission time.

---

### *4.1.3 Standby Redundant Systems*

A system is called a standby redundant system when some of its units remain idle until they are called for service by a sensing and switching device. For simplicity, let's consider a situation where only one unit operates actively and the others are in standby, as shown in Fig. 4.3.

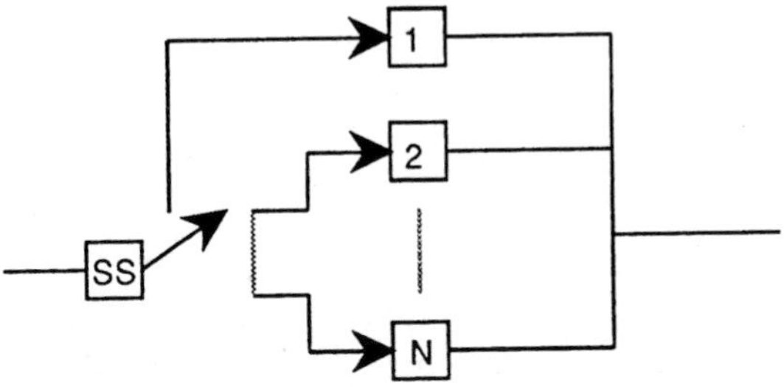


*Fig. 4.3 Standby Redundant System*

In this configuration, unit 1 operates constantly until it fails. The sensing and switching device recognizes a unit failure in the system and switches to another unit. This process continues until all standby units have failed, in which case the system is considered failed. Since units 2 to $N$ do not operate constantly (as is the case in active parallel systems), we would expect them to fail at a much slower rate. This is because the failure rate for components is usually lower when the components are operating than when they are idle or dormant.

It is clear that system reliability is totally dependent on the reliability of the sensing and switching device. The reliability of a redundant standby system is the reliability of unit 1 over the mission time $t$ (i.e., the probability that it succeeds the whole mission time) plus the probability that unit 1 fails at time $t_1$ prior to $t$ **and** the probability that the sensing and switching unit does not fail by $t_1$ **and** the probability that standby unit 2 does not fail by $t_1$ (in the standby mode) **and** the probability that standby unit 2 successfully functions for the remainder of the mission in an active operation mode, and so on.

Mathematically, the reliability function for a two unit standby device according to this definition can be obtained from

$$R_s(t) = R_1(t) + \int_0^t f_1(t_1)dt_1 \cdot R_{ss}(t_1) \cdot R_2'(t_1) \cdot R_2(t - t_1), \qquad (4.15)$$

where $f_1(t)$ is the *pdf* for the time to failure of unit 1, $R_{ss}(t_1)$ is the reliability of the sensing and switching device, $R'_2(t)$ is the reliability of unit 2 in the standby mode of operation, and $R_2(t-t_1)$ is the reliability of unit 2 after it started to operate at time $t_1$. Let's consider a case where time to failure of all units follows an exponential distribution.

$$\begin{aligned}R_s(t) &= \exp(-\lambda_1 t) + \int_0^t [\lambda_1 \exp(-\lambda_1 t_1)]\,[\exp(-\lambda_{ss} t_1)]\,[\exp(-\lambda'_2 t_1)] \\ &\quad [\exp(-\lambda_2(t-t_1))]\,dt_1. \\ &= \exp(-\lambda_1 t) \\ &\quad + \frac{\lambda_1 \exp(-\lambda_2 t)}{\lambda_1 + \lambda_{ss} + \lambda'_2 - \lambda_2}[1 - \exp - [(\lambda_1 + \lambda_{ss} + \lambda'_2 - \lambda_2)t]]. \end{aligned} \tag{4.16}$$

For the special case of perfect sensing and switching and no standby failures, $\lambda_{ss} = \lambda'_2 = 0$,

$$\begin{aligned}R_s(t) &= \exp(-\lambda_1 t) + \frac{\lambda_1 \exp(-\lambda_2 t)}{\lambda_1 - \lambda_2}[1 - \exp(-(\lambda_1 - \lambda_2)t)] \\ &= \exp(-\lambda_1 t) + \frac{\lambda_1}{\lambda_1 - \lambda_2}\left[\exp(-\lambda_2 t) - \exp(-\lambda_1 t)\right]. \end{aligned} \tag{4.17}$$

If the two units are identical, then $\lambda_1 = \lambda_2 = \lambda$,

$$R_s(t) = \exp(-\lambda t) + \lambda t \exp(-\lambda t) = (1 + \lambda t)\exp(-\lambda t). \tag{4.18}$$

In the case of perfect switching, a standby system poses the same characteristic as the shock model. That is one can assume that the $N^{th}$ shock (i.e., the $N^{th}$ unit failure) causes the system to fail. Thus, a gamma distribution can represent the time to failure of the system such that

$$\begin{aligned}R_s(t) &= 1 - \int_0^t \frac{\lambda^N}{\Gamma(N)} x^{N-1} \exp(-\lambda x)dx \\ &= \exp(-\lambda t)\left[1 + \lambda t + \frac{(\lambda t)^2}{2!} + \ldots + \frac{(\lambda t)^{N-1}}{(N-1)!}\right]. \end{aligned} \tag{4.19}$$

Accordingly, the *MTTF* of the above system is given by

$$MTTF_s = \frac{N}{\lambda}, \tag{4.20}$$

which is $N$ times the $MTTF$ of a single unit. Expression (4.20) explains why high reliability can be achieved through a standby system when the switching is perfect and no failure occurs during standby.

When more than two units are in standby, the equation becomes somewhat difficult, but the concept is almost the same. For example, for three units with perfect switching,

$$\begin{aligned} R_s(t) = R_1(t) &+ \int_{t_1=0}^{t} f_1(t_1)dt_1.R_2(t-t_1) \\ &+ \int_0^t f_1(t_1)dt_1 \int_0^{t-t_1} f_2(t_2)R_3(t-t_1-t_2)dt_2. \end{aligned} \tag{4.21}$$

If the sensing and switching device is not perfect, appropriate terms can be added to (4.21) to allow for its unreliability—similar to (4.15).

---

*Example 4.4*

Consider two identical independent units with $\lambda = 0.01\text{hr}^{-1}$. Mission time $t = 24$ hours. Compare the reliability of a system made of these units if they are placed in:

a) Parallel configuration
b) Series configuration
c) Standby configuration with perfect switching
d) Standby configuration with imperfect switching and standby failure rates of $\lambda_{ss} = 1 \times 10^{-6}\text{hr}^{-1}$ and $\lambda' = 1 \times 10^{-5}$, respectively.

*Solution:*

Let's assume an exponential time to failure model for each unit $R(t) = \exp(-\lambda t) = \exp(-0.01 \times 24) = 0.7866$.

a. For the parallel system, using (4.12), $R_s(24) = 1-[1-0.7866]^2 = 0.9544$.
b. For the series system, using (4.1) $R_s(24) = 0.7866 \times 0.7866 = 0.6187$.
c. For the standby system with perfect switches, using (4.18)

$$R_s(24) = (1 + 0.24)\exp(-0.01 \times 24) = 0.9748.$$

d. For the standby system with imperfect switching and standby failure rate using (4.16),

$$R_s(24) = 0.7866 + \frac{(0.01)(0.7866)}{1.1 \times 10^{-5}}[1 - \exp(-1.1 \times 10^{-5} \times 24)]$$
$$= 0.9754.$$

---

*4.1.4 Shared Load Systems*

A shared load system refers to a parallel system whose units equally share the system function. For example, if a set of two parallel pumps delivers $x$ gpm of water to a reservoir, each pump delivers $x/2$ gpm. If a minimum of $x$ gpm is required at all times, and one of the pumps fails at a given time $t_1$, then the other pump's speed should be increased to provide $x$ gpm alone. Other examples of load sharing are multiple load-bearing units (such as those in a bridge), and load-sharing multi-unit electric power plants. In these cases, when one of the units fails, the others should carry its load. Since these other units would then be working under a more stressful conditions, they would experience a higher rate of failure.

Let's assume that two units share a load (i.e., each unit carries half the load), and the time-to-failure distribution for both units to work is $f_h(t)$. When one unit fails (i.e., one unit carries the full load), the time to failure distribution is $f_f(t)$. Let's also assume that the corresponding reliability functions during full-load and half-load operation are $R_f(t)$ and $R_h(t)$ respectively. The system will succeed if both units carry half the load, or if unit 1 fails at time $t_1$ and unit 2 carries a full load thereafter, or if unit 2 fails at time $t_1$ and unit 1 carries the full load thereafter. Accordingly, the system reliability function $R_s(t)$ can be obtained from

$$R_s(t) = [R_h(t)]^2 + 2\int_0^t f_h(t_1)R_h(t_1)R_f(t - t_1)dt_1. \tag{4.22}$$

In (4.22), the first term shows the contribution from both units working successfully, with each carrying a half load; the second term represents the two equal probabilities that unit 1 fails first and unit 2 takes the full load at time $t_1$, or vice versa.

If there are switching or control mechanisms involved to shift the total load to the unfailed unit when one unit fails, then similar to (4.15), the reliability of the switching mechanism can be incorporated into (4.22).

In the special situation where an exponential time to failure model with failure rates $\lambda_f$ and $\lambda_h$ can be used for the two units for full and half load cases, respectively, then (4.22) can be simplified to

$$R_s(t) = \exp(-2\lambda_h t) + \frac{2\lambda_h \exp[-\lambda_f t]}{(2\lambda_h - \lambda_f)}\{\exp[-(2\lambda_h - \lambda_f)t]\}. \tag{4.23}$$

### *4.1.5 Complex Systems*

Most practical systems are not parallel or series, but exhibit some hybrid combination of the two. These systems are often referred to as **parallel-series system**. Fig. 4.4 shows an example of such a system.

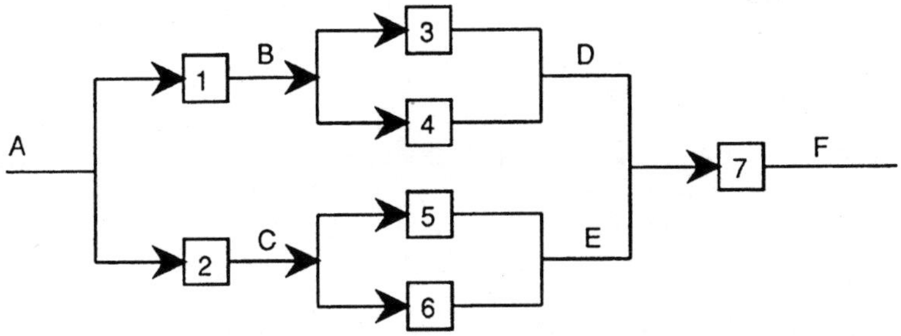


*Fig. 4.4 Complex Parallel-Series System*

Another type of complex system is one that is neither series nor parallel alone, nor parallel-series. Fig. 4.5 shows an example of such a system.

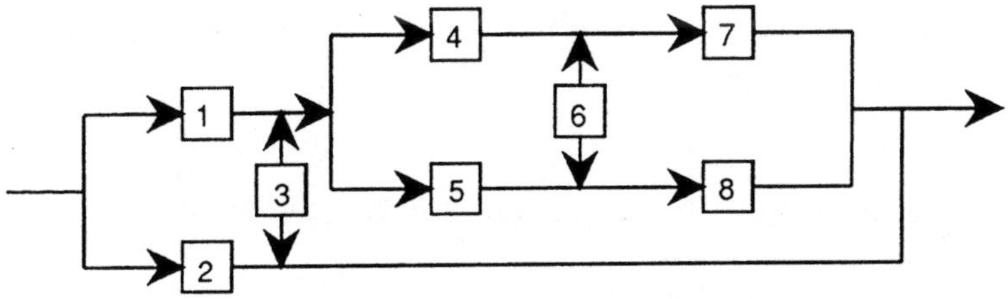


*Fig. 4.5 Complex Nonparallel-Series System*

A parallel-series system can be analyzed by dividing it into its basic parallel and series modules and then determining the reliability function for each module separately. The process can be continued until a reliability function for the whole system is determined. For the analysis of all types of complex systems, Shooman (1968) describes several analytical methods for complex systems. These are the **inspection method**, **event space method**, **path-tracing method**, and **decomposition**. These methods are good only when there are not a lot of units in the system. For analysis of large number of units, fault trees would be more appropriate. In the following, we discuss the decomposition and path-tracing methods.

In the decomposition method, Bayes' theorem is used to decompose the system. The reliability of a system is equal to the reliability of the system given that a chosen unit (e.g., unit 3 in Fig. 4.5) is good times the reliability of unit 3, plus the reliability of the system given unit 3 is bad times the unreliability of unit 3.

$$R_s(t) = R_s(t \mid \text{unit 3 good}) \cdot R_3(t) + R_s(t \mid \text{unit 3 bad})[1 - R_3(t)]. \quad (4.24)$$

If (4.24) is applied to all units that make the system a nonparallel series (such as units 3 and 6 in Fig. 4.5), the system would reduce to a simple parallel-series system. Thus, for Fig. 4.5 and for the conditional reliability terms in (4.24), it follows that

$$\begin{aligned} R_s(t|\text{units 3 good}) = {} & R_s(t \mid \text{unit 6 good} \cap \text{unit 3 good})R_6(t) \\ & + R_s(t \mid \text{unit 6 bad} \cap \text{unit 3 good})[1 - R_6(t)], \quad (4.25) \end{aligned}$$

or

$$\begin{aligned} R_s(t|\text{units 3 bad}) = {} & R_s(t \mid \text{unit 6 good} \cap \text{unit 3 bad})R_6(t) \\ & + R_s(t \mid \text{unit 6 bad} \cap \text{unit 3 bad})[1 - R_6(t)]. \quad (4.26) \end{aligned}$$

Each of the conditional reliability terms in (4.25) and (4.26) represents a purely parallel-series system, the reliability determination of which is simple. For example, $R_s(t \mid \text{unit 6 good} \cap \text{unit 3 bad})$ corresponds to a reliability block diagram shown in Fig. 4.6. The combination of (4.24) through (4.26) results in an expression for $R(s)$.

A more computationally intensive method for determining the reliability of a complex system involves the use of **path set** and **cut set** methods (path-tracing methods). A **path set** (or tie set) is a set of units that form a connection between input and output when

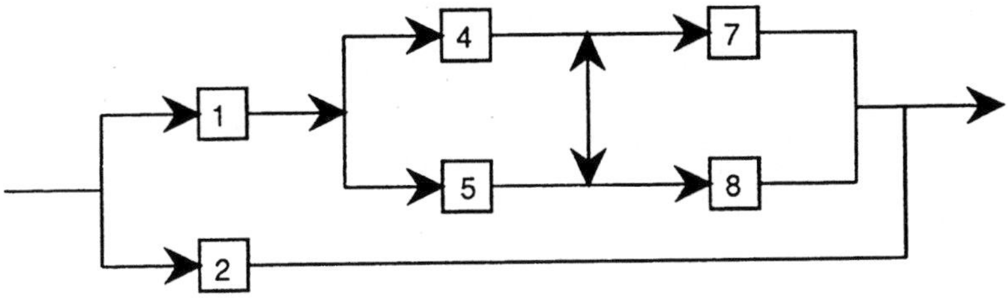


*Fig. 4.6 Representation of* $R_s(t \mid \text{unit 6 good} \cap \text{unit 3 bad})$

traversed in the direction of the reliability block diagram arrows. Thus, a path set merely represents a "path" through the graph. A **minimal path set** (or minimal tie set) is a path set containing the minimum number of units needed to guarantee a connection between the input and output points. For example, in Fig. 4.5, path set $P_1 = (1, 3)$ is a minimal path set, but $P_2 = (1, 3, 6)$ is not since units 1 and 3 are sufficient to guarantee a path.

A **cut set** is a set of units that interrupt all possible connections between the input and output points. A **minimal cut set** is the smallest set of units needed to guarantee an interruption of flow. In practice, minimal cut sets show a combination of unit failures that cause a system to fail. For example, in Fig. 4.5, the minimal path sets are: $P_1 = (2)$, $P_2 = (1, 3)$, $P_3 = (1, 4, 7)$, $P_4 = (1, 5, 8)$, $P_5 = (1, 4, 6, 8)$, $P_6 = (1, 5, 6, 7)$. The minimal cut sets are: $C_1 = (1, 2)$, $C_2 = (4, 5, 3, 2)$, $C_3 = (7, 8, 3, 2)$, $C_4 = (4, 6, 8, 3, 2)$, $C_5 = (5, 6, 7, 3, 2)$. If a system has $m$ minimal path sets denoted by $P_1, P_2, \ldots, P_m$, then the system reliability is given by

$$R_s(t) = \Pr(P_1 \cup P_2 \cup \ldots \cup P_m), \tag{4.27}$$

where each path set $P_i$ represents the event that units in the path set survive during the mission time $t$. This guarantees the success of the system. Since many path sets may exist, the union of all these sets gives all possible events for successful operation of the system. The probability of this union clearly represents the reliability of the system. It should be noted here that in practice, the path sets $P_i$s are not disjoint. This poses a problem for determining the left-hand side of (4.27). In Section 4.2, we will explain formal methods to deal with this problem. However, a useful upper bound on the system reliability may be obtained by assuming that the $P_i$s are highly disjoint. Thus,

$$R_s(t) \leq \Pr(P_1) + \Pr(P_2) + \ldots + \Pr(P_m). \tag{4.28}$$

Expression (4.28) yields better answers when we deal with small reliability values. Since this is not usually the case, (4.28) is not a good bound for use in practical applications.

Similarly, system reliability can be determined through minimal cut sets. If the system has $n$ minimal cut sets denoted by $C_1, C_2, \ldots, C_n$, then the system reliability is obtained from

$$R_s(t) = 1 - \Pr(C_1 \cup C_2 \cup \ldots \cup C_n), \tag{4.29}$$

where $C_i$ represents the event that units in the cut set fail sometime before the mission time $t$. This guarantees system failure. The $\Pr(\cdot)$ term on the right hand side of (4.29) shows the probability that at least one of all possible minimal cut sets exists before time $t$. Thus it represents the probability that the system fails sometimes before $t$. By subtracting this probability from 1, the reliability of the system is obtained. Similar to the union of path sets, the union of cut sets are not usually disjoint. Again, (4.29) can be written in the form of its lower bound, which is a much simpler expression given by

$$R_s(t) \geq 1 - [\Pr(C_1) + \Pr(C2) + \ldots + \Pr(C_n)]. \tag{4.30}$$

Notice that each element of a path set represents the **success** of a unit, whereas each element of a cut set represents the **failure** of a unit. Thus, for probabilistic evaluations, the reliability function of each unit should be used in connection with path set evaluations, i.e., (4.28), while the unreliability function should be used in connection with cut set evaluations, i.e., in (4.30).

The bounding technique used in (4.30), in practice, yields a much better representation of the reliability of the system than (4.28) because most engineering units have reliabilities greater than 0.9 over their mission time, making the use of (4.30) appropriate.

---

*Example 4.5*

For the reliability block diagram in Fig. 4.5. Determine the lower bound of the system reliability function if the hazard rates of each unit are constant and are $\lambda_1, \lambda_2, \ldots, \lambda_3$.

*Solution:*

Using the system cut sets discussed earlier and (4.30),

$$R_s(t) \geq 1 - [\Pr(C_1) + \Pr(C_2) + \ldots + \Pr(C_5)],$$

assuming $C_1$ and $C_2$ are independent, and

$$\Pr(C_1) = (1 - \exp(-\lambda_1 t))(1 - \exp(-\lambda_2 t)),$$

and so on. Therefore,

$$R_s(t) \geq 1-[(1-e^{-\lambda_1 t})(1-e^{-\lambda_2 t})+(1-e^{-\lambda_2 t})(1-e^{-\lambda_3 t})(1-e^{-\lambda_4 t})$$
$$(1-e^{-\lambda_5 t})+(1-e^{-\lambda_2 t})(1-e^{\lambda_3 t})(1-e^{-\lambda_7 t})(1-e^{-\lambda_8 t})$$
$$+(1-e^{-\lambda_2 t})(1-e^{-\lambda_3 t})(1-e^{-\lambda_4 t})(1-e^{-\lambda_6 t})(1-e^{-\lambda_8 t})$$
$$+(1-e^{-\lambda_2 t})(1-e^{-\lambda_3 t})(1-e^{-\lambda_5 t})(1-e^{-\lambda_6 t})(1-e^{-\lambda_7 t})].$$

For some typical values of $\lambda$, the lower bound for $R_s(t)$ can be plotted and compared to the exact value of $R_s(t)$. "Exact" means the cut sets are not assumed disjoint. For example, for $\lambda_1 = 1\times10^{-6}\text{hr}^{-1}$, $\lambda_2 = 1\times10^{-5}\text{hr}^{-1}$, $\lambda_3 = 2\times10^{-5}\text{hr}^{-1}$, and $\lambda_4 = \lambda_5 = \lambda_6 = \lambda_7 = \lambda_8 = 1\times10^{-4}\text{hr}^{-1}$. The plots in Fig. 4.7 can be obtained.

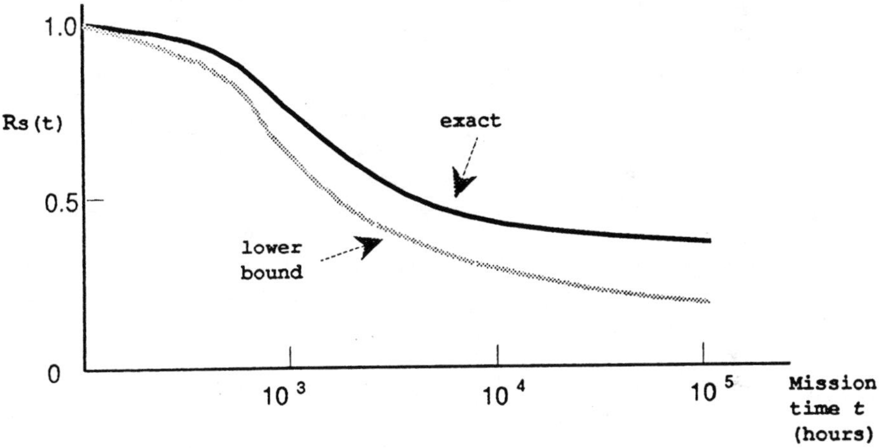


*Fig. 4.7 Comparison of the Reliability Functions for the Lower Bound and Exact Values*

It is evident from the graph that as time increases, the reliability of the system decreases (unit failure probability increases), causing (4.30) to yield a poor approximation. At this point, it is more appropriate to use (4.28). Again, notice that (4.28) and (4.30) assume the path sets and cut sets are disjoint.

---

In cases where we deal with very complex systems that have multiple failures modes for each unit and complex physical and operational interactions, the use of reliability block diagrams is difficult.

The method of fault tree and success tree analysis is more appropriate in this context, especially if the role of humans in the operation of the system needs to be modeled. We elaborate on this topic in the next section.

## 4.2 Fault Tree and Success Tree Methods

The operation of a system can be considered from two opposite viewpoints: the various ways that a system completely fails, or the various ways that a system completely succeeds. To logically and deductively model the failure or success of a system, the fault tree or success tree methods can be used, respectively. Most of the construction and analysis methods used are, in essence, the same for both fault trees and success trees. First we will discuss the fault tree method, and then describe the success the tree method by pointing out the differences between the two.

### *4.2.1 Fault Tree Method*

The fault tree approach is a deductive process whereby an undesirable event, called the **top event**, is postulated, and the possible means for this event to occur are systematically deduced. For example, a typical top event looks like "failure of control circuit A to send a signal when it should." The deduction process is performed so the fault tree embodies all component failures that contribute to the occurrence of the top event. It is also possible to include individual failure modes of each component as well as human errors during the system operation. The fault tree itself is a graphical representation of the various parallel and sequential combinations of faults that lead to the occurrence of the top event.

A fault tree does not necessarily contain all possible failure modes of the components (or units) of the system. Only those failure modes whose existence contribute to the existence of the top event are modeled. For example, consider a failed safe control circuit. If loss of the dc power to the circuit causes the circuit to open a contact, which in turn sends a signal to another system for operation, a top event of "control circuit fails to generate a safety signal" would not include the "failure of dc power source" as one its events, even though the dc power source (e.g., batteries) is part of the control circuit. This is because the top event would not occur due to the loss of the dc power source.

The postulated fault events that appear on the fault tree structure may not be exhaustive. Only those events considered important can be included. However, it should be noted that the decision for inclusion of fault events is not arbitrary; it is influenced by the fault tree construction procedure, system design and operation, operating history, available failure data, and the experience of the analyst. At each intermediate point, the postulated events represent the **immediate, necessary,** and **sufficient** causes for the occurrence of the intermediate (or top) events.

The fault tree itself is a logical representation model, and thus represents the qualitative characterization of the system logic. There are, however, many quantitative algorithms to evaluate fault trees. For example, the concept of **cut sets** discussed earlier can also be applied to fault trees by using the Boolean algebra method. By using $\Pr(C_1 \cup C_2 \ldots \cup C_m)$, the probability of occurrence of the top event can be determined using (4.29).

To understand the symbology of fault trees, consider Fig. 4.8. In essence, there are three types of symbols: **events**, **gates**, and **transfers**. Basic events, undeveloped events, condition events, and external events are sometimes referred to as **primary events**. When postulating events in the fault tree, it is important to include not only the undesired component states (e.g., applicable failure modes), but also the time they occur. It is also important to properly consider fault events.

To understand the fault tree concept better, let's consider the complex block diagram shown in Fig. 4.4. To better represent this system, let's assume that the block diagram is a circuit in which the arrows show the direction of electric current. A top event of "no current at point F" is selected, and all events that cause this top event are deductively postulated. Fig. 4.9 shows the results. For a more detailed discussion of the construction and evaluation of fault trees, refer to the Fault Tree Handbook (1981).

As another example, consider the pumping system shown in Fig. 4.10. Sufficient water is delivered from the water source $T_1$ when only one of the two pumps, P-1 or P-2, works. All the valves V-1 through V-5 are normally open. The sensing and control system $S$ senses the demand for the pumping system and automatically starts both P-1 and P-2, (If one of the two pumps fails to start or fails during operation, the mission is still considered successful if the other pump functions properly.) The two pumps and the sensing and control system use the same ac power source AC. Assume the water content in $T_1$ is sufficient and available, there are no human errors, and no failure in the pipe connections is considered important.

It is clear that the system's mission is to deliver sufficient water when needed. Therefore, the top event of the fault tree for this

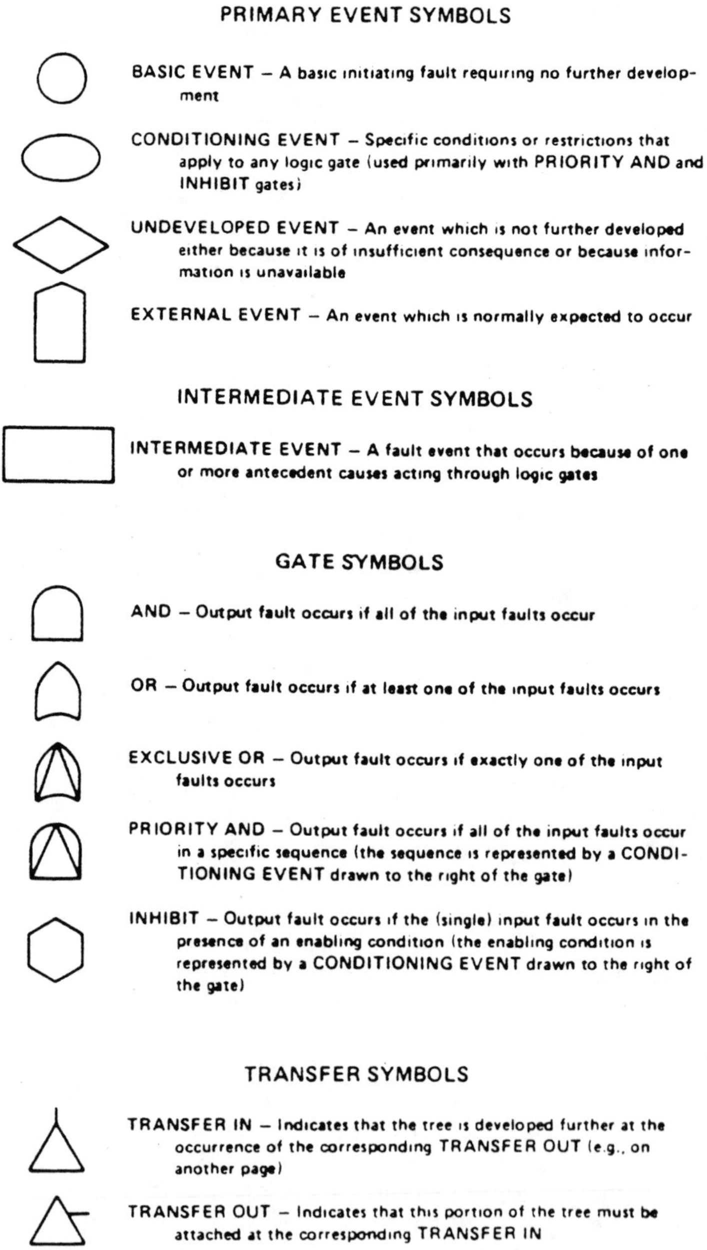
PRIMARY EVENT SYMBOLS
BASIC EVENT – A basic initiating fault requiring no further development
CONDITIONING EVENT – Specific conditions or restrictions that apply to any logic gate (used primarily with PRIORITY AND and INHIBIT gates)
UNDEVELOPED EVENT – An event which is not further developed either because it is of insufficient consequence or because information is unavailable
EXTERNAL EVENT – An event which is normally expected to occur
INTERMEDIATE EVENT SYMBOLS
INTERMEDIATE EVENT – A fault event that occurs because of one or more antecedent causes acting through logic gates
GATE SYMBOLS
AND – Output fault occurs if all of the input faults occur
OR – Output fault occurs if at least one of the input faults occurs
EXCLUSIVE OR – Output fault occurs if exactly one of the input faults occurs
PRIORITY AND – Output fault occurs if all of the input faults occur in a specific sequence (the sequence is represented by a CONDITIONING EVENT drawn to the right of the gate)
INHIBIT – Output fault occurs if the (single) input fault occurs in the presence of an enabling condition (the enabling condition is represented by a CONDITIONING EVENT drawn to the right of the gate)
TRANSFER SYMBOLS
TRANSFER IN – Indicates that the tree is developed further at the occurrence of the corresponding TRANSFER OUT (e.g., on another page)
TRANSFER OUT – Indicates that this portion of the tree must be attached at the corresponding TRANSFER IN

*Fig. 4.8 Primary Event, Gate, and Transfer Symbols*

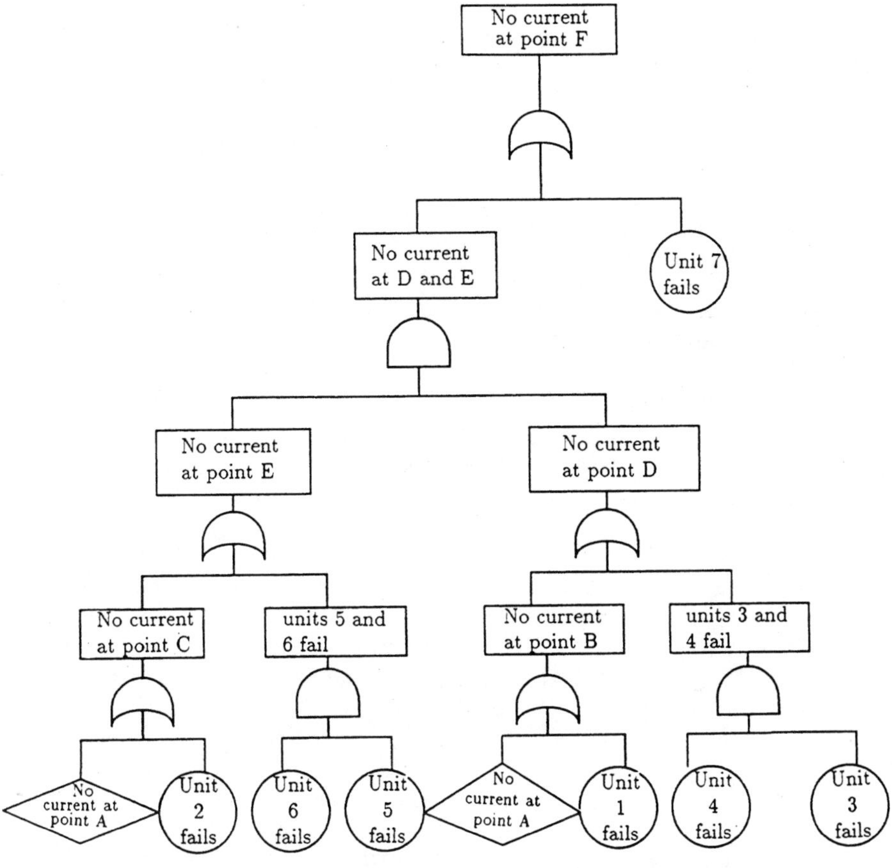


*Fig. 4.9 Fault Tree for the Complex Parallel-Series System in Fig. 4.4*

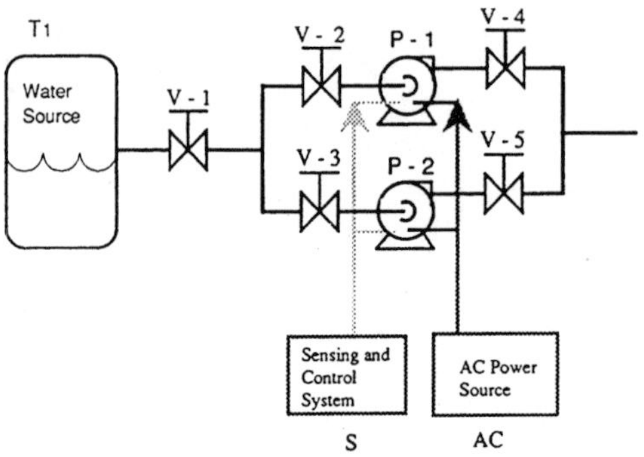


*Fig. 4.10 An Example of a Pumping System*

system should be "no water is delivered when needed." Fig. 4.11 shows the fault tree for this example. In Fig. 4.11, the failures of AC and $S$ are shown with **undeveloped events**. This is because one can further expand the fault tree if one knows what makes up the failures of AC and $S$, in which case these events will be **intermediate events**. However, since enough information (e.g., failure characteristics and probabilities) about these events is known, we have stopped their further development at this stage. Although the development of the fault tree in Fig. 4.11 is based on a strict deductive procedure (i.e., systematic decomposition of failures starting from "sink" and deductively proceeding toward "source"), one can rearrange it to the more concise and compact equivalent form shown in Fig. 4.12. While the development of the fault tree in Fig. 4.11 requires only a minimum understanding of the overall functionality and logic of the system, direct development of more compact versions requires a much better understanding of the overall system logic. If more complex logical relationships is required, other logical representations can be described by combining the two basic AND and OR gates. For example, the **K-out-of-N** and **exclusive OR** logics can be described, as shown in Fig. 4.13.

### Evaluation of Fault Trees

The evaluation of fault trees involves two distinct aspects: **logical** or **qualitative evaluation** and **probabilistic** or **quantitative evaluation**. Qualitative evaluation involves the determination of the fault tree cut sets or logical evaluations to rearrange the fault tree logic for computational efficiency (similar to the rearrangement presented in Fig. 4.12). Determining the fault tree cut sets involves

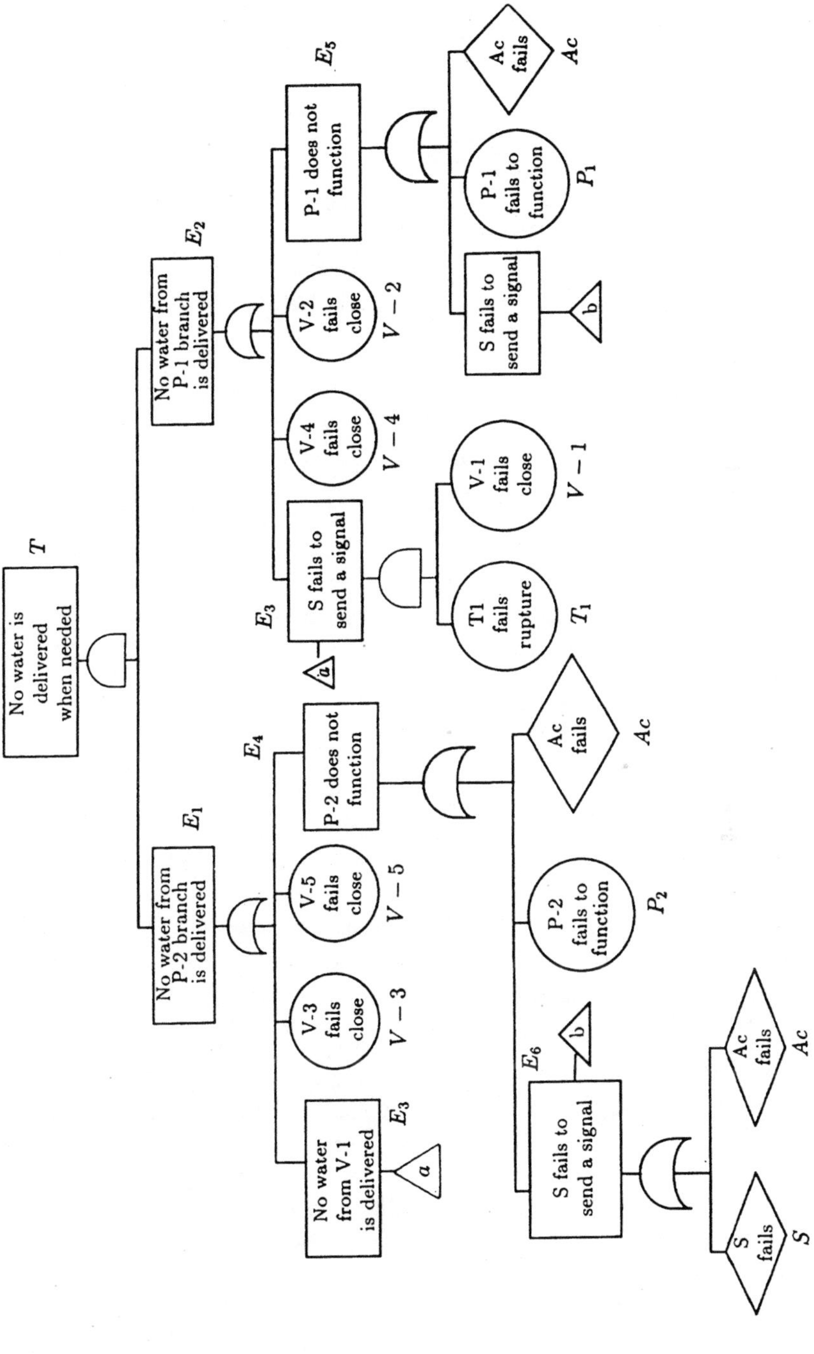


*Fig. 4.11 Fault Tree for the Pumping System in Fig. 4.1*

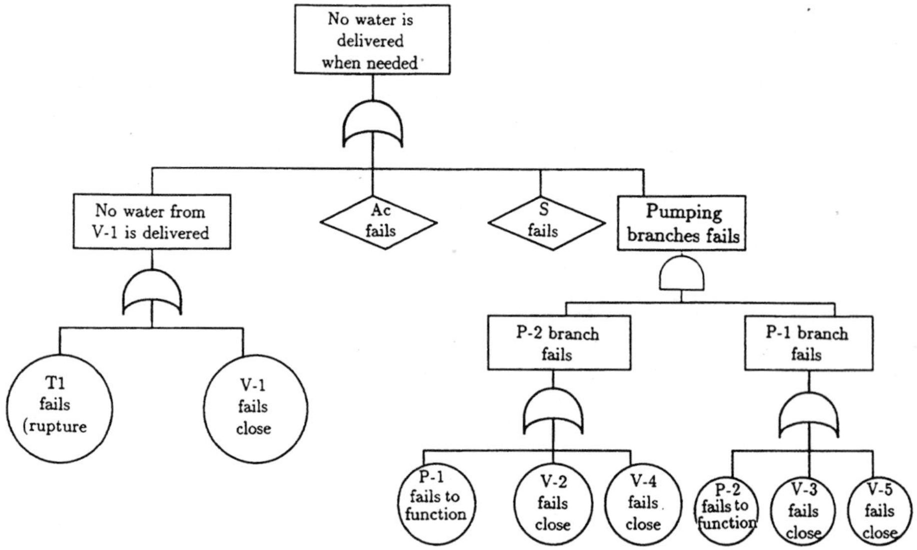


*Fig. 4.12 More Compact Form of the Fault Tree in Fig. 4.11*

some straightforward Boolean manipulation of events that we describe here. However, there are many types of logical rearrangements and evaluations, such as fault tree modularization, that are beyond the scope of this book. The reader is referred to the Fault Tree Handbook (1981) for a more detail discussion of this topic.

The quantitative evaluation of fault trees involves the determination of the probability of the occurrence of the top event. Accordingly, unreliability or reliability associated with the top event can also be determined.

The qualitative evaluation of fault trees through the generation of cut sets is conceptually very simple. The fault tree OR-gate logic represents the **union** of the events. That is, all the input events must occur to cause the output event to occur. For example, an OR

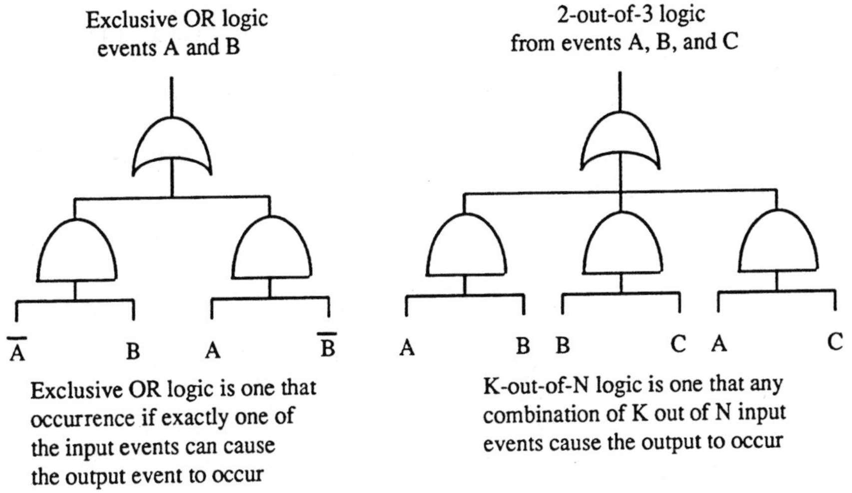
Exclusive OR logic
events A and B
2-out-of-3 logic
from events A, B, and C
Ā
B
A
B̄
A
B
B
C
A
C
Exclusive OR logic is one that occurrence if exactly one of the input events can cause the output event to occur
K-out-of-N logic is one that any combination of K out of N input events cause the output to occur

*Fig. 4.13 Exclusive OR and K-out-of-N Logics*

gate with two input events A and B and the output event Q can be represented by its equivalent Boolean expression, $Q = A \cup B$. Either A or B or both must occur for the output event Q to occur. Instead of the union symbol $\cup$, and for simplicity, the equivalent "+" symbol is often used. Thus, $Q = A + B$. Similarly an OR gate with $n$ input, $Q = A_1 + A_2 + \ldots + A_n$. The AND gate can be represented by the intersect logic. Therefore, the Boolean equivalent of an AND gate with two inputs A and B would be $Q = A \cap B$ (or $Q = A \cdot B$).

Determination of cut sets using the above expressions is possible through several algorithms. These algorithms include the **top-down** or **bottom-up** successive substitution method, the **modularization approach**, and **monte-carlo** simulation. The Fault Tree Handbook (1981) describes the underlying principles of these qualitative evaluation algorithms. The most widely used and straightforward algorithm is the successive substitution method. In this approach, the equivalent Boolean representation of each gate in the fault tree is determined and so that only primary events remain. Various Boolean identities are applied to reduce the Boolean expression to its most compact form, which represents the minimal cut sets of the fault tree. The substitution process can proceed from the top of the fault tree to the bottom or vice versa. Depending on the logic of the fault tree and its complexity, either the former or the latter approach, or a combination of the two, can be used.

As an example, let's consider the fault tree shown in Fig. 4.11. The step-by-step, top-down substitution of the top event is presented below.

Step 1: $T = E_1 \cdot E_2$.

Step 2: $E_1 = E_3 + V_3 + V_5 + E_4$,

$E_2 = E_3 + V_4 + V_2 + E_5$,

$T = E_3 + V_3 \cdot V_4 + V_3 \cdot V_2 + V_5 \cdot V_4 + V_5 \cdot V_2 + E_4 \cdot V_4 + E_4 \cdot V_2 + E_4 \cdot E_5 + V_3 \cdot E_5 + V_5 \cdot E_5$.

($T$ has been reduced by using the Boolean identities $E_3 \cdot E_3 = E_3$, $E_3 + E_3 \cdot X = E_3$, and $E_3 + E_3 = E_3$.)

Step 3: $E_3 = T_1 + V_1$,

$E_4 = E_6 + P_2 + AC$,

$E_5 = E_6 + P_1 + AC$,

$T = T_1 + V_1 + AC + V_3 \cdot V_4 + V_3 \cdot V_2 + V_5 \cdot V_4 + V_5 \cdot V_2 + V_4 \cdot P_2 + P_2 \cdot V_2 + E_6 + P_2 \cdot P_1 + V_3 \cdot P_1 + V_5 \cdot P_1$.

(Again, identities such as $AC + AC = AC$ and $E_6 + V_3 \cdot E_6 = E_6$ have been used to reduce $T$.)

Step 4: $E_6 = AC + S$,

$T = S + AC + T_1 + V_1 + V_3 \cdot V_4 + V_3 \cdot V_2 + V_5 \cdot V_4 + V_5 \cdot V_2 + V_4 \cdot P_2 + P_2 \cdot V_2 + P_2 \cdot P_1 + V_3 \cdot P_1 + V_5 \cdot P_1$.

The Boolean expression obtained in Step 4 represents four cut sets with one element (cut set size 1), and nine cut sets with two elements (cut set size 2). The size 1 cut sets are $S$, $AC$, $T_1$, $V_1$. The size 2 cut sets are $V_3$ and $V_4$; $V_3$ and $V_2$; $V_5$ and $V_4$; $V_5$ and $V_2$; $V_4$ and $P_2$; $P_2$ and $V_2$; $P_2$ and $P_1$; $V_3$ and $P_1$; and $V_5$ and $P_1$. A simple examination of each cut set shows that its occurrence guarantees the occurrence of the top event (failure of the system). For example, the cut set $V_5$ and $P_1$, which represents simultaneous failure of valve $V_2$ and pump $P_1$, causes the two flow branches of the system to be lost, which in turn incapacitates the system.

It is clear from this example that the evaluation of a large fault tree by hand can be a formidable job. A number of computer based programs are available for the analysis of fault trees. The PRA Procedures Guide (1982) elaborate on the important characteristics of these programs. Appendix B describes some of these programs

Quantitative evaluation of the cut sets has already been discussed under the context of the reliability block diagram. Expression (4.29) forms the basis for quantitative evaluation of the cut sets.

That is, the probability that the top event occurs in a mission time $t$ is

$$\Pr(T) = \Pr(C_1 \cup C_2 \cup \ldots C_n). \tag{4.31}$$

$\Pr(T)$ in a system reliability framework can be thought of as the unreliability of the system. To understand the complexities discussed earlier for the determination of $\Pr(T)$, let's consider the following situations:

$$C_1 = A \cdot B,$$
$$C_2 = A \cdot C.$$

Then,

$$\Pr(T) = \Pr(A \cdot B + A \cdot C). \tag{4.32}$$

According to (4.7), $\Pr(T) = \Pr(A \cdot B) + \Pr(A \cdot C) - \Pr(A \cdot B \cdot A \cdot C) = \Pr(A \cdot B) + \Pr(A \cdot C) - \Pr(A \cdot B \cdot C)$. If $A, B,$ and $C$ are independent, then

$$\Pr(T) = \Pr(A) \cdot \Pr(B) + \Pr(A) \cdot \Pr(C) - \Pr(A) \cdot \Pr(B) \cdot \Pr(C). \tag{4.33}$$

The determination of the cross-product terms, such as $\Pr(A)\Pr(B)\Pr(C)$ in (4.33), poses a dilemma in the quantitative evaluation of cut sets, especially when the number of the cut sets is large. In general, there are $2^n - 1$ of such terms in cut sets. For example, in the 13 cut sets generated for the pumping example, there are 8,191 such terms. For larger size problems, this can be a formidable job even for powerful mainframe computers.

Fortunately, evaluation of these cross product terms is often not necessary, and the boundary approach shown in (4.30) is quite adequate. As discussed earlier, this is true whenever we are dealing with small probabilities. In these cases, e.g., in (4.33), $\Pr(A) \cdot \Pr(B) \cdot \Pr(C)$ is substantially smaller than $\Pr(A) \cdot \Pr(B)$ and $\Pr(A) \cdot \Pr(C)$. Thus the boundary result can also be used as an approximation of the true reliability or unreliability value of the system. This is often called the **rare event approximation**. Let's assume, that $P(A) = \Pr(B) = \Pr(C) = 0.1$. Then $\Pr(A)\Pr(B) = \Pr(A) \cdot \Pr(C) = 0.01$, and $\Pr(A) \cdot \Pr(B) \cdot \Pr(C) = 0.001$. The latter is smaller than the former by an order of magnitude. Although $\Pr(T) = 0.019$, the rare event approximation yields $\Pr(T) \approx 0.02$. Obviously, the smaller the probabilities of the events, the better the approximation.

### *4.2.2 Success Tree Method*

The success tree method is conceptually the same as the fault tree method. By defining the **desirable** top event, all intermediate and primary events that guarantee the occurrence of this desirable event are deductively postulated. Therefore, if the logical complement of the top event of a fault tree is used as the top event of a success tree, the Boolean structure represented by the fault tree is the Boolean complement of the success tree. Thus, the success tree, which shows the various combinations of success events that guarantee the occurrence of the top event, can be logically represented by **path sets** instead of **cut sets**.

To better understand this problem, consider the simple block diagram shown in Fig. 4.14a. The fault tree for this system is shown in Fig. 4.14b and the success tree in Fig. 4.14c. Fig. 4.15 shows an equivalent representation of Fig. 4.14c.

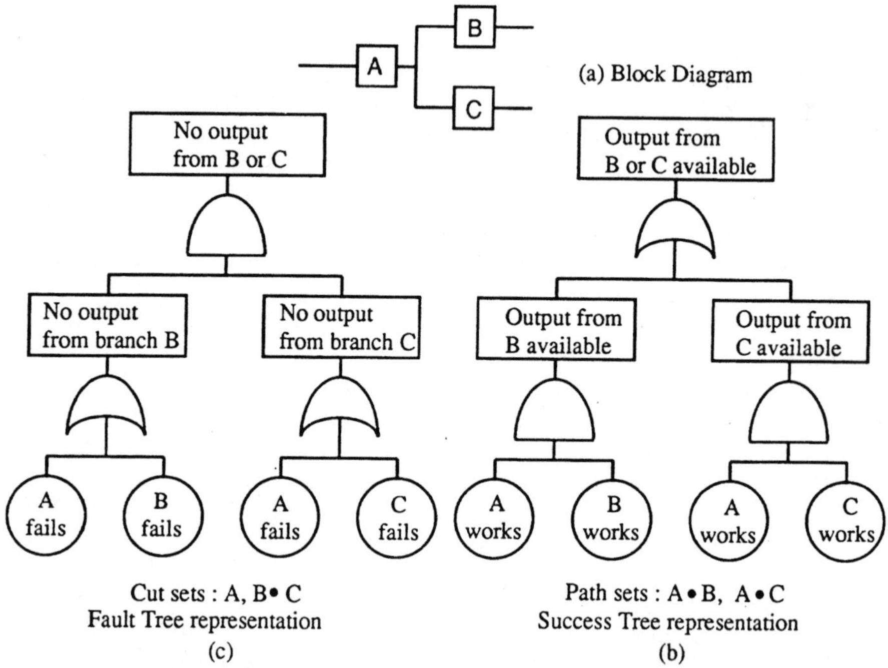


*Fig. 4.14 Representation of a Fault Tree and Success Tree for a System*

By inspecting Fig. 4.14b and Fig. 4.14c, it is easy to see that changing the logic of one tree (changing AND gates to OR gates and vice versa) and changing all primary and intermediate events to their logical complements yields the other tree. This is also true for cut sets and path sets. That is, the logical complement of the cut sets of the fault tree yields the path sets of the equivalent success tree. This can easily be seen in Fig. 4.14. The complement of cut sets is

$$\overline{A + B \cdot C} = (DeMorganTheorem)$$

$$\overline{A} \cdot \overline{B \cdot C} = (DeMorganTheorem)$$

$$\overline{A} \cdot (\overline{B} + \overline{C}) = A \cdot \overline{B} + \overline{A} \cdot \overline{C},$$

which are the path sets.

Qualitative and quantitative evaluations of success paths are mechanistically the same as those of fault trees. For example, the top-down successive substitution of the gates and reduction of the resulting Boolean expression yield the minimal path sets. Accordingly, the use of (4.27), or its lower bound (4.28), allows the determination of the top-event probability (in this case, reliability). As noted earlier, (4.27) poses a computational problem. In this context of using path sets, Wang and Modarres (1990) have described several options for efficiently dealing with this problem.

A convenient way to reduce complex Boolean equations, especially the paths sets, is to use the following expressions:

$$\begin{aligned}\Pr(T) &= \Pr(P_1 \cup P_2 \cup \ldots \cup P_n)\\ &= \Pr(P_1) + \Pr(\bar{P}_1 \cap P_2) + \Pr(\bar{P}_1 \cap \bar{P}_2 \cap P_3) + \ldots\\ &+ \Pr(\bar{P}_1 \cap \bar{P}_2 \cap \ldots \cap \bar{P}_{n-1} \cap P_n).\end{aligned} \tag{4.34}$$

For further discussions in applying (4.34), see Fong and Buzacoot (1987).

Success trees, as opposed to fault trees, provide a better understanding and display of how a system function successfully. While this is important for designers and operators of complex systems, fault trees are more powerful for analyzing of the failures associated with systems and determining the causes of system failures. The minimal path sets of a system show a system user how a system operate successfully. A collection of events in a minimal path set is sometimes referred to as a **success path**. A logical equivalent of a success tree can also be represented by using the top event as an output to an OR gate in which input to the gate would show the success

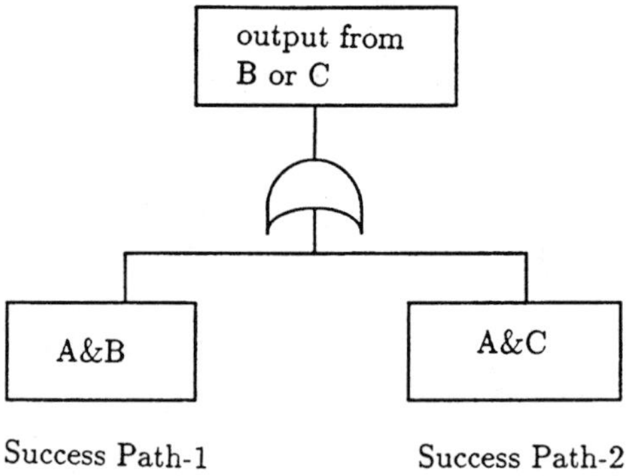


*Fig. 4.15 Equivalent Representation of a Success Tree*

paths. For example, Fig. 4.15 shows the equivalent representation the success tree in Fig. 4.14(c).

In complex systems, the type of representation given in Fig. 4.15 is useful for efficient system operation.

## 4.3 Event Tree Method

If successful operation of a system depends on an approximately chronological, but discrete, operation of its units or subsystems (e.g., units should work in a defined sequence for operational success), then an event tree is appropriate. This may not always be the case for a simple system, but it is often the case for complex systems, such as nuclear power plants where the subsystems should work according to a given sequence of events to achieve a desirable outcome. Event trees are particularly useful in these situations.

### *4.3.1 Construction of Event Trees*

Let's consider the event tree built for a nuclear power plant and shown in Fig. 4.16. The event trees are horizontally built structures that start on the left, where the **initiating event** is modeled. This event describes a situation where a legitimate demand for the operation of a system(s) occurs. Development of the tree proceeds chronologically, with the demand on each unit or subsystem being postulated. The first unit demanded appears first, as shown on the top of the structure. In Fig. 4.16, the events (referred to as **event tree headings**) are as follows:

RP = Operation of the reactor-protection system to shutdown the reactor
ECA = Injection of emergency coolant water by pump A
ECB = Injection of emergency coolant water by pump B
LHR = Long-term heat removal

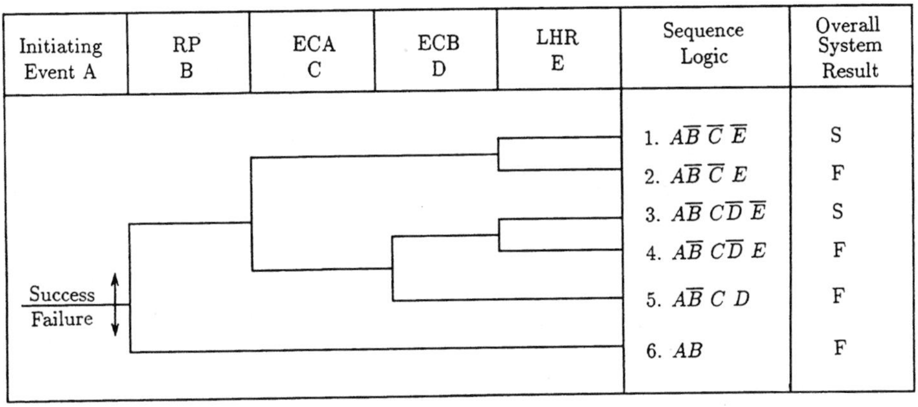


*Fig. 4.16 Simple Event Tree*

At a branch point, the upper branches of an event shows the **success** of the event heading and the lower branch shows its **failure**. In Fig. 4.16, following the occurrence of the initiating event A, unit RP needs to work (event B). If unit RP does not work (it fails), the overall system will fail (as shown by the lower branch of event B). If RP works, then it is important to know whether ECB functions or not. If ECB does not function, even though RP has worked, the overall system would still fail. However, if ECB functions properly, it is important for LHR to function. Successful operation of LHR leads the system to a successful operating state, and failure of LHR (event E) leads the overall system to a failed state. Likewise, if ECA functions, it is important that it be followed by a proper operation of LHR. If LHR fails, the overall system would be in a failed state. If LHR operates successfully, the overall system would be in a success state. It is obvious that certain subsystems (or units) may not be necessary given the Occurrence of some preceding events. For example, if ECA operates successfully it does not matter for the overall system success whether or not ECB operates.

The outcome of each of the sequences of events is shown at the end of each sequence. This outcome, in essence, describes the final outcome of each sequence, whether the overall system succeeds, fails, initially succeeds but fails at a later time, or vice versa. The logical representation of each sequence can also be shown in the form of a Boolean expression. For example, for sequence 5 in Fig. 4.16, events A, C, and D have occurred, but event B has not occurred (shown by $\overline{B}$).

The event trees are usually developed in a binary format; i.e., the heading events are assumed to either occur or not occur. In cases where a spectrum of outcomes is possible, the branching process can proceed with more than two outcomes. In these cases, the qualitative representation of the event tree branches in a Boolean sense would not be possible.

The development of an event tree, although deductive in principle, requires a good deal of inductive thinking by the analyst. To demonstrate this issue and further understand the concept of event tree development, let's consider the system shown in Fig. 4.10. One can think of a situation where the sensing and control system device (S) initiates one of the two pumps. At the same time, the ac power source (AC) should always exist to allow S and pumps P-1 and P-2 to operate. Thus, if we define three distinct events S, AC, and pumping system (PS) for a sequence of events starting with the initiating event, an event tree that includes these three events as its event headings can be constructed. Clearly if AC fails, both PS and S fail; if S fails, only PS fails. This would lead to placing AC as the first event tree heading followed by S and PS. This event tree is illustrated in Fig. 4.17.

Event headings represent discrete states of the systems. These states can be represented by fault trees. This way the event tree sequences and the logical combinations of events can be considered. This is a powerful aspect of event trees. If the event tree headings represent complex subsystems or units, using of a fault tree for each event tree heading can conveniently model the logic. Clearly, other system analysis models, such as reliability block diagrams and logical representations in terms of cut sets or path sets, can also be used.

### *4.3.2 Evaluation of Event Trees*

Qualitative evaluation of event trees is straightforward. The logical representation of each event tree heading, and ultimately each

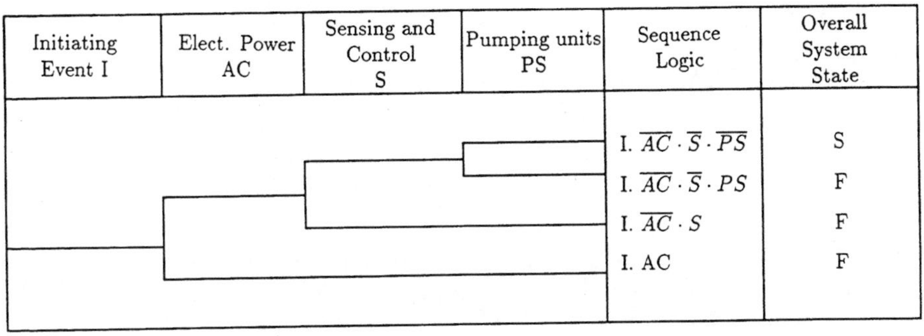

Initiating Event I
Elect. Power AC
Sensing and Control S
Pumping units PS
Sequence Logic
Overall System State
I. $\overline{AC}\cdot\overline{S}\cdot\overline{PS}$
I. $\overline{AC}\cdot\overline{S}\cdot PS$
I. $\overline{AC}\cdot S$
I. AC
S
F
F
F


*Fig. 4.17 Event Tree for the Pumping System*

event tree sequence, is obtained and then reduced through the use of Boolean algebra rules. For example, in sequence 5 of Fig. 4.16, if events $B$, $C$, and $D$ are represented by the following Boolean expressions, the reduced Boolean expression of the sequence can be obtained.

$$A = a,$$
$$B = b + c \cdot d,$$
$$C = e + d,$$
$$D = c + e \cdot h.$$

The simultaneous Boolean expression and reduction proceeds as follows:

$$\begin{aligned} A \cdot \overline{B} \cdot C \cdot D &= a \cdot (\overline{b + c \cdot d}) \cdot (e + d) \cdot (c + e \cdot h) \\ &= a \cdot (\overline{b} \cdot \overline{c} + \overline{b} \cdot \overline{d}) \cdot (e \cdot c + e \cdot h + d \cdot c) \\ &= a \cdot \overline{b} \cdot \overline{c} \cdot e \cdot h + a \cdot \overline{b} \cdot c \cdot \overline{d} \cdot e. \end{aligned}$$

If an expression explaining all failed states is desired, the union of the reduced Boolean equations for each sequence that leads to failure should be obtained and reduced.

Quantitative evaluation of event trees is similar to the quantitative evaluation of fault trees. For example, to determine the probability associated with a $A \cdot \overline{B} \cdot C \cdot D$ sequence,

$$\begin{aligned}\Pr(A \cdot \overline{B} \cdot C \cdot D) &= \Pr(a \cdot \bar{b} \cdot \bar{c} \cdot e \cdot h + a \cdot \bar{b} \cdot c \cdot \bar{d} \cdot e) \\ &= \Pr(a \cdot \bar{b} \cdot \bar{c} \cdot e \cdot h) + \Pr(a \cdot \bar{b} \cdot c \cdot \bar{d} \cdot e) \\ &= \Pr(a)[1 - \Pr(b)][1 - \Pr(c)] \cdot \Pr(e) \cdot \Pr(h) \\ &\quad + \Pr(a) \cdot [1 - \Pr(b)] \Pr(c) \cdot [1 - \Pr(d)] \cdot \Pr(e).\end{aligned}$$

Since the two terms are disjoint, the above probability is exact. However, if the terms are not disjoint, the rare event approximation can be used here.

## 4.4 Failure Mode and Effect Analysis

Failure mode and effect analysis (FMEA) is a powerful technique for reliability analysis. This method is inductive in nature and, in practice, is used in all aspects of system failure analysis from concept to development. FMEA analysis attempts to predict possible sequences of events that lead to system failure, determine their consequences, and devise methods to minimize their occurrence or reoccurrence.

The FMEA procedure consists of a sequence of steps starting with the analysis at one level or a combination of levels of abstraction, such as system functions, subsystems, or components. The analysis assumes a failure mode occurs and causes a failure. The effect of the failure is then determined as well as the causative agent for the failure, which is called the **failure mechanism**. The effect of the failure can be determined at various levels of abstraction starting at the component level. A criticality rating can also be determined for each failure mode and its resulting effect. The rating is normally based on the probability of the failure mode occurrence, the severity of its effect(s), and its detectability. Failures that score high in this rating can potentially be the source of system unreliability, and their causes should be corrected. Clearly, correct usage of FMEA results in improved reliability of an item by identifying potential areas of failures and providing strict documentation on how system failure occurs.

Although FMEA is an essential reliability task for many types of system design and development, it provides very limited insight into probabilistic representation of system reliability. Another limitation is that FMEA is performed for only one failure at a time. This may not be adequate for systems in which multiple failure modes can occur, with reasonable likelihood, at the same time. (Deductive methods are very powerful for identifying these kind of failures.) However, FMEA provides a lot of very valuable qualitative information about the system design and operation. An extension of FMEA is called Failure Mode and Effect Criticality Analysis (FMECA), which provides more quantitative treatment of failures. The methods of FMEA and FMECA are discussed in this section. For more detailed information, the readers are referred to MIL-STD-1629A (1978).

### *4.4.1 FMEA Procedure*

The procedure is straightforward. A table similar to Table 4.1 should be filled out for the level of detail desired. The information required for the columns of this table is obtained from the steps described below.

#### (a) System Description and Block Diagrams

It is important to first describe the system in a manner that allows the FMEA to be performed efficiently and understood by others. This description can be done in different levels of abstraction. For example, at the highest level (i.e., the functional level), the system can be represented by a **functional block diagram**. The functional block diagram is different from the **reliability block diagram** discussed earlier in this chapter. Functional block diagrams illustrate the operation, interrelationship, and interdependence of the functional entities of a system. For example, the pumping system of Fig. 4.10 can be represented by its functional block diagram, as shown in Fig. 4.18. In this figure, the components that support each function are also described.

If the failures are postulated at a lower abstraction level, such levels should be shown. The lowest level is the part level. The diagram used at this level is essentially more detailed than the one shown in Fig. 4.10 (even if the electric power and the sensing and control system are further decomposed to more basic components). The lower the abstraction level, the greater the level of detail required for the analysis. This step provides necessary information for the **identification number, functional identification (nomenclature)**, and **function** columns in the FMEA (Table 4.1).

#### (b) Failure Modes and Causes

The manner of failure of the function, subsystem, component, or part identified in the second column of the table is called the failure-mode and is listed in the **failure mode and causes** column of the

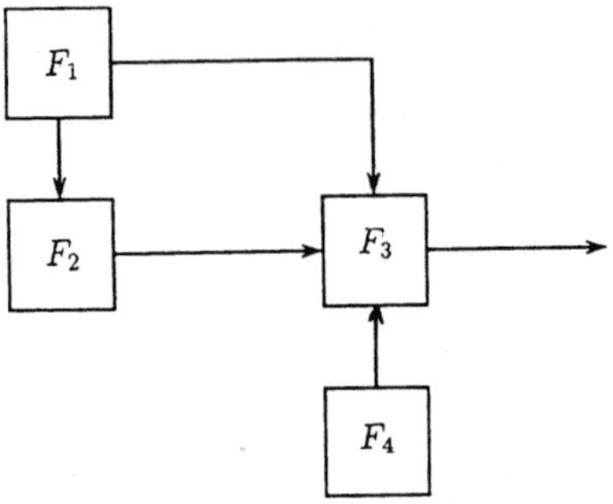


| Function | Functional Description | Components Involved |
|---|---|---|
| $F_1$ | AC Power | AC |
| $F_2$ | Sensing and Control | S |
| $F_3$ | Pumping | V-2, V-3, V-4 V-5, P-1, P-2 |
| $F_4$ | Source | T1, V-1 |

*Fig. 4.18 Functional Block Diagram for the Pumping System*

FMEA table. The causes (a failure mode can have more than one cause) of each failure mode should also be identified and listed in this column. The failure modes applicable to components and parts are often broadly known. For example, Table 4.2 shows failure modes for certain electronic components. Table 3.3 listed some representative failure modes for mechanical component. However, depending on the specific system under analysis, the environmental design, and other factors, only certain failure modes may apply. This should be known and specified by the analyst. Failure modes associated with individual functions of each functional block diagram are obtained or postulated from the stated requirements of the system.

### (c) Effect(s) of Failure

The consequences of each failure mode on the item's operation should be carefully examined and recorded in the column labeled **failure effects**. The effects can be distinguished at three levels: **local**, **next higher abstraction level**, and **end effect**. Local effects specifically show the impact of the postulated failure mode on the operation and function of the item under consideration. The consequence of each failure mode on the operation and functionality of an item under consideration is described as its local effect. It should be noted that sometime no local effects can be described beyond the failure mode itself. However, the consequences of each postulated failure on the **output** of the item should be described

*Table 4.1 FMEA Worksheet Format*

FAILURE MODE AND EFFECTS ANALYSIS

SYSTEM ____________ DATE ____________
INDENTURE LEVEL ____________ SHEET ____ OF ____
REFERENCE DRAWING ____________ COMPILED BY ____________
MISSION ____________ APPROVED BY ____________

| IDENTIFICATION NUMBER | ITEM/FUNCTIONAL IDENTIFICATION (NOMENCLATURE) | FUNCTION | FAILURE MODES AND CAUSES | MISSION PHASE/ OPERATIONAL MODE | FAILURE EFFECTS | | | FAILURE DETECTION METHOD | COMPENSATING PROVISIONS | SEVERITY CLASS | REMARKS |
|---|---|---|---|---|---|---|---|---|---|---|---|
| | | | | | LOCAL EFFECTS | NEXT HIGHER LEVEL | END EFFECTS | | | | |
| | | | | | | | | | | | |

*Table 4.2 Typical Modes of Failure for electrical components*

| *Component* | *FailureMode* |
|---|---|
| Transistors | Intermittent<br>High $I_{CE}$<br>B-C Short<br>Low $BV_{EBO}$<br>High $H_{FE}$ |
| Relays | Open or shorted (closed) contacts<br>Open or shorted coils<br>High contact resistance |
| Transformers | Open windings<br>Shorted windings |
| Resistors | Open<br>Short<br>Drift |
| Connectors | Misalignment<br>Corrosion |
| Capacitors | Short<br>Open<br>Leakage |

along with second order effects. End-effect analysis describes the effect of postulated failure on the operation, function, and status of the next higher abstraction level and ultimately on the system itself. The end effect shown in this column may be the result of multiple failures. For example, the failure of a supporting subsystem in a system can be catastrophic if it occurs along with another local failure. These cases should be clearly recognized and discussed in the end-effect column.

### (d) Failure Detections and Compensation

Failure detection features for each failure mode should be described. For example, previously known symptoms can be used based on the item behavior pattern(s) indicating that a failure has occurred. The described symptom can cover the operation of a component under consideration (logical symptom) or can cover both the component and the overall system, or equipment evidence of failure.

A detected failure should be corrected so as to eliminate its propagation to the whole system and to thus maximize reliability. Therefore, at each abstraction level provisions that will alleviate the effect of a malfunction or failure should be identified. These provisions include such items as a) redundant elements for continued and safe operation, b) safety devices, and c) alternative modes of operation, such as backup and standby units. Any action that may require operator action, should be clearly described. These findings should be described in the columns entitled **failure detection method** and **compensating provisions.**

### (e) Severity Classification

Severity classification is used to provide a qualitative indicator of the worst potential effect resulting from the failure mode. For FMEA purposes, MIL-STD-1629A classifies severity levels in the following categories:

| | |
|---|---|
| Category I, Catastrophic: | A failure mode that may cause death or complete mission loss. |
| Category II, Critical: | A failure mode that may cause severe injury or major system degradation, damage, or reduction in mission performance. |
| Category III, Marginal: | A failure that may cause minor injury or degradation in system or mission performance. |
| Category IV, Minor: | A failure that does not cause injury or system degradation but may result in system failure and unscheduled maintenance or repair. |

The classification selected for each failure mode should be described in the column labeled **severity class.**

### (f) Remarks

Any pertinent information, clarifying items, or notes should be entered in the column labeled **remarks.**

### *4.4.2 FMECA Procedure: Criticality Analysis*

The FMECA procedure is the same as the FMEA procedure in all respects, with one exception: a **criticality analysis** is also performed. Criticality analysis is the combination of a probabilistic determination of the occurrence of the failure modes and the determination of the impact of a failure mode on the reliability of the system (or lower abstraction level). Table 4.3 shows a sample of a criticality analysis (CA) worksheet format. Criticality analysis aspects of this table are explained below.

### (a) Failure Effect Probability $\beta$

The $\beta$ value represents the conditional probability that the failure effect with the specified criticality classification will occur given that the failure mode occurs. For complex systems, $\beta$ is difficult to calculate and thus becomes primarily a matter of judgment, meaning it is greatly driven by the analyst's prior experience. The general guidelines shown in Table 4.4 can be used for determining $\beta$. A similar table based on the severity of a failure mode can be used to determine $\beta$, as shown in Table 4.5. Table 4.5 has $\beta$ values for two types of systems, linear and nonlinear. The value of $\beta$ selected for a failure mode should be recorded in the **failure effect probability** ($\beta$) column of Table 4.3.

*Table 4.4 Failure Effect Probabilities for Various Failure Effects*

| Failure Effect | $\beta$ Value |
|---|---|
| Actual loss | 1.00 |
| Probable loss | $0.1 < \beta \leq 1.0$ |
| Possible loss | $0 < \beta \leq 0.1$ |
| No effect | 0 |

*Table 4.5 Failure Effect Probabilities for Various Severities*

| Severity Class | I | II | III | IV |
|---|---|---|---|---|
| Nonlinear System | 1 | $0.1 < \beta \leq 1$ | $0.05 < \beta \leq 0.1$ | $0 < \beta \leq 0.05$ |
| Linear System | 1 | 0.75 | 0.5 | 0.25 |

### (b) Failure Mode Ratio $\alpha$

The fraction of the item (component, part, etc.) failure rate $\lambda$ related to the particular failure mode under consideration is evaluated and recorded in the **failure mode ratio** ($\alpha$) column. The failure

*Table 4.3 CA Worksheet Format*

CRITICALITY ANALYSIS

SYSTEM ______ DATE ______
INDENTURE LEVEL ______ SHEET ___ OF ______
REFERENCE ______ COMPILED BY ______
MISSION ______ APPROVED BY ______

| IDENTIFICATION NUMBER | ITEM/FUNCTIONAL IDENTIFICAITON (NOMENCLATURE) | FUNCTION | FAILURE MODES AND CAUSES | MISSION PHASE/ OPERATIONAL MODE | SEVERITY CLASS | FAILURE PROBABILITY FAILURE RATE DATA SOURCE | FAILURE EFFECT PROBABILITY ($\beta$) | FAILURE MODE RATIO ($\alpha$) | FAILURE RATE ($\lambda_p$) | OPERATING TIME (T) | FAILURE MODE CRIT # $C_m=\beta\alpha\lambda_p t$ | ITEM CRIT # $C_r=\Sigma(C_m)$ | REMARKS |
|---|---|---|---|---|---|---|---|---|---|---|---|---|---|
| | | | | | | | | | | | | | |

mode ratio is the probability, expressed as a decimal fraction, that the item will fail in the identified mode of failure. If all potential failure modes of an item are listed, the sum of their corresponding $\alpha$ values should be equal to 1. The values of $\alpha$ should normally be available from the data source. However, if not available, the values can be judged by the analyst.

### (c) Failure Rate ($\lambda$)

The generic or specific failure rate for each failure mode of the item should be obtained and recorded in the **failure rate** ($\lambda$) column. This value can be obtained from statistical observations of actual failures during a test or in the field, or from generic sources of failure rates discussed in Section 3.7. Appropriate, applicable stress factors discussed in Section 3.7, such as **environmental** and **design factors**, should be applied to adjust the failure rate. For documentation purposes, the adjusting stress factors often need to be listed next to the $\lambda$ value computed.

### (d) Operating Time $T$

The operating time, in hours, or the number of operating cycles of the item should be listed in the corresponding column.

### (e) Failure Mode Criticality Number $C_r$

This value is an indicator of the number of expected system failures of a specified type (severity) due to the item's failure modes. For a particular severity classification, the $C_r$ for an item is the sum of the failure mode criticality numbers $C_m$ that have the same severity classification.

Thus,

$$C_r = \sum_{i=1}^{n} (C_m)_i, \quad \textit{and} \qquad (4.35)$$

and

$$C_m = \beta\alpha\lambda T,$$

where

$C_m$ = criticality number for a failure mode of an item with a known severity classification,

$n$ = all failure modes of the item having the same severity classification.

If an item's failure modes exhibit more than one severity classification, we would accordingly expect more than one $C_r$ for an item.

The values of $C_m$ and $C_r$ are recorded in the FMECA table columns labeled **failure mode criticality number** and **item criticality number**, respectively.

Based on the criticality number, a criticality matrix can be developed that provides a visual means of identifying and comparing each failure mode to all other failures with respect to severity. For example, Fig. 4.19 shows an example of such a matrix. This matrix can also be used for a qualitative criticality analysis in a FMEA-type study. On the vertical scale of the matrix, the probability of occurrence level (subjectively estimated by the analyst in a FMEA study) or the criticality number $C_r$ (calculated in a FMECA study) is entered. The criticality numbers or probability levels increase from bottom to top. On the horizontal scale the severity classification of an effect is entered. The severity increases from left to right. Each item on the FMEA or FMECA would be represented by one or more points on this matrix. If the item's failure modes correspond to more than one severity effect at the level under consideration, each severity effect will correspond to a different point on the matrix. Clearly those severities that fall in the upper-right quadrant of the matrix deserve immediate attention for reliability or design improvements. These are items that exhibit high probability of failure (or a high criticality number) and whose consequences are severe.

### *4.4.3 Role of FMEA and FMECA in System Reliability Analysis*

The FMECA can be a useful tool for system reliability analysis. To understand this concept better, let's consider a system of two amplifiers in parallel shown in Fig. 4.20. If this system, in a given mission, should function for a period of 72 hours, one can easily determine a FMECA table for this system. The result of this analysis, in summary form, is displayed in Table 4.6.

One can draw the following conclusions for this mission of the system.

1. The system will be expected to critically fail with a probability of $0.0036 + 0.0036 = 0.0072$.
2. The system will experience a critical failure due to system degradation with a probability of $0.0335 \times 10^{-3} \times 2$, which is $6.7 \times 10^{-5}$.
3. The system will experience a critical failure due to "open" circuit failure mode with a probability of $4.47 \times 10^{-3} \times 2 = 8.94 \times 10^{-3}$.

The above approximate probabilities can only hold true if $\alpha\beta\lambda T$ is small (e.g., $< 0.1$). Normally, criticality numbers are used as a

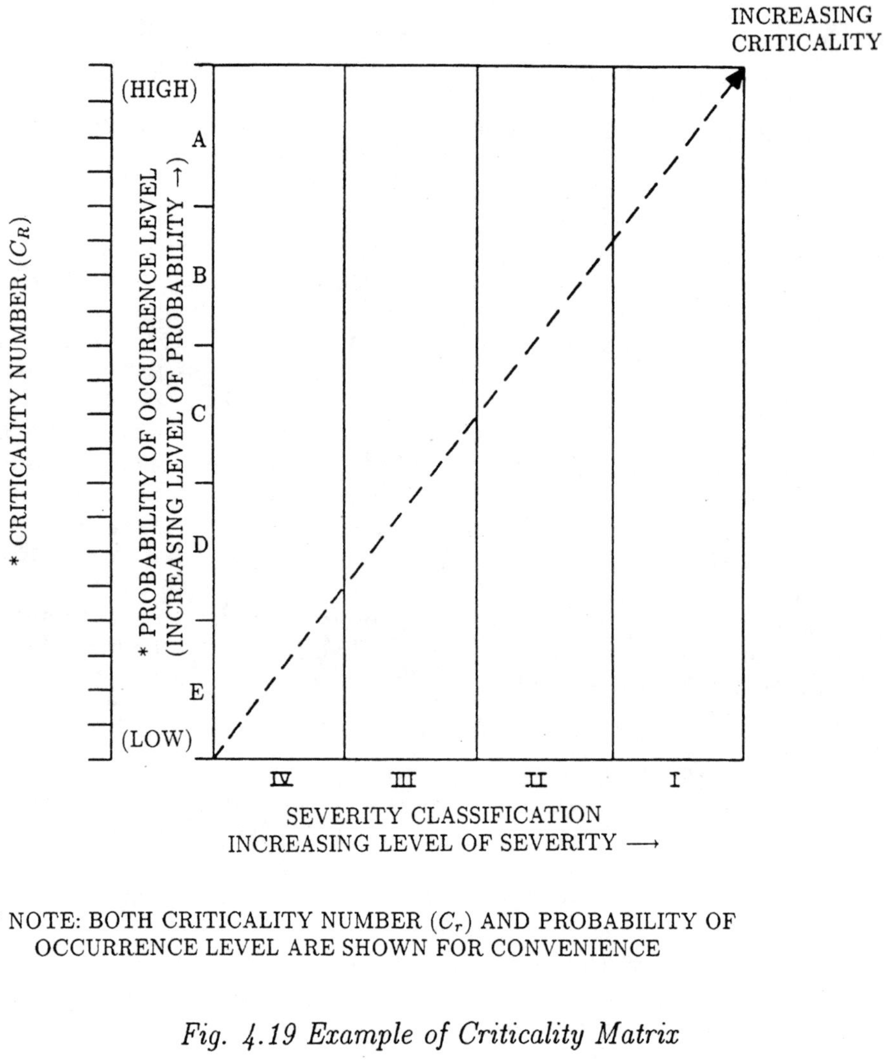


*Fig. 4.19 Example of Criticality Matrix*

measure of index of severity and not as a prediction of system reliability. Therefore, the most effective design would allocate more engineering resources on areas with high criticality numbers, and on minimizing the Class I and Class II severity failure modes.

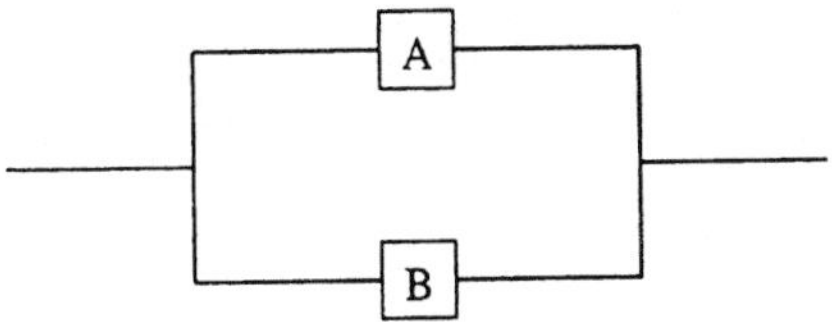


*Fig. 4.20 A System of Two Amplifiers in Parallel*

Conclusion 1 indicates that the failed-short failure mode is the most serious challenge to the system, and it occurs with a probability of 0.0072 (accordingly, $R \approx 0.9928$). Sometimes the failure probability values in FMEA or FMECA tables are used to perform the so called **part count** method. Here, the probability of a given part failure is summed for all items considered in the FMEA or FMECA table. The summation is used for determining the unreliability of the system. However, this is done without regard to the logic of the system. Clearly when redundancy exists, the resulting system unreliability is conservative. The result is an upper bound of the system unreliability that can be calculated with minimal effort. The part count technique therefore should be used only when actual reliability is not desired, and an upper bound only is adequate or when the system is being screened to determine areas where further detailed reliability evaluation should be performed.

The discussion above clearly demonstrates that probabilistic treatments offered by FMEA and FMECA are very limited. However, FMEA studies are very powerful for nonprobabilistic (deterministic) analysis of system design and operation.

## 4.5 Master Logic Diagram

For very complex systems such as a nuclear power plant, modelling for reliability analysis or risk assessment may become very difficult. In complex systems, there are always several subsystems that interact with each other, each of which can be modeled separately. However, it is necessary to find a logical representation of the overall system interactions with respect to the individual subsystems. The master logic diagram (MLD) is such as model.

Consider a functional block diagram of a complex system in which all of the functions modeled are necessary in one way or another to achieve a desired objective. For example, in the context of a nuclear power plant, the independent functions of heat generation, normal heat transport, emergency heat transport, reactor

*Table 4.6 FMECA for the Amplifier System*

| INDENT NO. | CKT NAME | FAILURE MODE | EFFECTS | | SEVERITY CLASS | $\beta$ | $\alpha$ | $\lambda$ | MISSION TIME (HOURS) | CRITICALITY NO. CM | REMARK |
|---|---|---|---|---|---|---|---|---|---|---|---|
| | | | LOCAL | SYSTEM | | | | | | | |
| 1. | A. | a) Open | A ckt failure | Degraded | III | $0.069^{(1)}$ | 0.90 | | | $4.47\times10^{-3}$ | |
| | | b) Short | Both A & B ckt failure | Failure | II | 1.00 | 0.05 | $1\times10^{-3}$ | 72 | $3.6\times10^{-3}$ | May cause secondary failure. |
| | | c) Other | A lost | Degraded | IV | $9.3\times10^{-3(2)}$ | 0.05 | | | $3.35\times10^{-5}$ | |
| 2. | B. | a) Open | B ckt failure | Degraded | III | 0.069 | 0.90 | | | $4.47\times10^{-3}$ | |
| | | b) Short | Both A & B ckt failure | Failure | II | 1.00 | 0.05 | $1\times10^{-3}$ | 72 | $3.6\times10^{-3}$ | May cause secondary failure |
| | | c) Other | B lost | Degraded | IV | $9.3\times10^{-3}$ | 0.05 | | | $3.35\times10^{-5}$ | |

$\Sigma\lambda_{II}=1.8\times10^{-3}$ , $\Sigma\lambda_{III}=1\times10^{-4}$, $\Sigma\lambda_{IV}=1\times10^{-4}$

$\Sigma\lambda = 2\times10^{-3}$ $C_r=\Sigma C_m=16.21\times10^{-3}$

Note: (1) Pr (System failure | A open) $= \beta$ for "A" open mode of failure $= 1 - R_B(72) = 1 - \text{EXP}(-1\times10^{-3} \times 72)$
$= 1 - 0.931 = 0.069$

(2) Assume failure rate doubles due to degradation: $R_{AD} = \text{EXP}(-2\times10^{-3} \times 72)=0.866$

Pr (System failure | A degraded) $= 1-[0.931+0.866-(0.931)(0.866)]= 1-0.99075 = 9.25\times10^{-3}$

shutdown, heat to mechanical conversion, and mechanical to electrical conversion collectively achieve the goal of safely generating electric power. Each of these functions, in turn, is achieved through the design and operating function from others. For example, emergency heat transport may require internal cooling, which is obtained from other so-called support functions.

The MLD clearly show the interrelationships among the independent functions (or systems) and the independent support functions. The MLD (in success space) can show the manner in which various functions, subfunctions, and hardware interact to achieve the overall system objective. On the other hand, an MLD in failed space can show the logical representation of the causes for failure of functions (or systems), and can easily map the propagation of the effect of failures.

In essence, the hierarchy of the MLD is displayed by the dependency matrix. For each function, subfunction, subsystem, and hardware item shown on the MLD, the effect of failure or success of all combinations of items modeled on the MLD is established and explicitly shown by •.

Consider the MLD shown (in success space) in Fig. 4.21. In this diagram, there are two major functions (or systems), $F_1$ and $F_2$. Together, they achieve the desired system objective. Each of these functions, because of reliability concerns is further divided into two identical subfunctions, each of which can achieve the respective parent functions. This means that both subfunctions of $F_1$ or $F_2$ must be lost for $F_1$ or $F_2$ to be lost. Suppose the development of the subfunctions (or systems) can be represented by their respective hardware, which interface with other support functions (or support systems) $S_1$, $S_2$, and $S_3$. However, function (or system) $S_1$ can be divided into two independent subfunctions (or systems) ($S_{1-1}$ and $S_{1-2}$), each of which can interact independently with the subfunctions (or systems) of $F_1$ and $F_2$. The dependency matrix is established by reviewing the design specifications or operating manuals that describe the relationship between the items shown in the MLD, in which the dependencies are explicitly shown by. For instance, the dependency matrix shows that failure of $S_3$ leads directly to failure of $S_2$ , which in turn results in failures of $F_{1-1}$, $F_{2-2}$, and $F_{2-1}$. This failure is highlighted on the MLD in Fig. 4.21.

A key element in the development of an MLD is assurance that the items for which the dependency matrix is developed (e.g., $S_{1-1}$, $S_{1-2}$, $S_3$, $F_{2-1}$, $F_{1-2}$, and $F_{2-2}$) are all physically independent. Physically independent mean they do not share any other system hardware or equipment. Sometimes it is difficult, a priori, to distinguish between main functions and supporting functions. In these situations, the dependency matrix can be developed irrespective of the main and supporting functions. Fig. 4.22 shows an example of such a development. However, the main functions can be identified easily by examining the resulting MLD; they are those functions that appear hierarchically, at the top of the MLD model and do not support other items.

The analysis of an MLD is straightforward. One must determine all possible $2^n$ combinations of failures of independent supporting items in the MLD and propagate their effects in the MLD structure. In practice, for example, in a nuclear plant, n is less than or about 50. However a large number of these combinations are eliminated (e.g. based on low probability and only several hundred to several thousand combinations require system unreliability.) Table 4.7 shows such combinations for Fig. 4.21. For reliability calculations, one can combine those end-states that are identical into a state that sometimes is referred to as an impact vector. Suppose independent items (e.g., systems or subsystems) $S_{1-1}$, $S_{1-2}$ $S_2$, and $S_3$ have a failure probability of 0.01 for a given mission, and the probability of independent failure of $F_{1-1}$, $F_{1-2}$, $F_{2-1}$, and $F_{2-2}$ is also 0.01. Table 4.8 shows the resulting combination of identical end-states. If needed, calculation of failure probabilities for the independent MLD items(e.g., subsystems) can proceed independent of the MLD, through one of the conventional system reliability analysis methods.(e.g., fault free analysis). Accordingly,

$$\Pr(\overline{S}) = \sum_i \Pr(e_i) \cdot \Pr(\overline{S}|e_i), \tag{4.36}$$

where

$\Pr(\overline{S})$ = system failure probability (unreliability),
$\Pr(e_i)$ = probability of the end state i,
$\Pr(\overline{S} \mid e_i)$ = probability of system failure given end state $i$.

The result of applying (4.36) to the results of Table 4.8 are summarized in Table 4.9. The reliability of the system for this mission is 0.9997. In certain complex systems, the major functions (front-line functions) must fail in a given sequence of events or in a given chronological sequence to cause system failure. In these cases, $\Pr(\overline{S} \mid e_i)$ should be determined in connection with an event tree model of the major functions.

It is clear that two of the combined end states are dominant (i.e., $F_{1-1} \times F_2$ and $F_1 \times F_2$). One can inspect these states, backward, to

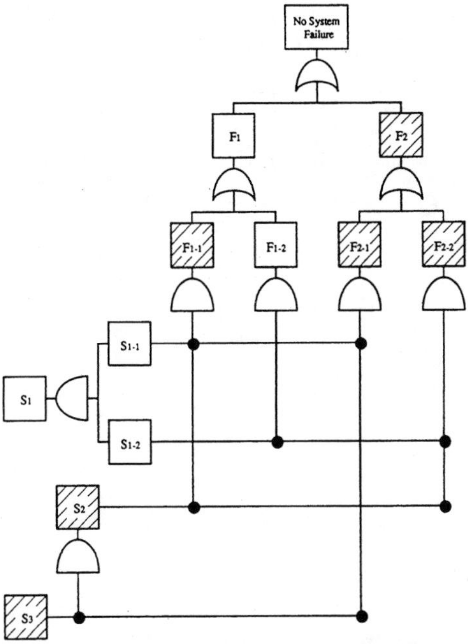


*Fig. 4.21 Master Logic diagram*

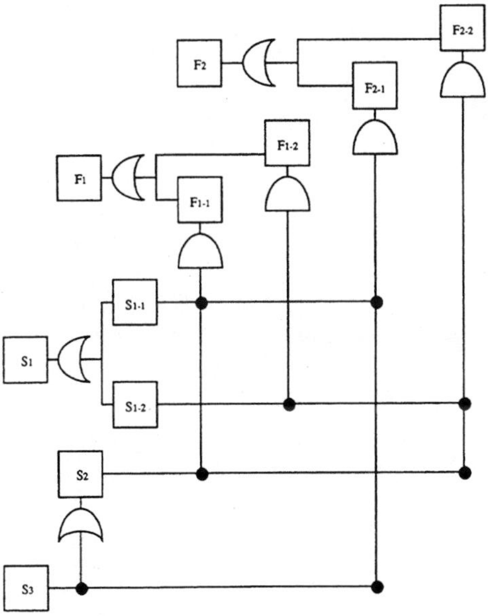


*Fig. 4.22 MLD With All System Functions Treated Similarly*

*Table 4.7 Combination of Support Function Failures and Their Respective Probabilities*

| Combination No. (i) | Failed Support Function ( or Equipment ) | Support Functions ( or Equip. ) Failed Because of Dependency | Probability $Pr(e_i)$ | End State |
|---|---|---|---|---|
| 1 * | $S_{1-1}$ | -- | $9.7 \times 10^{-3}$ ** | $F_{1-1}, F_{2-1}$ |
| 2 | $S_{1-2}$ | -- | $9.7 \times 10^{-3}$ | $F_{1-2}, F_{2-2}$ |
| 3 | $S_2$ | -- | $9.7 \times 10^{-3}$ | $F_{1-1}, F_{2-2}$ |
| 4 | $S_3$ | $S_2$ | $9.8 \times 10^{-3}$ *** | $F_{1-1}, F_2$ |
| 5 | $S_{1-1}, S_{1-2}$ | -- | $1.0 \times 10^{-4}$ **** | $F_1, F_2$ |
| 6 | $S_{1-1}, S_2$ | -- | $9.8 \times 10^{-5}$ | $F_{1-1}, F_2$ |
| 7 | $S_{1-1}, S_3$ | $S_2$ | $9.9 \times 10^{-5}$ | $F_{1-1}, F_2$ |
| 8 | $S_{1-2}, S_2$ | -- | $9.8 \times 10^{-5}$ | $F_1, F_{2-2}$ |
| 9 | $S_{1-2}, S_3$ | $S_2$ | $1.0 \times 10^{-4}$ | $F_1, F_2$ |
| 10 | $S_2, S_3$ | -- | $9.9 \times 10^{-5}$ | $F_{1-1}, F_2$ |
| 11 | $S_{1-1}, S_{1-2}, S_2$ | -- | $1.0 \times 10^{-6}$ | $F_1, F_2$ |
| 12 | $S_{1-1}, S_{1-2}, S_3$ | $S_2$ | $1.0 \times 10^{-6}$ | $F_1, F_2$ |
| 13 | $S_{1-1}, S_2, S_3$ | -- | $9.9 \times 10^{-7}$ | $F_{1-1}, F_2$ |
| 14 | $S_{1-2}, S_2, S_3$ | -- | $1.0 \times 10^{-6}$ | $F_1, F_2$ |
| 15 | $S_{1-1}, S_{1-2}, S_2$ $S_3$ | -- | $1.0 \times 10^{-8}$ | $F_1, F_2$ |
| 16 | No Failure | -- | 0.961 | No Failure |

* A total of $2^4$ combinations are expected

** $Pr(e_1) = Pr(S_{1-1}) \times Pr(\overline{S_{1-2}}) \times Pr(\overline{S_2}) \times Pr(\overline{S_3})$
$= 0.01 \times 0.99 \times 0.99 \times 0.99 = 9.7 \times 10^{-3}$

*** $Pr(e_4) = Pr(\overline{S_{1-1}}) \times Pr(\overline{S_{1-2}}) \times Pr(S_3)$
$= 0.99 \times 0.99 \times 0.01 = 9.8 \times 10^{-3}$

**** $Pr(e_5) = Pr(S_{1-1}) \times Pr(S_{1-2})$
$= 0.01 \times 0.01 = 1.0 \times 10^{-4}$ , since failure or success of other supporting items would not change the end state.

*Table 4.8 Conditional System Failure Probability for Combined States*

| State No. | Combined End State | $Pr(e_i)$ | No. of States Combined | $Pr(\bar{S} \mid e_i)$ |
|---|---|---|---|---|
| 1 | $F_{1\text{-}1}, F_{2\text{-}1}$ | $9.7 \times 10^{-3}$ | 1 | $1.0 \times 10^{-4}$ * |
| 2 | $F_{1\text{-}2}, F_{2\text{-}2}$ | $9.7 \times 10^{-3}$ | 1 | $1.0 \times 10^{-4}$ |
| 3 | $F_{1\text{-}1}, F_{2\text{-}2}$ | $9.7 \times 10^{-3}$ | 1 | $1.0 \times 10^{-4}$ |
| 4 | $F_{1\text{-}1}, F_2$ | $1.0 \times 10^{-2}$ | 5 | $1.0 \times 10^{-2}$ |
| 5 | $F_1, F_2$ | $2.0 \times 10^{-4}$ | 6 | 1.0 |
| 6 | $F_1, F_{2\text{-}2}$ | $9.8 \times 10^{-5}$ | 1 | $1.0 \times 10^{-2}$ |
| 7 | No Failures | 0.961 | 1 | $1.0 \times 10^{-8}$ |

* Obtained from $Pr(\bar{S} \mid e_i) = Pr(\text{failure of both } F_1 \text{ and } F_2 \mid e_i, \text{ failure of } F_{1\text{-}1} \text{ and } F_{2\text{-}1})$
$= Pr(F_{1\text{-}2}) \times Pr(F_{2\text{-}2}) = 1.0 \times 10^{-4}$.

*Table 4.9 System Failure Probability*

| Combined End State | System Failure for State i | Percent Contribution |
|---|---|---|
| $F_{1\text{-}1}, F_{2\text{-}1}$ | $9.7 \times 10^{-7}$ * | 0.32 |
| $F_{1\text{-}2}, F_{2\text{-}2}$ | $9.7 \times 10^{-7}$ | 0.32 |
| $F_{1\text{-}1}, F_{2\text{-}2}$ | $9.7 \times 10^{-7}$ | 0.32 |
| $F_{1\text{-}1}, F_2$ | $1.0 \times 10^{-4}$ | 32.91 |
| $F_1, F_2$ | $2.0 \times 10^{-4}$ | 65.80 |
| $F_1, F_{2\text{-}2}$ | $9.8 \times 10^{-7}$ | 0.33 |
| No Failures | $9.6 \times 10^{-9}$ | 0.00 |
| | $3.04 \times 10^{4}$ | 100 |

* $Pr(\bar{S}) = (9.7 \times 10^{-3})(1.0 \times 10^{-4}) = 9.7 \times 10^{-7}$.

identify areas that could be improved, or to identify human actions that could interrupt the progression of failure events or recover the sequence. For example, the state $F_{1-1} \times F_2$ is highly influenced by the failure of $S_3$ (i.e.,98% of the $F_{1-1} \times F_{1-2}$ state is due to the failure of $S_3$). Thus, failure of $S_3$, alone contributes 32.2% of the total yearly system failure probability. Clearly, if resources or efforts are being planned to improve system reliability, improvement in $S_3$ seems most plausible. For additional discussion about the MLD method see Modarres (1992).

---

*Example 4.6*

Consider the H-Coal process shown in Fig. 4.23. In case of an emergency, a shutdown device (SDD) is used to shutdown the hydrogen flow. If the reactor temperature is too high, an emergency cooling system (ECS) is also needed to reduce the reactor temperature. To protect the process plant when the reactor temperature becomes too high, both ECS and SDD must succeed. The SDD and ECS are actuated by a control device. If the control device fails, the emergency cooling system will not be able to work. However, an operator can manually operate (OA) the shutdown device and terminate the hydrogen flow. The power for the SDD, ECS, and control device comes from an outside electric company (off-site power-OSP). The failure data for these systems are listed in Table 4-10. Draw an MLD and use it to find the probability of losing both the SDD and ECS.

*Solution:*

The MLD is shown in Fig. 4.24. All combinations of independent failures of support system event sequences and their impacts on the SDD and ECS are listed in Table 4.11. The success parts of the support system sequences are not shown since the success probabilities are rather high (close to one). The combinations with the same impact on SDD and ECS are grouped together into end states and are listed in Table 4.12. The probability of losing both the ECS and SDD for each end state is calculated and listed in the third column of Table 4.12. Using (4.35), the probability of losing both systems can be calculated.

$$\begin{aligned}\Pr(\text{end state 1}) =& (\text{probability of combination1}) \\ &+ (\text{probability of combination 3}) \\ =& 9.7 \times 10^{-1} + 9.8 \times 10^{-3} = 0.98.\end{aligned}$$

Pr(SDD and ECS failure due to end state 1)

$$= \Pr(\text{end state 1}) \times \Pr(\text{SDD failure — end state}) \times \Pr(\text{ECS failure — end state 1}),$$

$$= (0.98) \times (1 \times 10^{-3}) \times (1 \times 10^{-3}) = 9.8 \times 10^{-7}.$$

Similarly, the above probabilities can be calculated for other end states. These are shown in Table 4.12. The total process failure probability would be the summation of all the probability contributions from each end state. Thus,

$$\begin{aligned} &\Pr(\text{total process failure given reactor temperature too high}) \\ &= 9.8 \times 10^{-7} + 9.8 \times 10^{-7} + 2.0 \times 10^{-2} \\ &\approx 2.0 \times 10^{-2}. \end{aligned}$$

Clearly, the primary contribution to failure comes from state 3. Thus, any anticipated improvement in system reliability should focus on improving failures that give rise to this state.

*Table 4.10 Failure Probability of Each System*

| System failure | Failure probability |
|---|---|
| OSP | $2.0 \times 10^{-2}$ |
| OA | $1.0 \times 10^{-2}$ |
| ACS | $1.0 \times 10^{-3}$ |
| SDD | $1.0 \times 10^{-3}$ |
| ECS | $1.0 \times 10^{-3}$ |

*Table 4.11 Support System End Sequence*

| Combination number | Failed support system | Failed front line (the impact vector) | Probability of this sequence |
|---|---|---|---|
| 1 | None | None | $9.7 \times 10^{-1}$ |
| 2 | ACS | ECS | $9.8 \times 10^{-4}$ |
| 3 | OA | None | $9.8 \times 10^{-3}$ |
| 4 | OA and ACS | ECS and SDD | $9.8 \times 10^{-6}$ |
| 5 | OSP(ACS will be lost) | ECS and SDD | $2.0 \times 10^{-2}$ |
| 6 | OSP and OA (ACS will be lost) | ECS and SDD | $2.0 \times 10^{-4}$ |

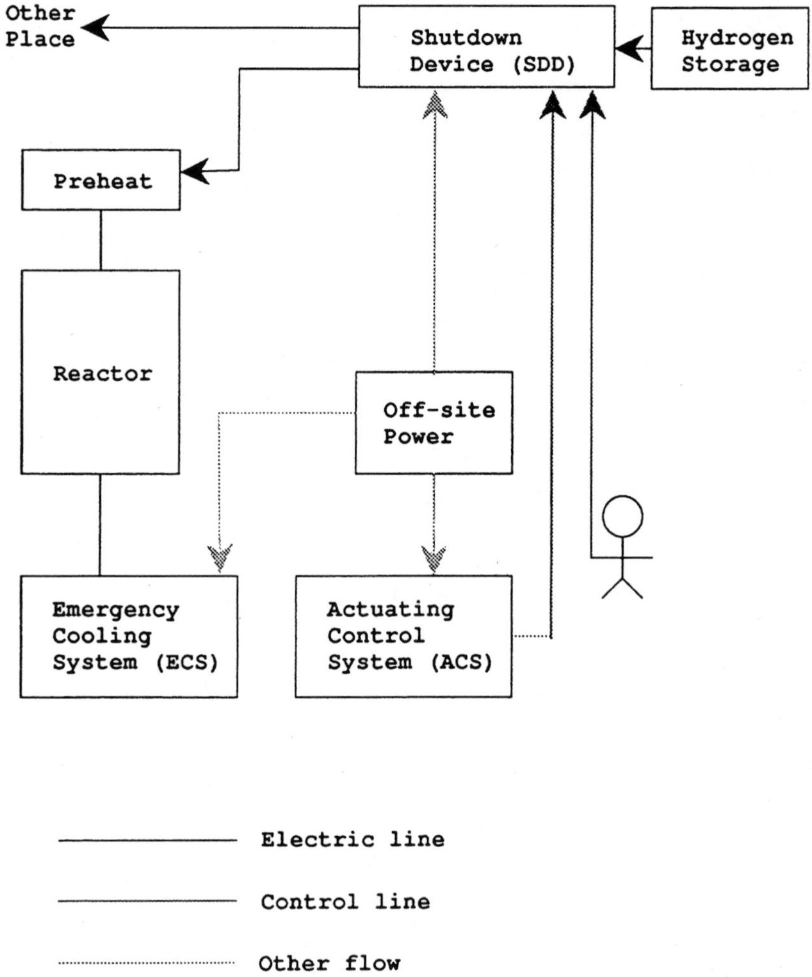


*Fig. 4.23 Simplified Diagram of the Safety Systems*

*Table 4.12 Support System End State and Probability of Losing Two Systems*

| State # | Contain sequence # | Impact | Probability of this state | Probability of losing two system |
|---|---|---|---|---|
| 1 | 1, 3 | None | $9.8 \times 10^{-1}$ | $9.8 \times 10^{-7}$ |
| 2 | 2 | ECS lost | $8.8 \times 10^{-4}$ | $9.8 \times 10^{-7}$ |
| 3 | 4, 5, 6 | All lost | $2.0 \times 10^{-2}$ | $2.0 \times 10^{-2}$ |

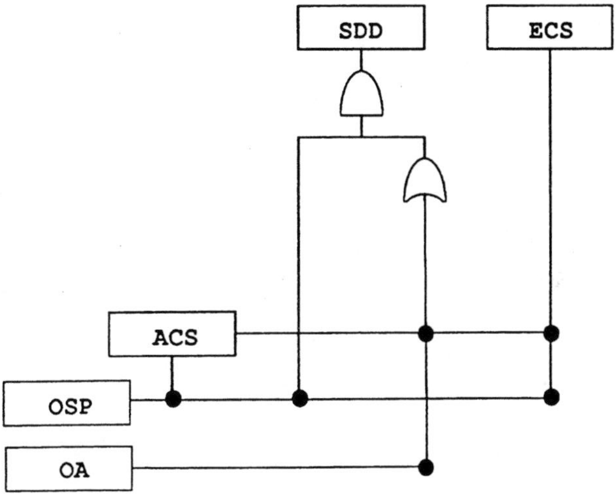


*Fig. 4.24 MLD for the Safety System in Fig. 4.23*

## BIBLIOGRAPHY

1. Fong, C.C. and J.A. Buzacoot (1987). "An Algorithm for Symbolic Reliability Combination with Path-Sets or Cut-Sets," *IEEE Trans. Reliability* , 36 (1), 34-37.

2. PRA Procedures Guide (1982). U.S. Nuclear regulatory Commission, Washington, D.C..
3. MIL-STD-1629A (1980). *Procedure for Performing a Failure Mode, Effects, and Criticality Analysis*, Department of Defense, Washington, D.C..
4. Shooman, M.L. (1990). *Probabilistic Reliability: An Engineering Approach*, 2nd Ed., Kreiger, Melbourne, FL.
5. Vesely, W.E., Goldberg F., Roberts N. and Haasl D. (1981).*Fault Tree Handbook*, NUREG-0492, U.S. Nuclear Regulatory Commission, Washington, D.C..
6. Wang, J. and Modarres, M. (1990). "REX : An Intelligent Decision and Analysis Aid for Reliability and Risk Studies," *Rel. Eng. & Syst. Safety J.*, 30, 185-239.
7. Modarres, M. (1992). "Application of the Master Plant Logic Diagram in Risk-Based Evaluations," Amer. Nucl. Society Topical Mtg. on Risk Management, Boston, MA.

## Exercises

4.1 Consider the circuit below.

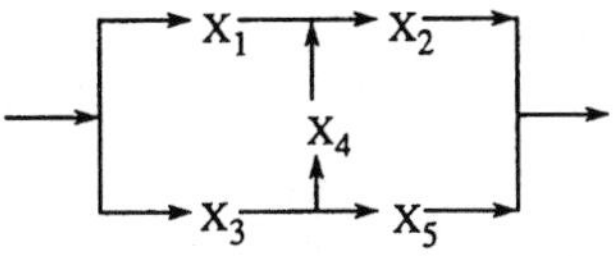


Assume the reliability of each circuit $R(x_i) = \exp(-\lambda_i t)$, and $\lambda_i = 2.0 \times 10^{-4}(\text{hr}^{-1})$ for all $i$. Find the following:

a) Minimal path sets
b) Minimal cut sets
c) MTTF
d) Reliability of the system at 1,000 hour using Bayes' theorem

4.2 Consider the circuit below:

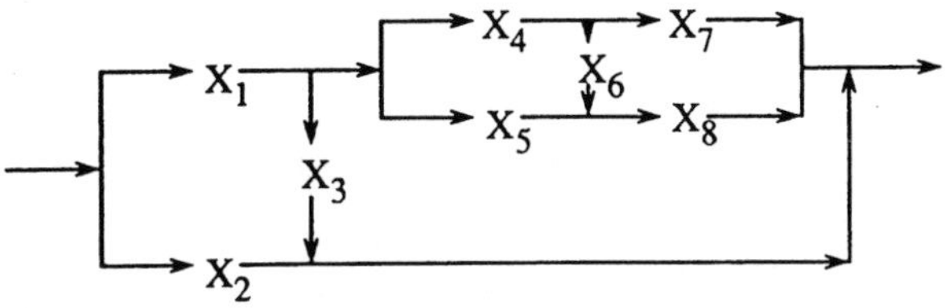


$$\Pr(R(x_i)) = a = \exp(-\lambda_i t), \quad i = 1, 2, \ldots, 8$$

$$\lambda_i = 0.1/1000 \text{ hour}, \quad \text{for all i}$$

Find the following:

a) Minimal path sets
b) Minimal cut sets
c) Reliability of the system at 1,000 hours
d) Probability of failure at 1,000 hours, using cut sets, to verify results from c)
e) Accuracy of the results of d) and/or c), using an approximate method
f) MTTF of the system

4.3 Calculate the reliability of the system shown in the figure below for a 1,000-hour mission. What is the MTTF for this system?

$$\lambda_1 = 1 \times 10^{-5}/\text{hr}, \qquad \lambda_2 = 10 \times 10^{-5}/\text{hr}^{-1}$$

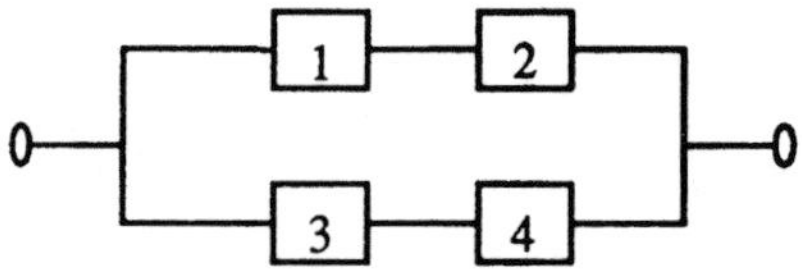


$\lambda_3 = 2 \times 10^{-5}/\text{hr}, \qquad \lambda_4 = 5 \times 10^{-5}/\text{hr}^{-1}$

4.4 Consider the piping system shown in the figure below. The purpose of the system is to pump water from point A to point B. All the valves and the pump can be represented by exponential distributions with failure rates $\lambda_v$ and $\lambda_p$ respectively.

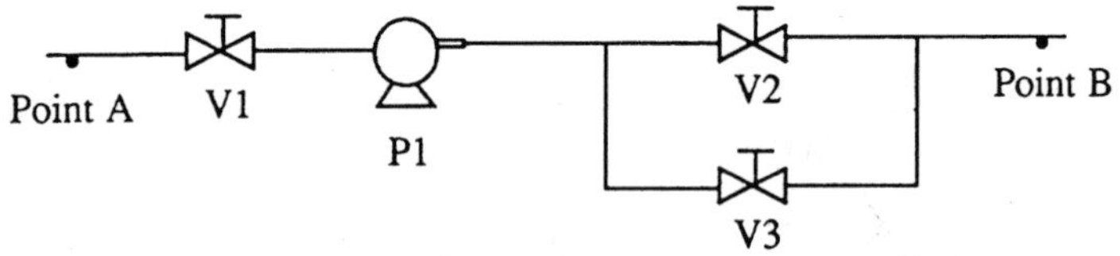


$(\lambda_{v1} = \lambda_{v2} = \lambda_{v3} = \lambda_v)$

a) Calculate the reliability function of the system. (One of the valves, V2 or V3, is sufficient for success).
b) If $\lambda_v = 10^{-3}(\text{hr}^{-1})$ and $\lambda_p = 2 \times 10^{-3}(\text{hr}^{-1})$, and the system has survived for 10 hours, what is the probability that it will survive another 10 hours?

4.5 Determine the reliability of the following reliability block diagram for a 2,500-hour mission. Assume a failure rate of $10^{-6}$/hour for each unit.

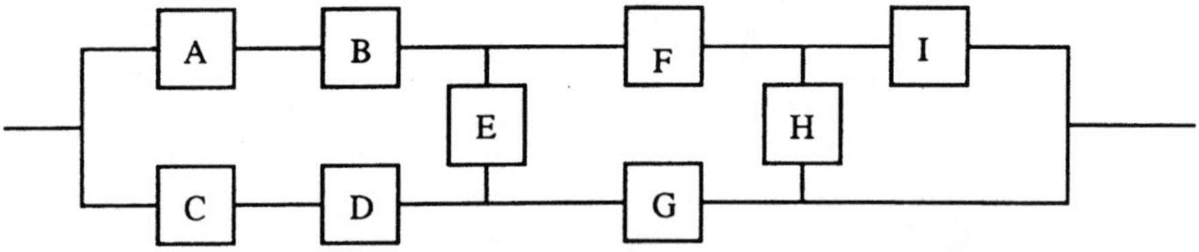


4.6 A containment spray system is used to scrub and cool the atmosphere around a nuclear reactor during an accident. Develop a fault tree using "No $H_2O$ spray" as the top event. Assume the following conditions:

- There are no secondary failures
- There is no test and maintenance
- There are no passive failures
- There are independent failures
- One of the two pumps and one of the two spray heads is sufficient to provide spray. (Only one train is enough)
- One of the valves sv1 or sv2 is opened after demand. However, sv3 and sv4 are always normally open.
- Valve sv5 is always in the closed position.
- There is no human error.
- SP1, SP2, sv1, sv2, sv3, and sv4 use the same power source P to operate

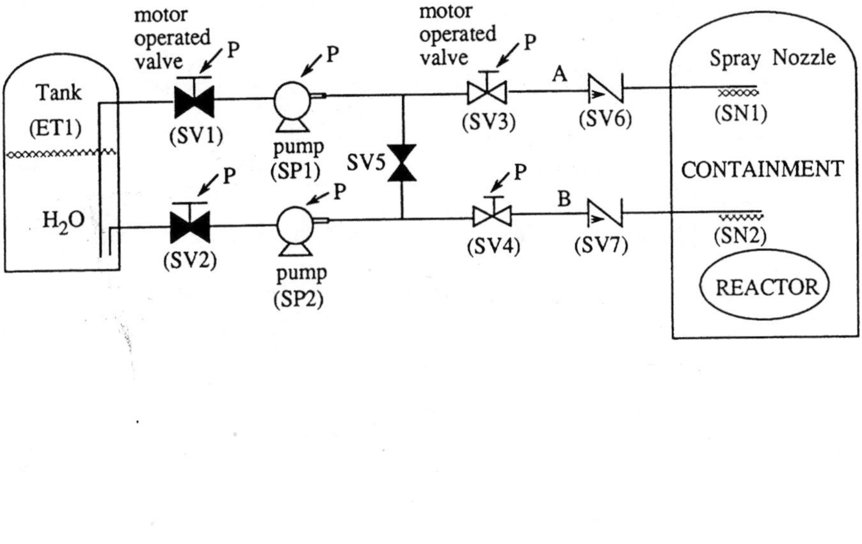

4.7 Consider the fault tree below.

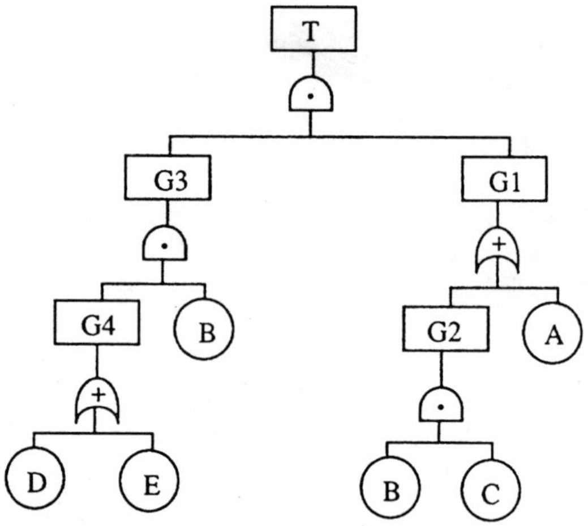


Find the following:

a) Minimal cut sets

b) Minimal path sets

c) Probability of the top event if the following probabilities apply:

$$P_1(A) = \Pr(C) = \Pr(E) = 0.01$$

$$\Pr(B) = \Pr(D) = 0.0092$$

4.8 Consider the pumping system below. System cycles every hour. Ten minutes are required to fill the tank. Timer is set to open contact 10 minutes after switch is closed. Operator opens switch or the tank emergency valve if he/she notices an overpressure alarm. Develop a fault tree for this system with the top event "Tank ruptures"

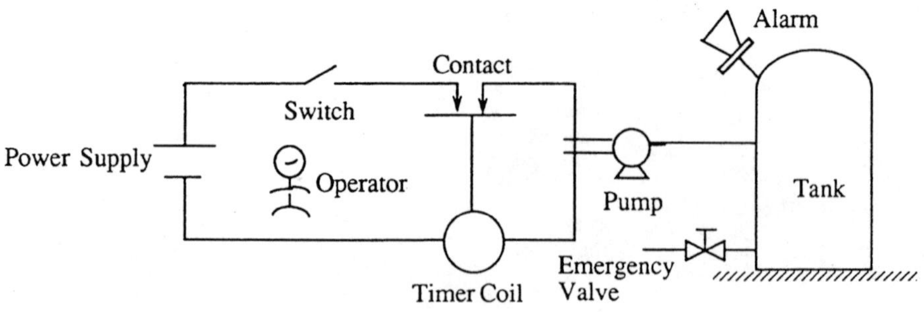

4.9 Find the cut sets and path sets of the fault tree shown below using the top-down substitution method.

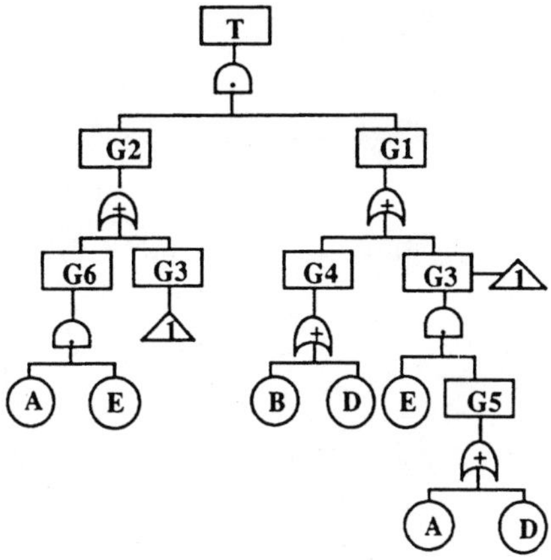


4.10 Compare Design 1 and Design 2 below.

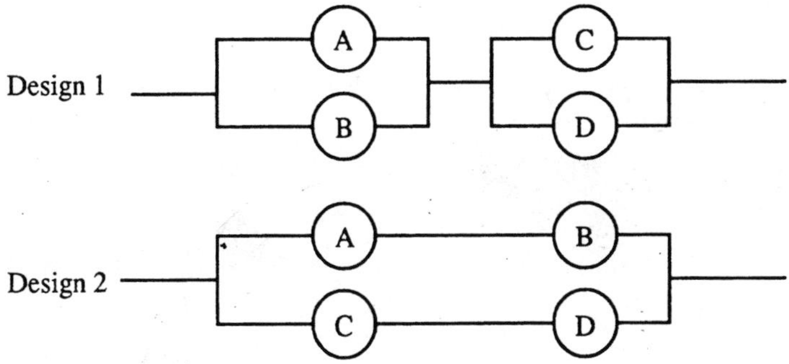

Failure rates ($hr^{-1}$)
$\lambda_A = 10^{-6}$ $\lambda_C = 10^{-3}$
$\lambda_B = 10^{-6}$ $\lambda_D = 10^{-3}$

a) Assume that the components are nonrepairable. Which is the better design?

b) Assume that our envisioned system failure probability cannot exceed $10^{-2}$. What is the operational life for Design 1 and for Design 2.

4.11 Consider the following simple electric circuit for providing emergency light during a blackout.

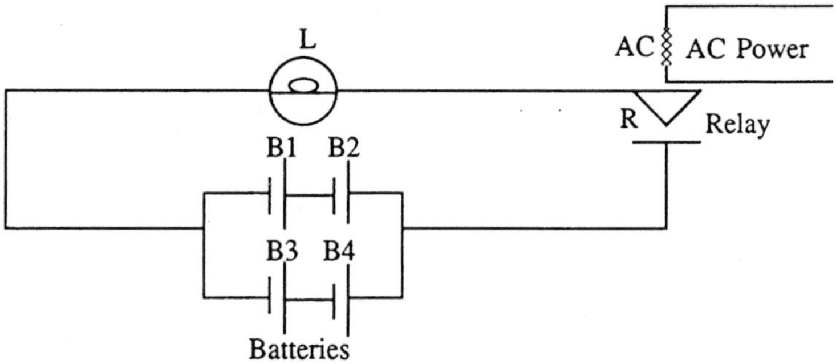


In this circuit, the relay is held open as long as AC power is available, and either of the four batteries is capable of supplying light power. Start with the top event "No Light When Needed."

a) Draw a fault tree for this system.

b) Find minimal cut sets of the system.

c) Find minimal path sets of the system.

4.12 Consider the following pumping system consisting of three identical parallel pumps, and a valve in series. The pumps have a constant failure rate of $\lambda_p(hr^{-1})$ and the valve has a constant failure rate of $\lambda_v(hr^{-1})$ (accidental closure).

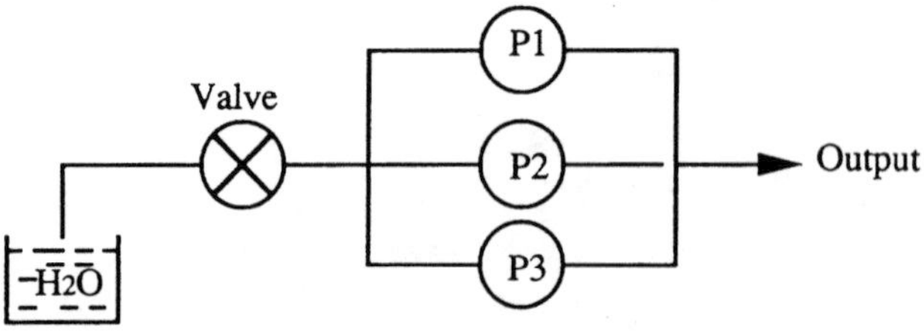

a) Develop an expression for the reliability of the system using the success tree. (Assume nonrepairable components.)
b) Find the average reliability of the system over the time period $T$.
c) Repeat questions (a) and (b) for the case that $\lambda t < 0.1$, and then approximate the reliability functions. Find the average reliability of the system when $\lambda = 0.001(\text{hr}^{-1})$ and $T = 10$ (hours).

4.13 In the following system, which uses active redundancy, what is the probability that there will be no failures in the first year of operation? Assume a constant hazard rate.

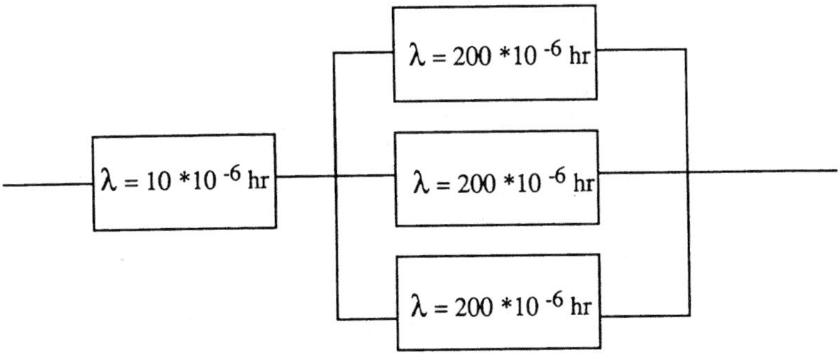


4.14 A filter system is composed of 30 elements, each with a failure rate of $2 \times 10^{-4}(\text{hr}^{-1})$. The system will operate satisfactorily with two elements failed. What is the probability that the system will operate satisfactorily for 1,000 hours?

4.15 Consider the reliability diagram below.
a) Find all minimal path sets (by inspection or any other method desired).
b) Find all minimal cut sets (by inspection or otherwise).
c) Assuming each component has a reliability of 0.90, compute the system reliability.

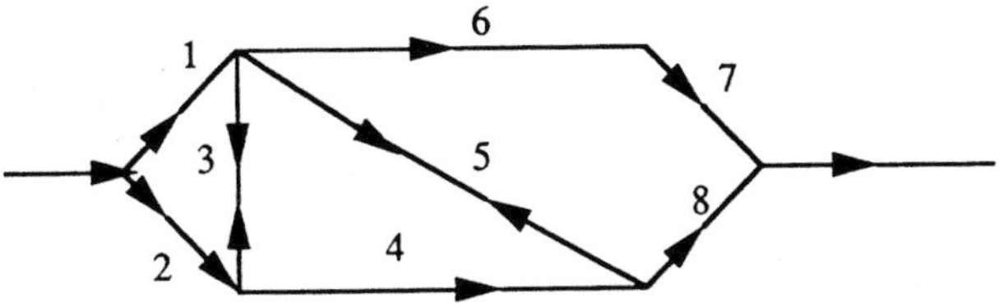

4.16 In the following fault tree, find all minimal cut sets and path sets. Assuming all component failure probabilities are 0.01, find the top event probability.

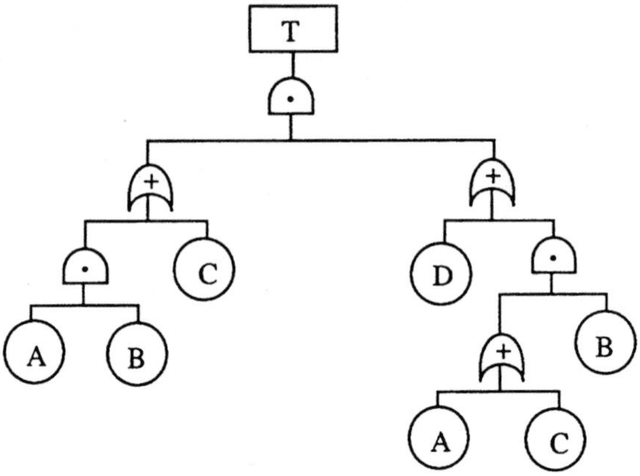


4.17 An event tree is used in reactor accident estimation as shown:

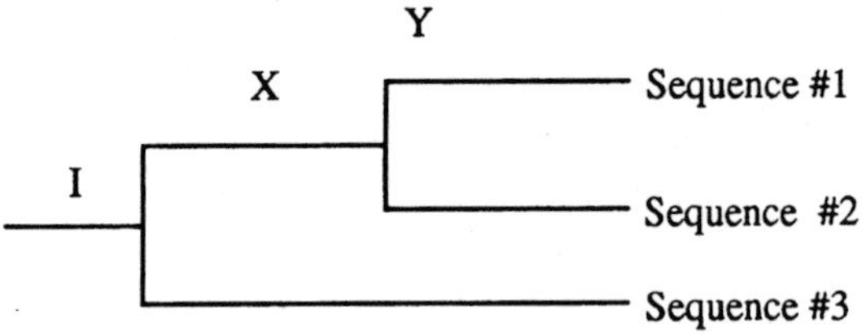


where sequence 1 is a success and sequences 2 and 3 are failures. The cut sets of system X and Y are

$$X = A \cdot B + A \cdot C + D, \quad \text{and}$$

$$Y = B \cdot D + E + A.$$

Find cut sets of sequence 2 and sequence 3.

4.18 In a cement production factory, a system such as the one shown below is used to provide cooling water to the outside of the furnace. Develop an MLD for this system.

| | |
|---|---|
| System bounds: | $S_{12}, S_{11}, S_9, S_8, S_7, S_6, S_5, S_4, S_3, S_2, S_1$ |
| Top event: | Cooling from legs 1 and 2 |
| Not-allowed events: | Passive failures and external failures |
| Assumptions: | Only one of the pumps or legs is sufficient to provide the necessary cooling. Only one of the tanks is necessary as a source. |

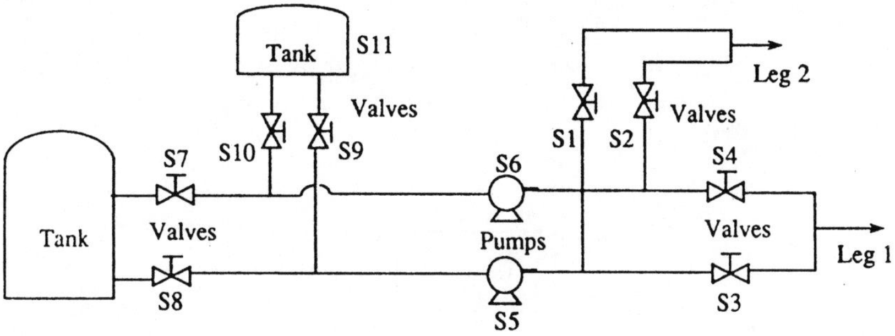

4.19 Develop an MLD model of the following system. Assume the following:

- One of the two product lines is sufficient for success.
- Control instruments feed the sensor values to the process-control computer, which calculates the position of the control valves.
- The plant computer controls the process-control computer.

  a. Develop a fault tree for the top event "inadequate product feed."

  b. Find all the cut sets of the top event.

  c. Find the probability of the top event occurring for the failure data provided

  d. Determine which components are critical to the design?

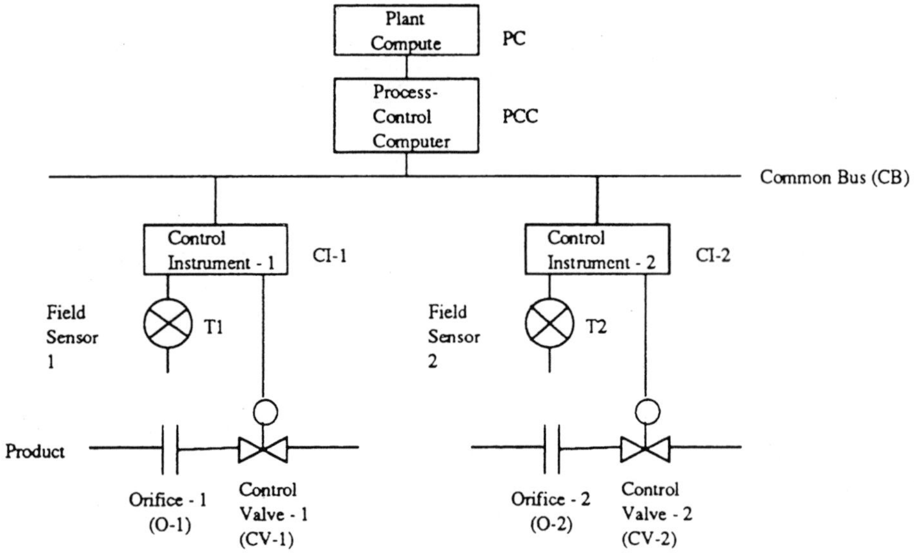


4.20 Perform a FMECA analysis for the system described in Exercise 4.19. Compare the results with part(d) of Exercise 4.19.

# 5

# Reliability and Availability of Repairable Items

When we perform reliability studies, it is important to distinguish between repairable and nonrepairable items. The reliability methods discussed in Chapters 3 and 4 are largely applicable to nonrepairable items. In this chapter, we examine the peculiar aspects of repairable systems and discuss methods used to determine the failure characteristics of these systems, as well as the methods for predicting their reliability and availability.

Nonrepairable items are those that are discarded and replaced with new ones when they fail. For example, light bulbs, transistors, contacts, unmanned satellites, missiles and small appliances are nonrepairable items. For these items, reliability is the **survival probability** of the item over its life, or a period of time during its life. Accordingly, the instantaneous probability of the first and only failure is referred to as the **hazard rate**. We elaborated on this topic in Chapter 3.

Repairable items are not replaced following the occurrence of a failure; rather, they are repaired and put into operation again. For these items, the probability that a failure will not occur over a period of time (when more than one failure can occur) is referred to as items reliability. This can be expressed in terms of the **rate of occurrence of failure** (ROCOF). Under the condition of constant failure rate, repairable item reliability can also be characterized by the mean time between failure (MTBF).

In this chapter, we are also interested in the notion of **availability**. items can be repaired, and repair activities take time. The probability that the item (e.g., system) is up (functioning) can be measured by a probability value called availability, which shows the probability that the system is up. (Conversely, the probability that the system is down is called unavailability.)

We will start with probabilistic and statistical models that are used in determining the failure characteristics of repairable items and their reliability. We will then define the concept of availability and explain availability evaluation methods for repairable items. Although the presentation of the material in chapter focuses on system reliability and availability, the methods are equally applicable to components.

## 5.1 Repairable System Reliability

Since repairable system failures occur as discrete random events, they should not be represented by continuous distributions. Rather, they should be represented by discrete distributions. These situations are also called **stochastic point processes**. To better understand this concept, consider the point estimate of the parameter $\lambda$ of an exponential distribution. The maximum likelihood estimator for $\lambda$ and confidence limits of $\lambda$ can be calculated using the statistical inference method [eq.(3.50)-(3.54)]. The main underlying assumption behind these estimates (i.e., point estimate and confidence interval), when applied to repairable systems, is that failure rate $\lambda$ is constant and will remain constant over all time intervals. Therefore, the data should be tested for potential increasing or decreasing trends.

The use of these estimators is justified only after it has been proven that the failure rate is reasonably constant, i.e., there is no evidence of an increasing or decreasing trend. An increasing trend is not necessarily due to random aging processes. Poor use of equipment, including poor testing, maintenance, and repair work, and out-of-spec (overstressed) operations, can lead to premature aging and be major contributions to increasing trends.

Fig. 5.1 which represents three cases of occurrences of failure in a repairable system, illustrates this point. The constant failure rate estimators give the same point estimate and confidence interval for each of the three situations shown in Fig. 5.1, since the number of failures and length of experience are the same for each. Clearly, Case 2 shows a decreasing failure rate, while Case 3 shows an increasing failure rate. We would therefore expect that, given a fixed length of time in the future, the system in Case 3 would be more likely to fail than the other two systems.

According to Ascher (1984) and O'Connor (1991), the following points should be considered in failure rate trend analyses:

1) Failure of a component may be partial, and repair work done on a failed component may be imperfect. Therefore, the time periods between successive failures are not necessarily independent. This is a major source of trend in the failure rate.

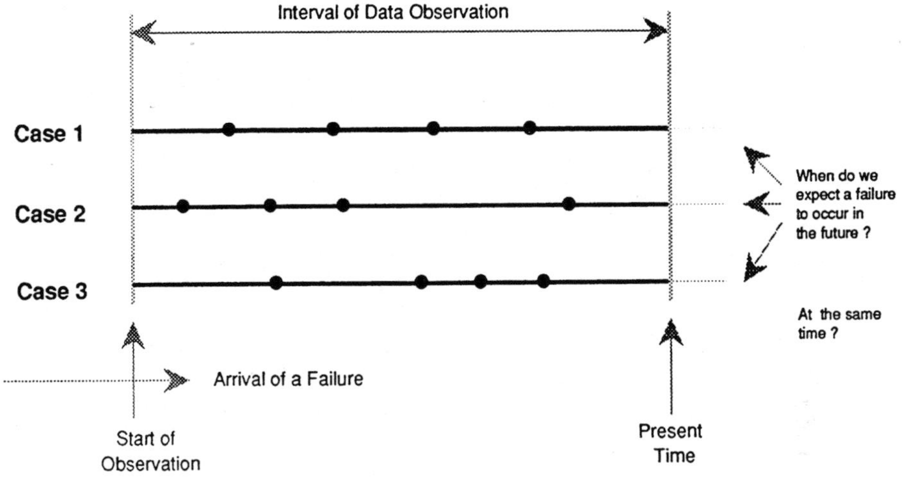


*Fig. 5.1 Three Cases of Failure Occurrence*

2) Imperfect repairs performed following failures do not renew the system, i.e., the component will not be **as good as new** following maintenance or repair. The constant failure rate assumption holds only if the component is assumed to be as good as new; only then can the statistical inference methods using a constant failure rate assumption be used.
3) Repairs made by adjusting, lubricating, or otherwise treating component parts that are wearing out provide only a small additional capability for further operation, and do not renew the component or system. These types of repair may result in a trend of increasing failure rates.
4) A component may fail more frequently due to aging and wearing out. In the remainder of this section, we provide a summary of a typical trend-analysis process, and discuss the subsequent calculation of unavailability estimates.

Situations in which failure events occur randomly in a continuum cannot be truly represented by a single continuous distribution function. Failures occurring in repairable systems (e.g., failure of safety pumps or the occurrence of accidents) are the result of discrete events occurring over time. These situations are often called stochastic point processes. For example, the Poisson distribution is used to represent a situation in which events occur randomly and at a constant average rate. This situation is known as a homogeneous Poisson process (HPP). An HPP is a stationary point process, since the distribution of the number of events in a fixed length of time does not vary. In a nonhomogenous Poisson process (NHPP), the point process is not stationary, i.e., the distribution of the number of events in a fixed length of time may change. Typically, the number of discrete events may increase or decrease over time due to trends in the observed data. An essential condition of any HPP is that the probability of events occurring in any period is independent of what has occurred in the preceding periods. Thus, an HPP describes a sequence of independently and identically distributed (IID) random variables. Conversely, an NHPP describes a sequence of random variables that is neither independently nor identically distributed. To determine whether a process is an HPP or NHPP, one must perform a trend analysis to determine whether an IID situation exists.

Several methods may be used to perform the trend analysis using the arrival values of failures of a component (or subsystem or system). For example, if failure times (failure arrivals) of a component (or system) are $t_1, t_2, \ldots, t_n$, the interarrival failure values would be $T_1, T_2, \ldots, T_n$. (Interarrivals are the time periods between successive failure events, i.e., $T_i = t_i - t_{i-1}$.) The Centroid test (sometimes called the Laplace test) can be used as one of several alternative tests to determine trends in the data. For this purpose, the statistic $U$, which approximates a standardized normal variate, is used for determining a trend such that

$$U = \frac{\frac{\sum_{i=1}^{n-1} t_i}{n-1} - \frac{t_o}{2}}{t_o\sqrt{1/[12(n-1)]}}, \tag{5.1}$$

where $t_o$ is the length of time during which failures are observed and $n$ is the total number of observed failure events. Equation (5.1) is used when failure observation is terminated at the occurrence of the last failure (therefore, $t_0$ is the time of the last failure). However, if the observation is terminated sometimes after the occurrence of the last failure, then $(n-1)$ values both in the numerator and denominator of (5.1), should be replaced by $n$. If $U = 0$ (or practically speaking close to 0), there is no evidence of trend in the data, and the process is assumed to be stationary (i.e., an HPP). If $U < 0$, the trend is decreasing , i.e., the interarrival values are becoming larger. If $U > 0$, the trend is increasing. For the latter two situations, the process is not stationary (i.e., it is an NHPP). It is recommended that a null hypothesis test for $U$ be performed by testing the value of $U$ against the values of the standard normal variate $z$. The Centroid test is good for this purpose when $n > 3$.

---

*Example 5.1*

Consider the failure arrival data for a motor-operated rotovalve in a process system. This valve is normally in standby mode, and is demanded when overheating occurs in the process. The only major failure mode is "failure to start upon demand." The arrival dates of this failure mode (in calendar time) are shown in the table below. Determine whether an increasing failure rate is justified. Assume that a total of 5,256 demands occurred between January 1, 1970 and August 12, 1986, and that demands occur at a constant rate. The last failure occurred on August 12,1986.

| Date | Failure Occurrence | Date | Failure Occurrence |
|---|---|---|---|
| 04-20-1970 | 1 | 05-04-1981 | 16 |
| 09-19-1970 | 2 | 05-05-1981 | 17 |
| 10-09-1975 | 3 | 08-31-1981 | 18 |
| 12-16-1975 | 4 | 09-04-1981 | 19 |
| 12-21-1975 | 5 | 12-02-1982 | 20 |
| 07-24-1977 | 6 | 03-23-1983 | 21 |
| 01-22-1978 | 7 | 12-16-1983 | 22 |
| 01-29-1978 | 8 | 03-28-1984 | 23 |
| 06-15-1978 | 9 | 06-06-1984 | 24 |
| 01-01-1979 | 10 | 07-19-1984 | 25 |
| 05-12-1979 | 11 | 06-23-1985 | 26 |
| 07-23-1979 | 12 | 07-01-1985 | 27 |
| 11-17-1979 | 13 | 01-08-1986 | 28 |
| 07-24-1980 | 14 | 04-18-1986 | 29 |
| 11-23-1980 | 15 | 08-12-1986 | 30 |

*Solution:*

Let's distribute the total number of demands (5,256) over the period of observation. Let's also calculate the interarrival time of failures (in months), the interarrival of demands (number of demands between two successive failures), and the arrival demand. These values are shown in Table 5.1. Since the observation ends at the last failure, the following results are obtained using (5.1):

$$\sum t_i = 95898,$$

$$n - 1 = 29,$$

$$\frac{\sum t_i}{n-1} = 3307,$$

$$\frac{t_o}{2} = 2628,$$

$$U = \frac{3307 - 2628}{5256\sqrt{1/(12 \times 29)}} = 2.41.$$

To test the null hypothesis that there is no trend in the data and the failure rate of rotovalves is constant, we would use Table A.1. When $U = 2.41$, $\Phi(z) = 0.992$. Therefore, we can reject the null hypothesis at the 1% statistical significance level.

---

The existence of a trend in the data in Example 5.1 indicates that the interarrivals of rotovalve failures are not IID, and thus the stationary process for evaluating reliability or availability of rotovalves is incorrect. Rather, the probability is described by a NHPP. The following equation represents the probability of a failure of a repairable system in any interval $(t_1, t_2)$:

$$F(t_1, t_2) = 1 - \exp\left[-\int_{t_1}^{t_2} \lambda(t)dt\right]. \tag{5.2}$$

Clearly, if no trend had been observed, then $\lambda(t) = \lambda$, and the above equation would be simplified to that used in Chapter 3. For small values of the integral, i.e., when $\lambda(t)dt < 0.1$, then $F(t_1, t_2) \approx \int_{t_1}^{t_2} \lambda(t)dt$. If a trend exists, the failure rate is a function of time. One form of $\lambda(t)$ proposed by Bassin (1969, 1973) and Crow (1974) is

$$\lambda(t) = \lambda\beta t^{\beta-1} (\lambda, \beta \geq 0, t \geq 0). \tag{5.3}$$

Expression (5.3) has the same form as (3.12) with $\lambda = (\alpha^{-\beta})$. Accordingly, using (5.2), the conditional reliability of the repairable systems can be obtained from

$$R(t_1 \mid t_1) = \exp\left[-|\lambda(t + t_1)^\beta - \lambda t^\beta|\right], \tag{5.4}$$

*Table 5.1 Arrival and Interarrival for the Rotovalve*

| Date | Interarrival Time (month) | Interarrival Demand | Arrival Demand |
|---|---|---|---|
| 04-20-1970 | 4 | 104 | 104 |
| 09-19-1970 | 5 | 131 | 235 |
| 10-09-1975 | 62 | 1597 | 1832 |
| 12-16-1975 | 2 | 59 | 1891 |
| 12-21-1975 | 0 | 4 | 1895 |
| 07-24-1977 | 19 | 503 | 2398 |
| 01-22-1978 | 6 | 157 | 2555 |
| 01-29-1978 | 0 | 6 | 2561 |
| 06-15-1978 | 5 | 118 | 2679 |
| 01-01-1979 | 7 | 173 | 2852 |
| 05-12-1979 | 4 | 113 | 2966 |
| 07-23-1979 | 2 | 62 | 3028 |
| 11-17-1979 | 4 | 101 | 3129 |
| 07-24-1980 | 8 | 216 | 3345 |
| 11-23-1980 | 4 | 106 | 3451 |
| 05-04-1981 | 5 | 140 | 3591 |
| 05-05-1981 | 0 | 1 | 3592 |
| 08-31-1981 | 4 | 102 | 3694 |
| 09-04-1981 | 0 | 3 | 3697 |
| 12-02-1982 | 15 | 393 | 4090 |
| 03-23-1983 | 4 | 96 | 4186 |
| 12-16-1983 | 9 | 232 | 4418 |
| 03-28-1984 | 3 | 89 | 4507 |
| 06-06-1984 | 2 | 61 | 4568 |
| 07-19-1984 | 1 | 37 | 4605 |
| 06-23-1985 | 11 | 293 | 4898 |
| 07-01-1985 | 0 | 7 | 4905 |
| 01-08-1986 | 6 | 165 | 5070 |
| 04-18-1986 | 3 | 86 | 5157 |
| 08-11-1986 | 4 | 99 | 5256 |

where $t$ = present age of the system, and $t + t_1$ is its age in $t_1$ units of time from the present.

Crow (1974) has shown that under the condition of a single system observed to its $n^{th}$ failure, the maximum likelihood estimator of $\beta$ and $\lambda$ are as follows:

$$\hat{\beta} = \frac{n}{\sum_{i=1}^{n-1} \ln \frac{t_o}{t_i}}, \tag{5.5}$$

$$\hat{\lambda} = \frac{n}{t_o^{\hat{\beta}}}. \tag{5.6}$$

The $1 - \alpha$ confidence limits for inferences on $\beta$ and $\lambda$ have been developed and discussed by Bain (1978).

---

*Example 5.2*

Using the information in Example 5.1, calculate the maximum likelihood estimator of $\beta$ and $\lambda$. Also, plot the probability of the rotovalve failure on demand as a function of time from 1971 to 1999.

*Solution:*

Using (5.5) and (5.6), we can calculate $\hat{\beta}, \hat{\lambda}$, as 1.59 and $3.71 \times 10^{-5}$ respectively.

Using $\hat{\beta}$ and $\hat{\lambda}$, the functional form of the demand failure rate can be obtained by using (5.2).

$$\lambda(d) = 3.71 \times 10^{-5} \times 1.59 d^{0.59},$$

where $d$ represents the demand number.

The functional form of the demand failure rate (NHPP) and its confidence bound are plotted as a function of calendar time for the rotovalve in Fig. 5.2. For comparison purposes, the constant demand failure rate function (IID case) are also shown. For the IID, the point estimate of $\lambda$ was obtained by dividing the number of failures by the number of demands. The upper and lower confidence intervals were obtained using the IID assumption.

---

*Example 5.3*

In a repairable system, the following six interarrival times between failures have been observed: 16, 32, 49, 60, 78, and 182 (in hours). Assuming the observation ends at the time when the last failure is observed.

a) What is the trend?
b) Given the data, what is the probability that the interarrival time for the seventh failure will be greater than 200 hours?

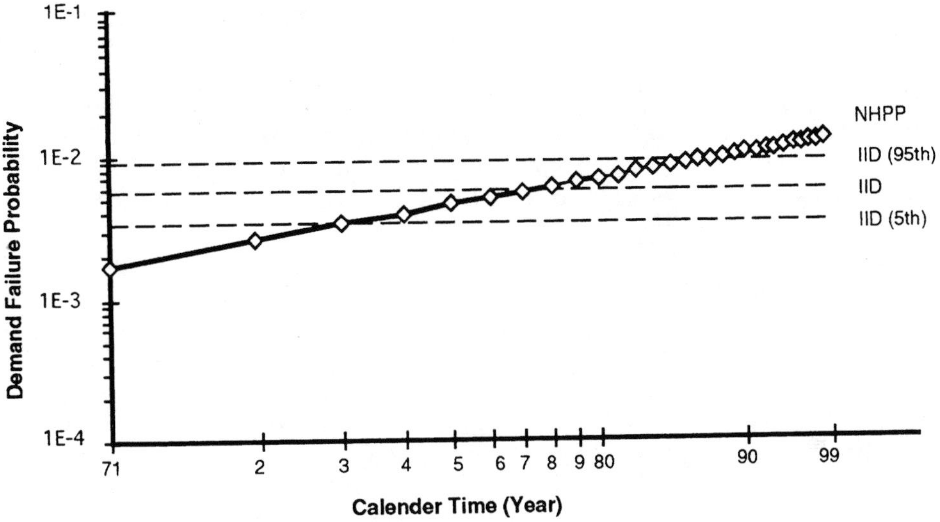


*Fig. 5.2 Comparison of NHPP and IID for Rotovalve Failure Estimates*

*Solution:*

From (5.1), $t_i = T_i + t_{i-1}$, and

$$U = \frac{\frac{16+(16+32)+\ldots}{5} - \frac{417}{2}}{417\sqrt{1/[12(5)]}} = -0.45.$$

Notice that $t_o = 417$.

Thus, the NHPP is applicable (the trend is decreasing). Using (5.5) and (5.6),

$$\hat{\beta} = \frac{6}{\ln\frac{417}{16} + \ln\frac{417}{16+32} + \ldots} = 0.712,$$

$$\hat{\lambda} = \frac{6}{(417)^{0.712}} = 0.0817, hr^{-1}.$$

Thus, $\hat{\lambda}(t) = 0.058t^{-0.288}$. From (5.2) with $t_1 = 200$,

$$Pr(\text{7th failure occurs within 200 hours}) = 1 - exp[-\lambda(|(t_o + t_1)^{\beta} - (t_o)^{\beta}|),$$
$$= 0.85.$$

The probability that the interarrival time is greater than 200 hours is 0.15.

---

Crow (1990) has expanded (5.2) through (5.6) to include cases where data originate from multi unit cases. Another form of $\lambda(t)$ for the NHPP model proposed by Cox and Lewis (1966) is $\lambda(t) = \exp(\alpha_o + \alpha_1 t)$. Lawless (1982) has applied this model to failure data from a set of similar air-conditioning units.

## 5.2 Availability of Repairable Systems

We defined reliability as the probability that a component or system will perform its required function for the time interval that it is required to do so. Similarly, we define availability as the probability that a system (or component) will function as planned when called upon to do so. It is possible to offer three different definitions for availability.

1) Instantaneous (pointwise) availability: $a(t)$ is the probability that the system (or component) is up at time $t$.
2) Limiting pointwise availability: $a$ is defined as

$$a = \lim_{t \to \infty} a(t). \tag{5.7}$$

3) Average availability: $\bar{a}$ is defined for a fixed period of time $T$ as

$$\bar{a} = \frac{1}{T} \int_0^T a(t)dt. \tag{5.8}$$

It is also possible to define the limiting average availability as

$$\bar{a}_l = \lim_{T \to \infty} \frac{1}{T} \int_0^T a(t)dt. \tag{5.9}$$

However, this definition has limited applications. We elaborate on each of the three definitions of availability in the remainder of this section.

### *5.2.1 Pointwise Availability*

The most frequently chosen model for pointwise availability is the exponential distribution. As defined earlier, availability would be represented by

$$a(t) = \exp\left[- \int_0^t \lambda(\theta)d\theta\right], \tag{5.10}$$

where $a(t)$ is the probability that the component will be in a success state by time $t$, given it is working at time $t = 0$. Conversely, the unavailability $q(t)$ can be defined as

$$q(t) = 1 - a(t). \tag{5.11}$$

It is important to understand the general types of systems (or components), defined below.

1) Time-independent systems. The failure probability and repair time (if any) are independent of time.
2) Nonrepairable systems. The model shown in (5.10) applies, in which $\lambda(\theta)$ represents the instantaneous rate of failure.
3) Repairable systems for which failure is immediately detected (revealed faults).
4. Repairable systems for which failure is detected upon inspection (sometimes referred to as periodically tested systems).

For Type 1 systems, the unavailability can be obtained from

$$a = \frac{T_u}{T_u + T_d}, \quad or$$
$$q = \frac{T_d}{T_u + T_d}, \tag{5.12}$$

where $T_u$ = up time and $T_d$ = downtime. This is clearly an ideal case and is rarely seen in practice. We will discuss an example of this type in Section 5.3.

For Type 2 systems, (5.10) represents the availability of the system.

For Type 3 systems, since the systems are repairable, the rate at which repair is performed enters the calculation of availability. In these cases $a(t)$ is obtained from the following set of ordinary differential equation:

$$\frac{da(t)}{dt} = -\lambda(t)a(t) + \mu(t)q(t),$$
$$\frac{dq(t)}{dt} = \lambda(t)q(t) - \mu(t)a(t), \tag{5.13}$$

where $\lambda(t)$ is the failure rate and $\mu(t)$ is the repair rate.

It is possible to show, by solving (5.13), that if no trend exists in the failure rate and repair rate, the availability of the system (or component) can be obtained from

$$a(t) = \frac{\mu}{\lambda + \mu} + \frac{\lambda}{\lambda + \mu} \exp[-(\lambda + \mu)t],$$
$$q(t) = \frac{\lambda}{\lambda + \mu} - \frac{\lambda}{\lambda + \mu} \exp[-(\lambda + \mu)t]. \tag{5.14}$$

Note that in (5.14), $\mu = \frac{1}{\tau}$, where $\tau$ is the average length of time per repair (sometimes referred to as mean time to repair-MTTR). Clearly, MTBF $= 1/\lambda$ in this case.

For Type 4 systems, the availability equations are rather involved. Caldorola (1977) presents a form of $a(t)$ for cases where no trend in the failure rate exists, and the inspection interval ($\eta$), duration of inspection ($\theta$), and duration of repair ($\tau$) are fixed. In these cases,

$$a(t) = \exp[-(t - m\eta)\lambda_{eff}] \cdot \left\{1 - \exp\left[-(\frac{t - m\eta}{\theta})^q\right]\right\},$$

where

$$\lambda_{eff} = \frac{\eta - \theta}{\eta}\left(\frac{n - \theta}{\eta} + \frac{2\tau}{\eta}\right) + 2\left[1 - \frac{\Gamma(1/q)}{q}\right]\frac{\theta}{\eta^2},$$
$$q = \ln[3 - \ln(\theta\lambda)], \quad \text{and} \tag{5.15}$$

$m =$ inspection interval number $= 1, 2, \ldots, n$. When $t > m\eta + \theta$, it is easy to show that $a(t) \approx \exp[-(t - m\eta)\lambda_{eff}]$.

---

*Example 5.4*

Find the unavailability, as a function of time, for a system that is inspected once a month. Duration of inspection is 1.5 hours. Any required repair takes an average of 19 hours. Assume the failure rate of the system is $3 \times 10^{-6}(hr^{-1})$.

*Solution:*

Using (5.15) and (5.11), for $\theta = 1.5, \tau = 19, \eta = 720, \lambda = 3 \times 10^{-6}$, we can get the plot of $q(t)$ as shown in Fig. 5.3.

---

For simplicity, the pointwise availability function can be represented in an approximate form. This simplifies availability calculations significantly. For example, in a periodically tested component, if the repair and test durations are very short in comparison to operation time, and the test and repair are assumed perfect, one can neglect their contribution to unavailability of the system. This can be shown using Taylor expansion of the unavailability equation (see Lofgren, 1985). In this case for each test interval $T$, the availability and unavailability functions are

$$a(t) \approx 1 - \lambda t,$$
$$q(t) \approx \lambda t. \tag{5.16}$$

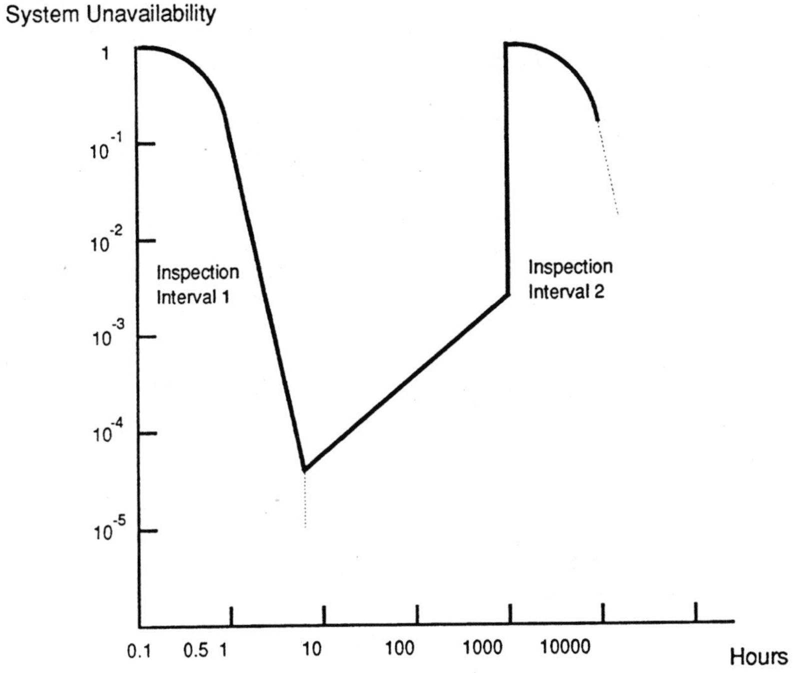


*Fig. 5.3 Unavailability of the System as a Function of Time*

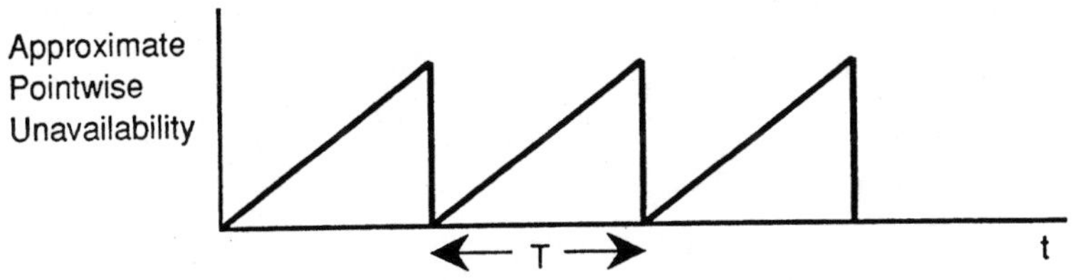


*Fig. 5.4 Approximate Pointwise Unavailability for a Periodically Tested Item*

The plot of the unavailability as a function of time, using (5.16), will take a shape similar to that in Fig. 5.4. Clearly if the test and repair durations are long, one must include their effect.

Vesely and Goldberg (1977) have used the approximate pointwise unavailability functions for this case. The functions and their plot are shown in Fig. 5.5. The average values of the approximate unavailability functions shown in Fig. 5.4 and 5.5 are discussed in Section 5.2.3 and are presented in Table 5.2.

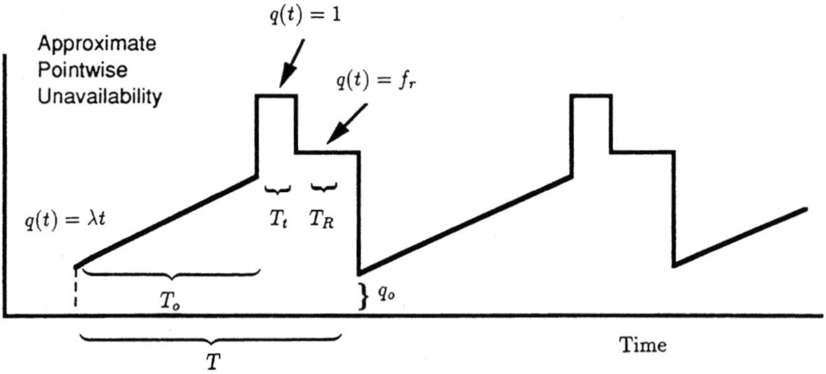


$T$ = Test interval, $T_R$= Average repair time (hr), $T_t$= Average test duration (hr), $f_r$= Frequency of repair, $q_o$= Residual unavailability.

*Fig. 5.5 Pointwise Unavailability for a Periodically Tested Item Including Test and Repair Outages*

It should be noted that, due to random imperfection in test and repair activities, it is possible that a residual unavailability $q$ would remain following a test and/or repair. Thus, unlike the unavailability function shown in Fig. 5.4, the unavailability function in Fig. 5.5 exhibits a residual unavailability $q_o$ due to these random imperfections.

*5.2.2 Limiting Pointwise Availability*

It is easy to see that some of the pointwise availability equations discussed in Section 5.2.1 have limiting values: For example, (5.14) has the following limiting value:

$$a = \lim_{t\to\infty} a(t),$$
$$= \frac{\mu}{\lambda+\mu}, \text{ or its equivalent}$$

$$a = \frac{MTBF}{MTBF + MTTR}. \tag{5.17}$$

Equation (5.17) sometimes is referred to as the asymptotic availability of a repairable system with a constant failure rate.

*5.2.3 Average Availability*

According to its definition, average availability is a constant measure of availability over a period of time $T$. For noninspected items, $T$ can be any value (preferably it should be the mission length). For inspected items, $T$ is normally the inspection (or test) interval or mission length $T_m$. Thus, for nonrepairable items, if the inspection interval is $T$, then the approximate expression for pointwise availability with constant $\lambda$ can be used. If we assume $\bar{a} \approx 1 - \lambda t$ (which in only true if $\lambda t < 0.1$), then

$$a = \frac{1}{T}\int_0^T (1 - \lambda t)dt = 1 - (1/2)\lambda T. \tag{5.18}$$

Accordingly, for all types of systems, one can generate such average availabilities. Vesely et al. (1981) have discussed the average unavailabilities for various types of systems. Table 5.2 shows these functions.

*Table 5.2 Average Availability Functions*

| Type of Item<br>Type of Item | Average<br>Unavailability | Average<br>Availability |
|---|---|---|
| Time Independent,<br>Constant | q | a |
| Nonrepairable | $\frac{1}{2}\lambda T_m$ | $1 - \frac{1}{2}\lambda T_m$ |
| Repairable<br>Revealed Fault | $\frac{\lambda\tau}{1+\lambda\tau}$ | $\frac{1}{1+\lambda\tau}$ |
| Repairable<br>Periodically Tested | $\frac{1}{2}\lambda T_o + f_r\frac{T_R}{T} + \frac{T_t}{T}$ | $1 - \frac{1}{2}\lambda T_o - f_r\frac{T_R}{T} - \frac{T_t}{T}$ |

$\lambda =$ constant failure rate $(hr^{-1})$,
$T_m =$ mission length (hr),
$\tau =$ average downtime or MTTR (hr),
$T =$ test interval (hr),

$T_R =$ average repair time (hr),
$T_t =$ average test duration (hr),
$f_r =$ frequency of repair per test intervals,
$T_o =$ operating time (up time) $= T - T_R - T_t$.

Equations in Table 5.2 also apply to standby equipment, with $\lambda$ representing the standby (or demand) failure rate, and the mission length or operating time being replaced by the time between two tests.

## 5.3 Use of Markovian Methods for Determining System Availability

Markovian methods are useful tools for evaluating systems that have multiple states (e.g., up, down, and degraded). For example, consider a system with the states shown in Fig. 5.6. In the Markovian models, the transitions between various states are characterized by constant **transition rates** (these rates may not necessarily be constant, as discussed in in Section 5.1).

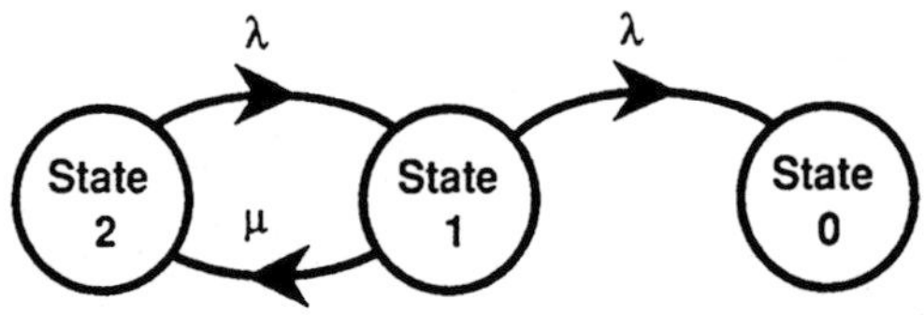


*Fig. 5.6 A Markovian Model for a System with Three Discrete States*

Consider a system with a number of discrete states, with the following definitions:

$Pr_i(t) =$ $Pr$ (the system is in state $i$ at time $t$),
$\rho_{ij}(t) =$ transition rate from state $i$ to state $j$.

By assuming that $Pr_i(t)$ is differentiable, it is possible to show that

$$\frac{d\,\mathrm{Pr}_i(t)}{dt} = -\Big(\sum_j \rho_{ij}(t)\Big) Pr_i(t) + \Big(\sum_j \rho_{j_i}(t) Pr_j(t)\Big). \qquad (5.19)$$

If a differential equation similar to (5.19) is written for each state, and the resulting set of differential equations is solved, one may obtain the time-dependent probability of each state. This can be seen better in Example 5.5.

---

*Example 5.5*

Consider a system with failure rate $\lambda$ and repair rate $\mu$ in a standby redundant configuration. When the system fails, its repair starts immediately, which puts it back into operation. The system has two states:

- State 0, when the system is down, and
- State 1, when the system is operating.

a) Determine probability of each state.
b) Determine the availability of this system.

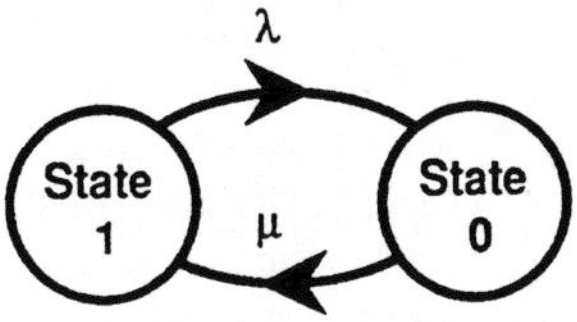


*Fig. 5.7 Markovian Model for Example 5.5*

*Solution:*

The governing differential equations according to (5.19) are

$$\frac{d\,\mathrm{Pr}_0(t)}{dt} = \lambda Pr_1(t) - \mu Pr_0(t),$$
$$\frac{d\,\mathrm{Pr}_1(t)}{dt} = -\lambda Pr_1(t) + \mu Pr_1(t).$$

By assuming that $Pr_1(0) = 1$ and $Pr_0(0) = 0$, the above equations can be solved either by using the Laplace transformation or by discretizing the equations. Below, we take the Laplace transform of both sides of the equations.

$$s\bar{P}_1(s) - 1 = \lambda\bar{P}_1(s) - \mu\bar{P}_0(s),$$
$$s\bar{P}_0(s) = -\lambda\bar{P}_1(s) + \mu\bar{P}_0(s).$$

For the above set of equations, $A = \begin{pmatrix} \lambda & -\mu \\ -\lambda & \mu \end{pmatrix}$ is referred to as the **transition matrix**. Therefore,

$$\bar{P}_1(s) = \frac{s+\mu}{s(s+\lambda+\mu)},$$

$$\bar{P}_0(s) = \frac{\lambda}{s(s+\lambda+\mu)}.$$

It follows that availability $a(t)$ is obtained from

$$a(t) = Pr_1(t) = \mathcal{L}^{-1}\{\frac{s+\mu}{s(s+\lambda+\mu)}\} = \frac{\mu}{\lambda+\mu} + \frac{\lambda}{\lambda+\mu}\exp[-(\lambda+\mu)t].$$

Accordingly, unavailability is

$$\bar{a}(t) = Pr_2(t) = 1 - A(t) = \frac{\lambda}{\lambda+\mu} - \frac{\lambda}{\mu+\lambda}\exp[-(\lambda+\mu)t].$$

For very large values of $t$, $\exp[-(\lambda+\mu)t] = 0$, and $a(t)$ asymptote can be obtained, as follows:

$$a(\infty) = \frac{\mu}{\lambda+\mu}.$$

---

*Example 5.6*

A system that consists of two cooling units has the three states shown in the Markovian model in Fig. 5.8. When one unit system fails, the other system takes over and repair on the first starts immediately. When both systems are down, there are two repair crews to simultaneously repair the two systems. The three states are as follows:

- State 0, when both systems are down,
- State 1, when one of the systems is operating and the other is down, and
- State 2, when the first system is operating and the second is in standby (in an operating ready condition).

a) Determine the probability of each state.
b) Determine the availability of the entire system.

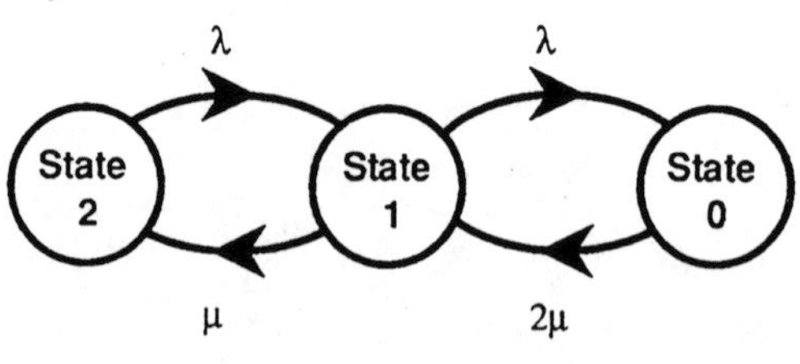


*Fig. 5.8 Markovian Model for Example 5.5*

*Solution:*

a) The governing differential equations are

$$\frac{d\text{Pr}_2(t)}{dt} = -\lambda Pr_2(t) + \mu Pr_1(t),$$
$$\frac{d\text{Pr}_1(t)}{dt} = \lambda Pr_2(t) - (\mu + \lambda)Pr_1(t) + 2\mu Pr_0(t),$$
$$\frac{d\text{Pr}_0(t)}{dt} = \lambda Pr_1(t) - 2\mu Pr_0(t).$$

Taking the Laplace transform of both sides of the equations yields the following:

$$sP_2(s) - Pr_2(0) = -\lambda P_2(s) + \mu P_1(s),$$
$$sP_1(s) - Pr_1(0) = \lambda P_2(s) - (\mu + \lambda)P_1(s) + 2\mu P_0(s),$$
$$sP_0(s) - Pr_0(0) = \lambda P_1(s) - 2\mu P_0(s).$$

$Pr_2(0) = 1$ and $Pr_1(0) = Pr_0(0) = 0$. By solving the above set of equations, $Pr_i(s)$ can be calculated.

$$P_2(s) = \frac{1}{\lambda + s} + \frac{\mu\lambda(2\mu + s)}{s(s + \lambda)(s - k_1)(s - k_2)},$$
$$P_1(s) = \frac{\lambda(2\mu + s)}{s(s - k_1)(s - k_2)},$$
$$P_0(s) = \frac{\lambda^2}{s(s - k_1)(s - k_2)},$$

where

$$k_1 = \frac{-2\lambda - 3\mu - \sqrt{4\lambda\mu + \mu^2}}{2},$$
$$k_2 = \frac{2\mu\lambda + \lambda^2 + 2\mu^2}{k_1}.$$

If the inverses of the above Laplace transforms are taken, the probability of each state can be determined.

$$Pr_2(t) = \exp[-\lambda t] + G_1 \exp[-\lambda t] + G_2 \exp[k_1 t] + G_3 \exp[k_2 t] + G_4,$$

where

$$G_1 = \frac{\mu(\lambda - 2\mu)}{(\lambda + k_1)(\lambda + k_2)}, \qquad G_2 = \frac{\mu\lambda(2\mu + k_1)}{(k_1 + \lambda)(k_1 - k_2)(k_1)},$$
$$G_3 = \frac{\mu\lambda(2\mu + k_1)}{k_2(k_2 - k_1)(k_2 + \lambda)}, \qquad G_4 = \frac{2\mu^2}{(2\mu\lambda + \lambda^2 + 2\mu^2)}.$$

And,

$$Pr_1(t) = A_1 \exp[k_1 t] + A_2 \exp[k_2 t] + A_3,$$

where

$$A_1 = \frac{2\mu\lambda}{(k_1 - k_2)k_1} + \frac{\lambda}{k_1 - k_2}, \quad A_2 = \frac{\lambda(2\mu + k_2)}{(k_1 - k_2)(k_2)}, \quad A_3 = \frac{2\mu\lambda}{2\mu\lambda + \lambda^2 + 2\mu^2}.$$

And,

$$Pr_0(t) = B_1 \exp[k_1 t] + B_2 \exp[k_2 t] + B_3$$

, where

$$B_1 = \frac{\lambda^2}{(k_1 - k_2)k_1}, \quad B_2 = \frac{\lambda^2}{(k_2 - k_1)k_1}, \quad \text{and} \quad B_3 = \frac{\lambda^2}{2\mu\lambda + \lambda^2 + 2\mu^2}.$$

b) The availability of the two units system, is $a(t) = Pr_2(t) + Pr_1(t)$, and the unavailability of the entire system is $q(t) = Pr_0(t)$.

---

It is possible to simply find the limiting pointwise availability from the governing equations of the system. For this purpose, consider the Markovian transition diagram shown in Fig. 5.9.

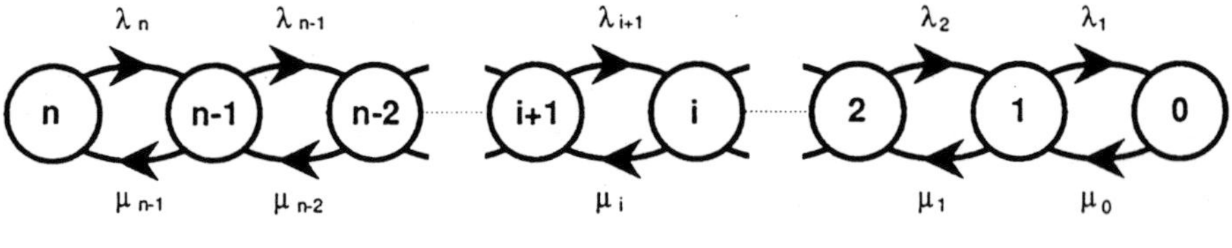


*Fig. 5.9 A Markovian Transition Diagram with n States*

Lamaire (1977) and others have shown that

$$Pr_{i+1}(\infty)\lambda_{i+1} = \mu_i Pr_i(\infty), \tag{5.20}$$

or,

$$Pr_i(\infty) = \frac{\lambda_{i+1} \cdot \lambda_{i+2} \cdots \lambda_n}{\mu_i \cdot \mu_{i+1} \cdots \mu_{n-1}} Pr_n(\infty). \tag{5.21}$$

Since $\sum_{i=0}^{n} Pr_i(\infty) = 1$, solving (5.21) for $Pr_n(\infty)$ yields

$$Pr_n(\infty) = \frac{1}{1 + \sum_{i=0}^{n-1} \frac{\lambda_{i+1} \cdot \lambda_{i+2} \cdots \lambda_n}{\mu_i \cdot \mu_{i+1} \cdots \mu_{n-1}}}. \tag{5.22}$$

Accordingly, the system's limiting pointwise unavailability (and similarly its availability) can be obtained.

$$q = Pr_0(\infty) = \frac{Pi_{i=1}^{n}\lambda_i}{Pi_{i=0}^{n-1}\mu_i} Pr_n(\infty). \tag{5.23}$$

If the system is unavailable when it is in any of the states $(0, 1, \ldots, r-1)$, then

$$q = \sum_{i=0}^{r-1} Pr_i(\infty) = Pr_n(\infty) \sum_{i=1}^{r-1} \frac{\lambda_{i+1} \cdot \lambda_{i+2} \ldots \lambda_n}{\mu_i \cdot \mu_{i+1} \ldots \mu_{n-1}}. \tag{5.24}$$

---

*Example 5.7*

For Example 5.6, determine the limiting pointwise unavailability from (5.23) and confirm it with the results obtained in that example.

*Solution:*

Since $\lambda_2 = \lambda_1 = \lambda, \quad \mu_1 = \mu, \quad \mu_0 = 2\mu$ from (5.20),

$$Pr_1(\infty) = \frac{2\mu}{\lambda} Pr_0(\infty),$$

and

$$Pr_2(\infty) = \frac{\mu}{\lambda} Pr_1(\infty) = \frac{2\mu^2}{\lambda^2} Pr_0(\infty).$$

Since

$$Pr_0(\infty) + Pr_1(\infty) + Pr_2(\infty) = 1 \quad \text{from (5.25)},$$

$$q = Pr_0(\infty) = \frac{\lambda^2}{2\mu^2 + 2\mu\lambda + \lambda^2}.$$

Accordingly,

$$a = Pr_1(\infty) + Pr_2(\infty) = \frac{2\mu^2 + 2\mu\lambda}{2\mu^2 + 2\mu\lambda + \lambda^2}.$$

This can be verified from the solution for $Pr_0(t)$. Since $k_1$ and $k_2$ are negative, the exponential terms approach zero, then

$$Pr_0(\infty) = B_3 = \frac{\lambda^2}{2\mu^2 + 2\mu\lambda + \lambda^2}.$$

Similarly,

$$Pr_1(\infty) = A_3, \quad \text{and} \quad Pr_2(\infty) = G_4.$$

Thus

$$a = \frac{2\mu^2}{2\mu^2 + 2\mu\lambda + \lambda^2} + \frac{2\mu\lambda}{2\mu^2 + 2\mu\lambda + \lambda^2}.$$

Therefore, the results obtained in Examples 5.5 and 5.6 are consistent.

---

It is clear that if a trend exists in the parameters that characterize system availability (e.g., failure rate and repair rate), one cannot use the Markovian method; only solutions to (5.19) can be used. Solutions to such equations may pose difficulty in systems with many states. However, with the emergence of efficient numerical algorithms and powerful computers, solutions to these equations are indeed possible.

## 5.4 Use of System Analysis Techniques in the Availability Calculations of Complex Systems

In Chapter 4, we discussed a number of methods for estimating the reliability of a system from the reliability of its individual components or units. The same concept applies here also. That is, one can use the availability (or unavailability) functions for each component of a complex system and use, for example, system cut sets to obtain system availability (or unavailability). The method of determining system availability in these cases is exactly similar to the system reliability estimation methods.

---

*Example 5.8*

Assume all components of the system shown in Fig. 4.4 are repairable (revealed fault) with a failure rate of $10^{-3}(\text{hour}^{-1})$ and a mean down time of 15 hours. Component 7 has a failure rate of $10^{-5}(\text{hour}^{-1})$, with a mean downtime of 10 hours. Calculate the average system unavailability.

*Solution:*

The cut sets are (7), (1, 2), (1, 5, 6), (2, 3, 4) and (3, 4, 5, 6). The unavailability of component 1 through 6, according to Table 5.2, is

$$q_{1-6} = \frac{\lambda\tau}{1 + \lambda\tau} = \frac{10^{-3} \times 15}{1 + 10^{-3} \times 15} = 9.85 \times 10^{-3}.$$

Similarly,

$$q_7 = \frac{10^{-5} \times 10}{1 + 10^{-5} \times 10} = 9.99 \times 10^{-5}.$$

Using the rare event approximation,

$$q_{sys} = q(\text{cut sets}) = q_7 + q_1 \cdot q_2 + q_1 \cdot q_5 \cdot q_6 + q_2 \cdot q_3 \cdot q_4 + q_3 \cdot q_4 \cdot q_5 \cdot q_6.$$

Thus,

$$\begin{aligned} q_{sys} &= 9.99 \times 10^{-5} + 9.70 \times 10^{-5} + 9.56 \times 10^{-7} + 9.56 \times 10^{-7} + 9.41 \times 10^{-9} \\ &= 1.99 \times 10^{-4}. \end{aligned}$$

---

*Example 5.9*

The auxiliary feedwater system in a pressurized water reactor (PWR) plant is used for emergency cooling of steam generators. The simplified piping and instrument diagram (P & ID) of a typical system like this is shown in Fig. 5.10(a). The reliability block diagram in Fig. 5.10(b) represents this P & ID. Calculate the system unavailability. Assume all of the components are in standby mode and are periodically tested with the following characteristics. (Characteristics are shown collectively for each block.)

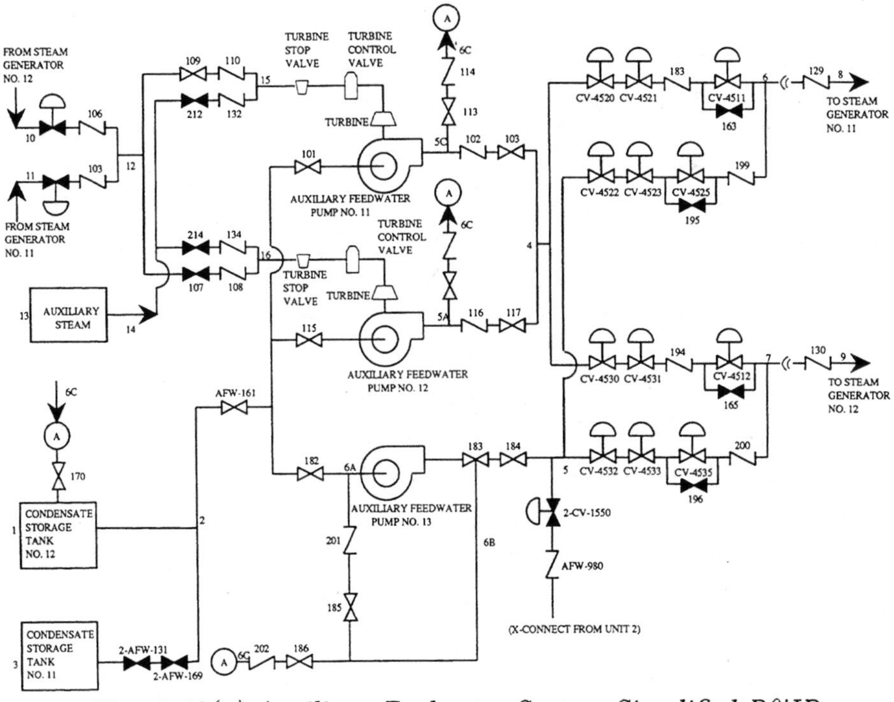


*Fig. 5.10(a) Auxiliary Feedwater System Simplified P&ID*

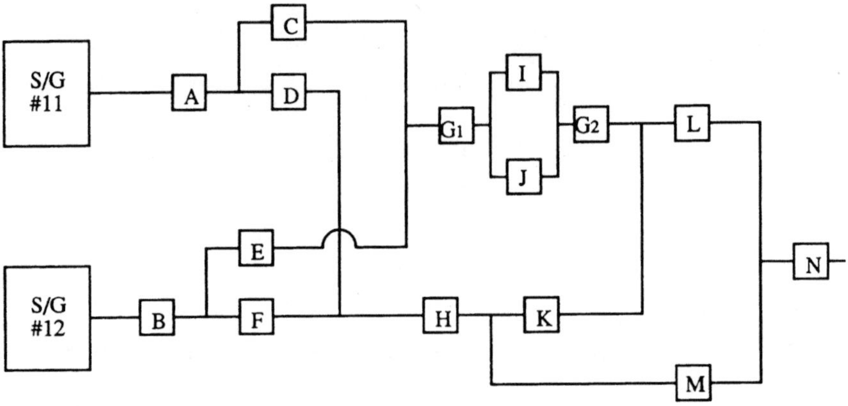


*Fig. 5.10(b) Simplified Auxiliary Feedwater System of a PWR*

| Block Name | Failure Rate ($hr^{-1}$) | Frequency of Repair | Average Test Duration (hours) | Average Repair Time (hours) | Test Interval (hours) |
|---|---|---|---|---|---|
| A | $1 \times 10^{-7}$ | $9.2 \times 10^{-3}$ | 0 | 5 | 720 |
| B | $1 \times 10^{-7}$ | $9.2 \times 10^{-3}$ | 0 | 5 | 720 |
| C | $1 \times 10^{-6}$ | $2.5 \times 10^{-2}$ | 0 | 10 | 720 |
| D | $1 \times 10^{-6}$ | $2.5 \times 10^{-2}$ | 0 | 10 | 720 |
| E | $1 \times 10^{-6}$ | $2.5 \times 10^{-2}$ | 0 | 10 | 720 |
| F | $1 \times 10^{-6}$ | $2.5 \times 10^{-2}$ | 0 | 10 | 720 |
| G($G_1$ and $G_2$) | $1 \times 10^{-7}$ | $7.7 \times 10^{-4}$ | 0 | 15 | 720 |
| H | $1 \times 10^{-7}$ | $1.8 \times 10^{-4}$ | 0 | 24 | 720 |
| I | $1 \times 10^{-4}$ | $6.8 \times 10^{-1}$ | 2 | 36 | 720 |
| J | $1 \times 10^{-4}$ | $6.8 \times 10^{-1}$ | 2 | 36 | 720 |
| K | $1 \times 10^{-5}$ | $5.5 \times 10^{-1}$ | 2 | 24 | 720 |
| L | $5 \times 10^{-7}$ | $4.3 \times 10^{-3}$ | 0 | 10 | 720 |
| M | $3 \times 10^{-4}$ | $1.5 \times 10^{-1}$ | 0 | 10 | 720 |
| N | $1 \times 10^{-7}$ | $5.8 \times 10^{-4}$ | 0 | 5 | 720 |

*Solution:*

According to Table (5.2), we can calculate the unavailability of each block.

| Block Name | Unavailability |
|---|---|
| A | 1.0E-4 |
| B | 1.0E-4 |
| C | 7.0E-4 |
| D | 7.0E-4 |
| E | 7.0E-4 |
| F | 7.0E-4 |
| G($G_1$ and $G_2$) | 5.2E-4 |
| H | 4.2E-4 |
| I | 7.3E-4 |
| J | 7.3E-4 |
| K | 1.4E-4 |
| L | 2.4E-4 |
| M | 1.1E-1 |
| N | 4.0E-5 |

The cut sets of the block diagram in Fig. 5.10(b) are as follows:

1) N
2) L M
3) H L
4) G H
5) A B
6) H J I
7) G K M
8) D F L
9) D G F
10) C E H
11) B D L
12) B D G
13) B C H
14) B C D
15) A F L
16) A E F
17) A E H
18) A G F
19) J I K M
20) D F J I
21) C E K M
22) C D E H
23) B D J I
24) B C K M
25) A F J I
26) A E K M

Using the same procedure as the one used in Example 5.8 and rare event approximation, we can easily compute the average system unavailability as

$$q_{sys} = 7.49 \times 10^{-5}.$$

---

One important point to recognize in the availability estimation of redundant systems with periodically tested components is that components whose simultaneous failures cause the system to fail (i.e., sets of components in each cut set of the system) should be tested in a **staggered** manner. This way the system would not become totally unavailable during the testing and repair of its components. For example, consider a system of two parallel units, each of which is periodically tested and has a pointwise unavailability behavior that can be approximated by the model shown in Fig. 5.5. If the components are not tested in a staggered manner, the system's pointwise unavailability exhibits the shape shown in Fig. 5.11. On the other hand, if the components are tested in a staggered manner, the system unavailability would exhibit the shape illustrated in Fig. 5.12. Clearly, the average unavailability in the case of staggered testing is lower. This subject is discussed in more detail by Vesely and Goldberg (1981) and Ginzburg and Vesely (1990). Also, to minimize unavailability, one can find an optimum value for test interval as well as the optimum degree of staggering.

Modarres (1984) has suggested a simple method for estimating approximate average system unavailability of a series-parallel system having a single input node and single output node, and repairable (revealed fault) components. In this method, it is assumed that the components or blocks are independent and $\lambda_i \tau_i << 1$ for each component or block of the system, where $\lambda_i$ is the constant failure rate (i.e., no failure rate trend is assumed), and $\tau_i$ is the component's mean downtime. In this method, series and parallel blocks of the system are systematically replaced with equivalent "super blocks." The equivalent failure rate (or occurrence rate) $\Lambda$ and mean downtime $\tau$ of the super blocks can be calculated from Table 5.3. Example 5.10 is an application of this method.

---

*Example 5.10*

Consider the series-parallel system shown in Fig. 5.13, with the component data shown in Table 5.4. This system is composed of two parallel blocks. Each block is composed of sub-block(s) and component(s).

a) Determine the approximate occurrence rate $\Lambda$ and mean downtime $\tau$ of this system.
b) Determine the approximate average unavailability of the system.

*Solution:*

Assuming independence between blocks and super-blocks:

a) The super-blocks are enclosed by dotted lines in Fig. 5.13. First, all of the blocks are resolved and their equivalent $\Lambda$ and $\tau$ are obtained. Next, their equivalent $\Lambda$ and $\tau$ are determined. Finally, the whole system is resolved. Equations in Table 5.3 are applied to the system along with the failure

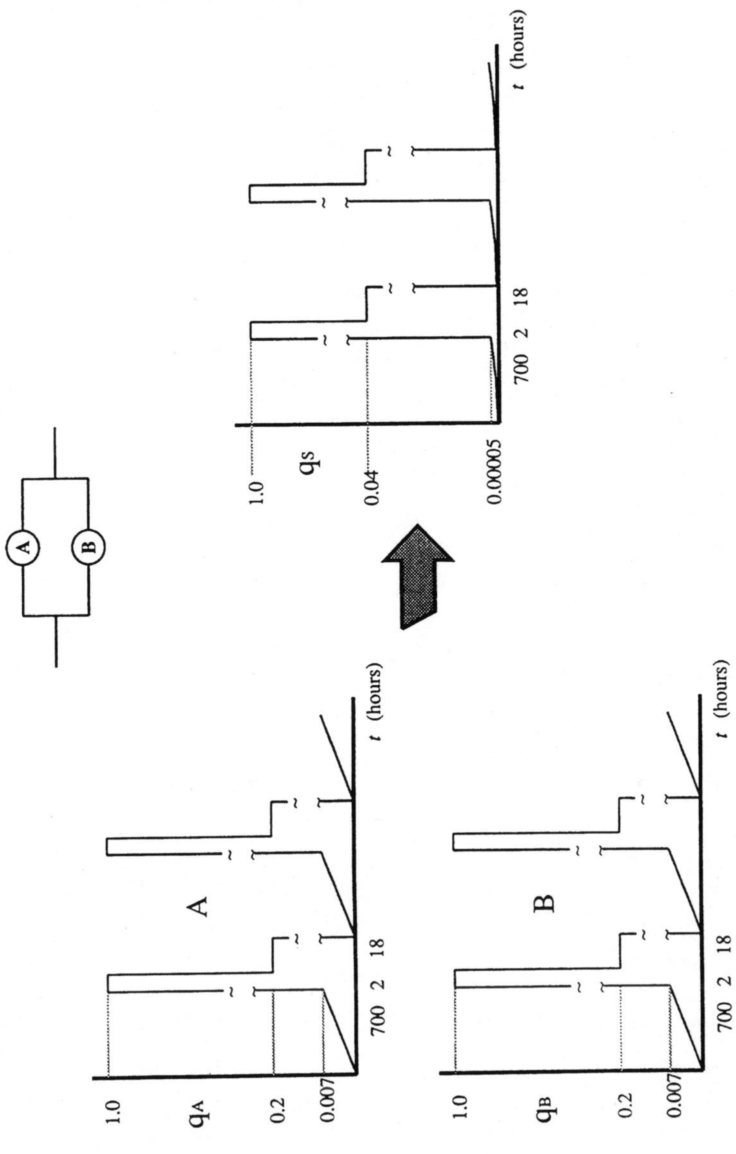


*Fig. 5.11 Unavailability of a Parallel System Using Nonstaggered Testing*

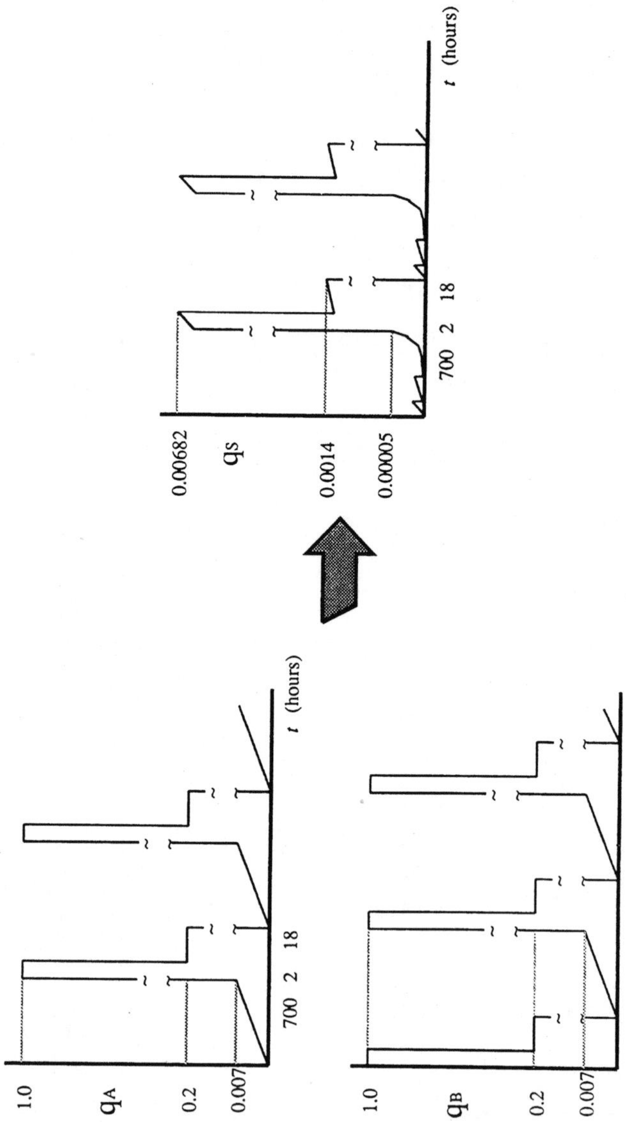


*Fig. 5.12 Unavailability of a Parallel System Using Staggered Testing*

*Table 5.3 Failure Characteristics for Parallel or Series Blocks*

| Type of Block | Block Failure Characteristic | |
|---|---|---|
| | Occurrence Rate $\Lambda$ | Mean Down Time $\tau$ |
| Parallel | $\left[\prod_{i=1}^{n}\lambda_i\tau_i\right]\sum_{i=1}^{n}\frac{1}{\tau_i}$ | $\frac{1}{\sum_{i=1}^{n}\frac{1}{\tau_i}}$ |
| Series | $\left(1-\sum_{i=1}^{n}\lambda_i\tau_i\right)\sum_{i=1}^{n}\lambda_i$ | $\frac{\sum_{i=1}^{n}\lambda_i\tau_i}{\left(1-\sum_{i=1}^{n}\lambda_i\tau_i\right)^2\sum_{i=1}^{n}\lambda_i}$ |

data summarized in Table 5.4 to obtain $\Lambda$ and $\tau$ values. The steps are illustrated in Fig. 5.14.

b) The approximate unavailability of the system can be calculated using $q = \Lambda\tau/(1+\Lambda\tau)$ from Table 5.2. Thus $q = 2.9\times10^{-5}\times2.3/(1+2.9\times10^{-5}\times2.3) = 6.67\times10^{-5}$. This can be compared with the direct calculation method using the cut set concept (similar to Examples 5.8 and 5.9), which yields the average system unavailability of $6.57\times10^{-5}$. The difference is due to the approximate nature of this approach and the assumption that the whole system's time to failure approximately follows an exponential distribution.

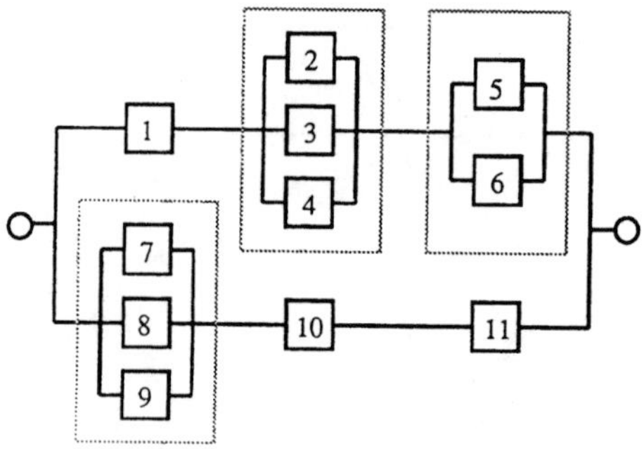


*Fig. 5.13 Sample Series-Parallel System*

*Table 5.4 Summary of Failure Data for the Components Shown in Fig. 5.12*

| Component Serial Number | Failure Rate $\lambda_i$ (per 1000 hour) | Mean Downtime $\tau_i$ (hour) |
|---|---|---|
| 1 | 1 | 5.0 |
| 2 | 10 | 7.5 |
| 3 | 10 | 7.5 |
| 4 | 10 | 7.5 |
| 5 | 5 | 6.0 |
| 6 | 5 | 6.0 |
| 7 | 10 | 7.5 |
| 8 | 10 | 7.5 |
| 9 | 10 | 7.5 |
| 10 | 10 | 5.0 |
| 11 | 10 | 5.0 |

**STEP 1**

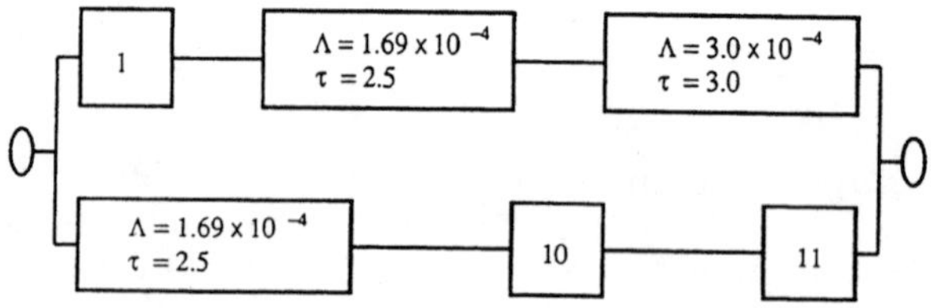


*Fig. 5.14 Step-by-Step Resolution of the System in Fig. 5.12*

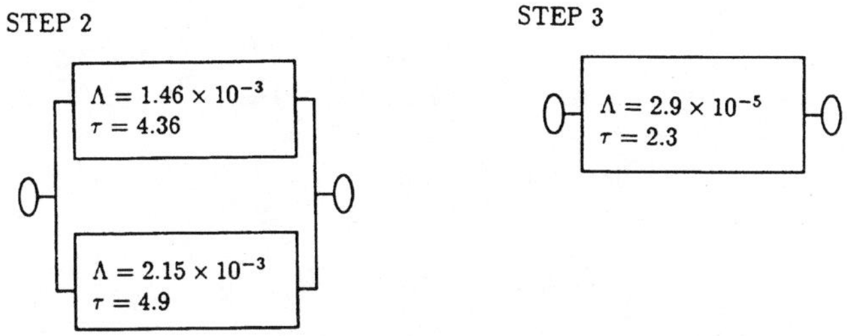


*Fig.5.14 Step-by-Step Resolution of the System in Fig. 5.12 (cont.)*

## BIBLIOGRAPHY


1. Ascher, H. and H. Feingold (1984). *Repairable Systems Reliability: Modeling and Inference, Misconception and Their Causes,* Marcel Dekker, New York.
2. Bain, L.J. (1978). *Statistical Analysis of reliability and Life-Testing Models Theory and Methods,* Marcel Dekker, New York.
3. Bassin, W.M. (1969). "Increasing Hazard Functions and Overhaul Policy," *ARMS IEEE-69C 8-R*, pp. 173–180.
4. Bassin, W.M. (1973). "A Bayesian Optimal Overhaul Interval Model for the Weibull Restoration Process," *J. Am. Stat. Soc.* 68, pp. 575–578 (1973).
5. Caldorola, G. (1977). "Unavailability and Failure Intensity of Components," Nuclear Engineering and Design J., 44, p. 147.
6. Cox, D.R. and P.A. Lewis (1978). *The Statistical Analysis of Series and Events,* Methuen, London.
7. Crow, L.H. (1974). "Reliability Analysis for Complex Repairable Systems," *Reliability and Biometry*, F. Proschan and R.J. Serfling, eds., SIAM, Philadelphia.
8. Crow, L.H. (1990). "Evaluating the Reliability of repairable Systems," Proc. of Ann. Rel. and Maint. Symp., IEEE, Orlando, FL.
9. Ginzburg, T. and W.E. Vesely (1990). "FRANTIC-ABC User's Manual: Time-Dependent Reliability Analysis and Risk Based Evaluation of Technical Specifications," Applied Biomathematics, Inc., Setauket, New York.
10. Lawless, J.F. (1982). *Statistical Models and Methods for Lifetime Data,* Wiley, New York.
11. Lofgren, E. (1985). "Probabilistic Risk Assessment Course Documentation," U.S. Nuclear Regulatory Commission, NUREG/CR-4350, Vol. 5–System Reliability and Analysis Techniques, Washington, DC.

12. Modarres, M. (1984). "A Method of Predicting Availability Characteristics of Series-Parallel Systems," *IEEE Transaction on Reliability*, R-33,4, pp. 309–312.
13. O'Connor, P. (1991). *Practical Reliability Engineering,* 3rd edition, Wiley, New York.
14. Vesely, W.E., F.F. Goldberg, J.T. Powers, J.M. Dickey, J.M. Smith and E. Hall (1981). "FRANTIC II–A Computer Code for Time-Dependent Unavailability Analysis," U.S. Nuclear Regulatory Commission, NUREG/CR-1924, Washington, DC.

## Exercises

5.1 The following table shows fire incidents during 6 equal time intervals of 22 chemical plants.

| Time Interval | 1 | 2 | 3 | 4 | 5 | 6 |
|---|---|---|---|---|---|---|
| No. of fires | 6 | 8 | 16 | 6 | 11 | 11 |

Do you believe the fire incidents are time-dependent? Prove your answer.

5.2 A simplified schematic of the electric power system at a nuclear power plant is shown in the figure below.

a) Draw a fault tree with the top event "Loss of Electric Power from Both Safty Load Buses."

b) Determine the unavailability of each event in the fault tree for 24 hours of operation.

c) Determine the top event probability.

Assume the following

* Either the main generator *or* one of the two diesel generators is sufficient.
* One battery is required to start the corresponding diesel generator.
* Normally, the main generator is used. If that is lost, one of the diesel generators provides the electric power on demand.

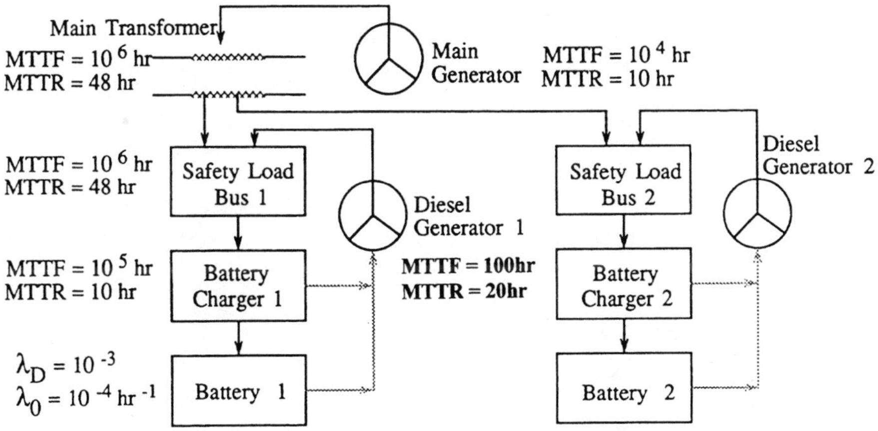

5.3 An operating system is repaired each time it has a failure and is put back into service as soon as possible (monitored system). During the first 10,000 hours of service, it fails five times and is out of service for repair during the following times:

| Out-of-Service |
|---|
| 1000 - 1050 hrs. |
| 3660 - 4000 hrs. |
| 4510 - 4540 hrs. |
| 6130 - 6170 hrs. |
| 8520 - 8560 hrs. |

a) Is there a trend in the data?
b) What is the reliability of the system 100 hours after the system is put into operation? What is the asymptotic availability?
c) If the system has been operating for 10 hours without a failure, what is the probability that it will continue to operate for the next 10 hours without a failure?
d) What is the 80% confidence interval for the mean time to repairs ($\tau = \frac{1}{\mu}$)?

# 6

# Selected Topics in Reliability Analysis

In this chapter, we discuss a number of topics important to reliability analysis. These topics are not significantly related to each other, nor are they presented in a particular order. Most of the topics are still the subject of current research; the methods presented represent a summary of the state of the art.

## 6.1 Analysis of Dependent Failures

Dependent failures are extremely important in reliability analysis and must be given adequate treatment so as to minimize gross overestimation of reliability. In general, dependent failures are defined as events in which the probability of each failure is dependent on the occurrence of other failures. According to (2.14), if a set of dependent events $\{E_1, E_2, \ldots, E_n\}$ exists, then the probability of each failure in the set depends on the occurrence of other failures in the set.

The probabilities of dependent events in the left-hand side of (2.14) are usually, but not always, greater than the corresponding independent probabilities. Determining the conditional probabilities in (2.14) is generally very difficult. However, there are parametric methods that can take into account the conditionality and generate the probabilities directly. These methods are discussed later in this section.

Generally, dependence among various events, e.g., failure events of two items, is either due to the internal environment of these systems or external environment (or events). The internal aspects can be divided into three categories: internal challenges, intersystem dependencies, and intercomponent dependencies. The external aspects are natural or human-made environmental events that make failures dependent. For example, the failure rates for items exposed to extreme heat, earthquakes, moisture, and flood will increase. The intersystem and intercomponent dependencies can be categorized into

*Table 6.1 Types of Dependent Events*

| Dependent Event Type | Dependent Event Category | Subcategory | Description |
|---|---|---|---|
| Internal | 1. Challenge | — | Internal transients or deviations from the normal operating envelope introduce a challenge to a number of items. |
| | 2. Intersystem | 1. Functional | Power to several independent systems is from the same source. |
| | (Failure between two or more systems) | 2. Shared equipment | The same equipment, e.g., a valve, is shared between otherwise independent systems. |
| | | 3. Physical | An extreme environment (e.g., high temperature) causes dependencies between independent systems. |
| | | 4. Human | Operator error causes failure of two or more independent systems. |
| | 3. Intercomponent | 1. Functional | A component in a system provides multifunctions. |
| | (Failure between two or more components ) | 2. Shared equipment | Two independent trains in a hydraulic system share the same common header. |
| | | 3. Physical | Same as system interdependency above. |
| | | 4. Human | Design errors in redundant pump controls cause dependency in a system. |
| External | — | — | Earthquake or fire fails a number of independent systems or components. |

four broad categories: functional, shared equipment, physical, and human caused dependencies. These are described in Table 6.1.

The major causes of dependence among a set of systems or components as described in Table 6.1 can be explicitly described and modeled, e.g., by system reliability analysis models, such as fault trees. However, the rest of the causes can be collectively modeled using the concept of common cause failures (CCFs). CCFs are considered as the collection of all sources of dependencies described in Table 6.1 (especially between components) that are not known, or are difficult to explicitly model in the system or component reliability analysis. For example, functional and shared equipment dependencies are often handled by explicitly modeling them in the system analysis, but other dependencies are considered collectively using CCF.

CCFs have been shown by many reliability studies to contribute significantly to the overall unavailability or unreliability of complex systems. There is no unique and universal definition for CCFs. However, a fairly general definition of CCF is given by Mosleh et al. (1988) as "... a subset of dependent events in which two or more component fault states exist at the same time, or in a short time interval, and are direct results of a shared cause."

To better understand CCFs, consider a system with three redundant components $A$, $B$, and $C$. The total failure probability of $A$ can be expressed in terms of its independent failure $A_I$ and dependent failures as follows:

$C_{AB} =$ Failure of components $A$ and $B$ (and not $C$) from common causes.

$C_{AC} =$ Failure of components $A$ and $C$ (and not $B$) from common causes.

$C_{ABC} =$ Failure of components $A$, $B$, and $C$ from common causes.

Component $A$ fails if any of the above events occur. The equivalent Boolean representation of total failure of component $A$ is $A_T = A_I + C_{AB} + C_{AC} + C_{ABC}$. Similar expressions can be used for components $B$ and $C$.

Now suppose that the success criteria for the system is 2-out-of-3 for components $A$, $B$, and $C$. Accordingly, the failure of the system can be represented by the following events (cut sets): $(A_I \cdot B_I)$, $(A_I \cdot C_I)$, $(B_I \cdot C_I)$, $C_{AB}$, $C_{AC}$, $C_{BC}$, $C_{ABC}$. Thus, the Boolean representation of the system failure will be

$$S = (A_I \cdot B_I) + (A_I \cdot C_I) + (B_I \cdot C_I) + C_{AB} + C_{AC} + C_{BC} + C_{ABC}.$$

It is evident that if only independence is assumed, the first three terms of the above Boolean expression are used, and the remaining terms are neglected. Applying the rare event approximation, the system failure probability is given by

$$Q_s \approx \Pr(A_I) \cdot \Pr(B_I) + \Pr(A_I) \cdot \Pr(C_I) + \Pr(B_I) \cdot \Pr(C_I) + \Pr(C_{AB}) + \Pr(C_{AC}) + \Pr(C_{BC}) + \Pr(C_{ABC}).$$

If components $A$, $B$, and $C$ are similar (which is often the case since common causes among different components has a much lower probability), then

$$\Pr(A_I) = \Pr(B_I) = \Pr(C_I) = Q_1,$$
$$\Pr(C_{AB}) = \Pr(C_{AC}) = \Pr(C_{BC}) = Q_2, \text{ and}$$
$$\Pr(C_{ABC}) = Q_3.$$

Therefore,

$$Q_s = 3(Q_1)^2 + 3Q_2 + Q_3.$$

In general, $Q_k$ represents the probability of CCF among $k$ specific components in a component group of size $m$, such that $1 \leq k \leq m$. There are several methods for calculating $Q_k$. These methods are generally based on parameters that can be estimated through the analysis of failure data. Table 6.2 describes these models. In this table, two general categories of models are presented: shock and nonshock. Moreover, with respect to the number of parameters required for modeling common cause events, there are two distinct categories: single parameter models, and multiple parameter models.

*Table 6.2 Key Characteristics of the Parametric Models, Mosleh (1991)*

| Estimation Approach | Model | Model Parameters | General Form for Multiple Component Failure Frequency |
|---|---|---|---|
| Nonshock models Single parameter | Beta factor | $Q_t, \beta$ | $Q_k = \begin{cases} (1-\beta)Q_t & k=1 \\ 0 & 1<k<m \\ \beta Q_t & k=m \end{cases}$ |
| Nonshock models Multiparameter | Multiple Greek letters | $Q_t, \underbrace{\beta, \gamma, \delta}_{m-1 \text{ parameters}}$ | $Q_k = \frac{1}{\binom{m-1}{k-1}}(1-\rho_{k+1})(\prod_{i=1}^{k}\rho_i)Q_t$ <br> $\rho_1 = 1, \rho_2 = \beta, \ldots, \rho_{m+1} = 0$ |
| | Alpha factor | $Q_t, \alpha_1, \alpha_2, \ldots, \alpha_m$ | $Q_k = \frac{k}{\binom{m-1}{k-1}} \frac{\alpha_k}{\alpha_t} Q_t, \; k = 1, \ldots, m$ <br> $\alpha_t = \sum_{k=1}^{m} k\alpha_k$ |
| Shock models | Binomial failure rate | $Q_1, \mu, \rho, w$ | $Q_k = \begin{cases} \mu\rho_k(1-\rho)^{m-k} & k \neq m \\ \mu\rho^m + w & k=m \end{cases}$ |

### 6.1.1 Single Parameter Models

Single parameter models are those that use one parameter in addition to the total component failure probability to calculate the CCF probabilities. One of the most commonly used single parameter models defined by Fleming (1975) is called the $\beta$-factor model. It is the first parametric model applied to CCF events in risk and reliability analysis. The sole parameter of the model ($\beta$) can be associated with that fraction of the component failure rate that is due to common cause events, which are shared by the other component in the system. That is,

$$\beta = \frac{\lambda_c}{\lambda_c + \lambda_I} = \frac{\lambda_c}{\lambda_t}, \tag{6.1}$$

where

$\lambda_c =$ failure rate due to common cause failures,
$\lambda_I =$ failure rate due to independent failures, and
$\lambda_t = \lambda_c + \lambda_I$.

An important assumption of this model is that whenever a common cause event occurs, all components of a redundant component system fail. In other words, if a CCF shock strikes a redundant system, all components are assumed to fail immediately without any delays.

Based on the $\beta$-factor model, for a system of $m$ components, the probabilities of basic events involving $k$ specific components ($Q_k$), where $1 \le k \le m$, are 0 except $Q_1$ and $Q_m$. These quantities are given as

$$\begin{aligned} Q_1 &= (1-\beta)Q_t, \\ Q_2 &= 0, \\ &\vdots \\ Q_{m-1} &= 0, \\ Q_m &= \beta Q_t, \qquad \text{with } m = 1, 2, \ldots \end{aligned} \tag{6.2}$$

In general, the estimate for the total failure rate of the component must be generated from generic sources of data, while the estimators of the corresponding $\beta$ factor do not explicitly depend on system success data, but require specific assumptions concerning the interpretation of the data. Some generic values of $\beta$ are given in Mosleh et al. (1988). It should be noted that although this model can be used with some degree of accuracy for two component redundancy, the results tend to be conservative for a higher level of redundancy. However, due to its simplicity, this model has been widely used in risk and reliability studies. To get a more reasonable result with regard to higher level of redundancy, more generic parametric models are used.

---

*Example 6.1*

Consider the following system with two redundant trains. Suppose each train is composed of a valve and a pump (each driven by a motor). The pump failure modes are "failure to start" and "failure to run following successful start". The valve failure mode is "failure to open." Develop an expression for the probability of system failure and consider CCF between identical units using the $\beta$-factor method.

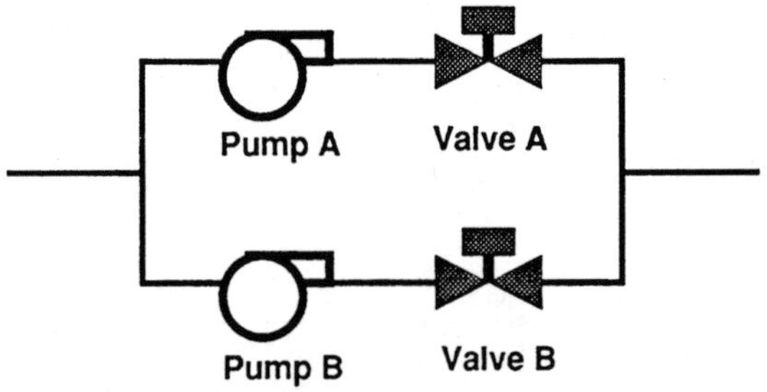


*Solution:*

Use a system fault tree and expand it to include CCFs.

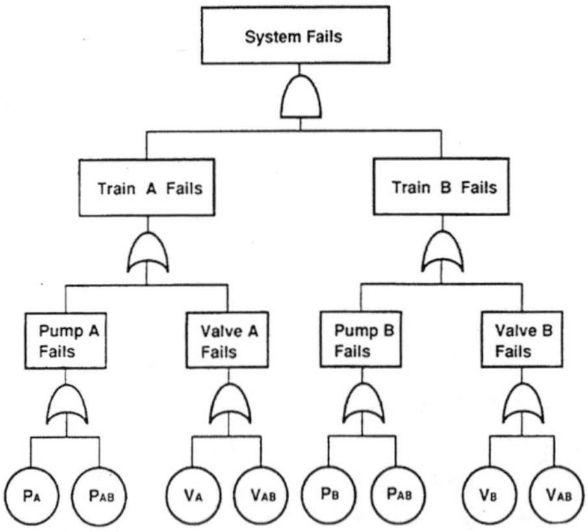


where,

$P_A =$ independent failure of pump A.

$P_B =$ independent failure of pump B.

$P_{AB} =$ dependent failure of pumps A and B.

$V_A =$ independent failure of valve A.
$V_B =$ independent failure of valve B.
$V_{AB} =$ dependent failure of valves A and B.

By solving the fault tree, the following cut sets can be identified:

$C_1 = (P_A, P_B)$, $C_2 = (P_{AB})$,
$C_3 = (V_A, V_B)$, $C_4 = (V_{AB})$,
$C_5 = (P_A, V_B)$, $C_6 = (P_B, V_A)$.

Use the $\beta$-factor method to calculate the probability of each cut set.

$$\begin{aligned}
\Pr(C_1) &= (1-\beta_{PS})^2(q_{PS})^2 + (1-\beta_{PR})^2(\lambda_{PR}t)^2,\\
\Pr(C_2) &= \beta_{PS}\, q_{PS} + \beta_{PR}(\lambda_{PR}t),\\
\Pr(C_3) &= (1-\beta_{VO})^2(q_{VO})^2,\\
\Pr(C_4) &= \beta_{VO}(q_{VO}),\\
\Pr(C_5) &= \Pr(C_6) = (q_{PS} + \lambda_{PR}t)(q_{VO}),
\end{aligned}$$

where,

$PS =$ Pump failure to start,
$PR =$ Pump failure to run,
$VO =$ Valve failure to open,
$q =$ Failure on demand,
$\lambda =$ Failure to run,
$t =$ Mission time,

System failure probability is calculated using rare event approximation, as follows:

$$\Pr(\text{sys. failure}) \cong \sum_{i=1}^{6} P(C_i).$$

---

### *6.1.2 Multiple Parameter Models*

Multiple parameter models are used to get a more accurate assessment of CCF frequencies in systems with a higher level of redundancy. These models have several parameters that are usually associated with different event characteristics. This category of models can be further divided into two subcategories, shock and nonshock models.

## Multiple Greek Letter Model

The Multiple Greek Letter (MGL) model introduced by Fleming et al. (1986) is a generalization of the $\beta$-factor model. Other parameters such as $\gamma$, $\delta$, etc., are used in addition to $\beta$ to distinguish among common cause events affecting different numbers of components in a higher level of redundancy. For a system of $m$ redundant components, and for each given failure mode, $m$ different parameters are defined. For example, for $m = 4$, we have the following

$Q_t =$ Total probability of failure of the item due to common cause and independent events.

$\beta =$ Conditional probability that the common cause of failure an item will be shared by one or more additional items.

$\gamma =$ Conditional probability that the common cause of an item failure that is shared by one or more items will be shared by two or more items in addition to the first.

$\delta =$ Conditional probability that the common cause of an item failure shared by two or more items will be shared by three or more items in addition to the first.

It should be noted that the $\beta$-factor model is a special case of the MGL model in which all other parameters excluding $\beta$ are equal to 1.

### Alpha Factor Model

The $\alpha$-factor model discussed by Mosleh and Siu (1987) develops CCF frequencies from a set of failure ratios and the total component failure rate. The parameters of the model are

$Q_t =$ Total failure probability of each component due to all independent and common cause events.

$\lambda_k =$ Fraction of the total probability of failure in the system that involves the failure of $k$ components due to a common cause.

Using these parameters, depending on the assumption regarding the way systems in the data base are tested, the probability of a common cause basic event involving failure of $k$ components in a system of $m$ components is calculated according to the equation given in Table 6.2. For example, the probabilities of the basic events of the three-component system described earlier will be

$Q_1 = (\alpha_1/\alpha_t)Q_t$,
$Q_2 = (\alpha_2/\alpha_t)Q_t$,
$Q_3 = (3\alpha_3/\alpha_t)Q_t$, where $\alpha_t = \alpha_1 + 2\alpha_2 + 3\alpha_3$.

Therefore, the system probability failure for the three redundant components discussed earlier can now be written as

$$Q_s = 3(\frac{\alpha_1}{\alpha_t}Q_t)^2 + 3(\frac{\alpha_2}{\alpha_t}Q_t) + (3\frac{\alpha_3}{\alpha_t}Q_t).$$

## Binomial Failure Rate Model

The Binomial Failure Rate (BFR) model discussed by Atwood (1983), unlike the $\alpha$-factor model and MGL model, is a shock dependent model. It estimates the frequency of failure of two or more components of a redundant system as the product of the CCF shock rate and the conditional failure of components given the shock. This model considers two types of shock: lethal and nonlethal. The assumption is that, given a nonlethal shock, components fail independently, each with a constant probability of $p$, whereas in the case of lethal shock, all components fail with a probability of 1. The expansion of this model is called the Multinominal Failure Rate (MFR) model. In this model, the conditional probability of failure of $k$ components is calculated directly from component failure statistics without any further assumptions. Therefore, the MFR model becomes essentially the same as the nonshock models, because the separation of the CCF frequency into the shock rate and conditional probability of failure given shock is, in general, a statistical rather than a physical modeling step. This is essentially a common characteristic that relates all the nonlethal shock models. The parameters of the BFR model if a lethal shock is present are as follows:

$Q_1 =$ Independent failure probability of each component.
$\mu =$ Nonlethal shock rate.
$\rho =$ Conditional probability of failure of each component given the occurrence of a nonlethal shock.
$w =$ Lethal shock rate.

Table 6.2 provides the probability of basic events in accordance with the BFR model. Due to its complexity and the lack of data to obtain its parameters, the BFR model is not widely used in practice.

### *6.1.3 Data Analysis for Common Cause Failure Parameters*

Despite the different forms and formulations, of the models described in Section 6.1.1, they have similar data requirements. Basically, most of these models need the same statistical information extracted from the raw data. Therefore, one should not expect to find significant differences among the numerical results of these models. The relative differences in the results may be attributed to the distributional aspects of the parameter estimates, which are related to the type of assumptions made in developing the parameter estimator, and to parameter dependence in CCF probability quantification. In the case of the BFR model, there is an additional requirement to classify events caused by lethal shocks and those caused by nonlethal shocks. A relatively sizable difference between the results based on the BFR model and those based on all other models can be obtained, depending on the number of events that can be caused by lethal shocks. This is because only the BFR model makes an extra assumption about the occurrence of CCFs. With the BFR model, if there are no lethal shocks present in the data base, the estimated frequencies could be significantly different from those obtained with other models.

The most important steps in the quantification of CCFs are gathering information from the raw data and selecting a model that can use most of the information. The parameter estimation approaches discussed in Chapters 2 and 3 are applied to estimate the parameters. If models rely on the same type of information in estimating the CCF probabilities, and similar assumptions regarding the mechanism of CCFs are used, comparable numerical results can be generated. It is important to use information extracted from the data base because most of the model parameters are estimated based on that information. Therefore, to get an accurate estimate of the CCF probability, efforts should be directed toward using the large amount of information gathered from the data base. Table 6.3 summarizes simple point estimators for parameters of various nonshock CCF models. In this table, $n_k$ is the total number of observed failure events involving failure of $k$ similar items due to a shared cause, $m$ is the total number of redundant items considered; and $N_D$ is the total number of system demands. If the item is normally operating (not standby), then $N_D$ can be replaced by $T$, the total test (operation) time. The estimators in Table 6.3 assume that in every system demand, all components and possible combination of components are challenged. Therefore, the estimators apply to systems whose tests are nonstaggered.

---

*Example 6.2*

For the system described in example 6.1, estimate the $\beta$ parameters, $\lambda$ and $q$, for the valves and pumps if the following data are observed:

| Failure Mode | Event Statistics | | |
|---|---|---|---|
| | $n_1$ | $n_2$ | $T$ (hrs) or $N_D$ |
| Pump fails to start ($PS$) | 10 | 1 | 500 (demands) |
| Pump fails to run ($PR$) | 50 | 2 | 10,000 (hours) |
| Valve fails to open ($VO$) | 15 | 1 | 10,000 (demands) |

*Table 6.3 Simple Point Estimators for Various Parametric Models*

| Method | Point Estimator |
|---|---|
| Beta Factor | $\hat{Q}_t = \frac{1}{mN_D}\sum_{i=1}^{m} kn_k$ <br> $\hat{\beta} = \left(\sum_{k=2}^{m} kn_k\right) / \left(\sum_{k=1}^{m} kn_k\right)$ |
| Multiple Greek Letters | $\hat{Q}_t = \frac{1}{mN_D}\sum_{k=1}^{m} kn_k$ <br> $\hat{\beta} = \left(\sum_{k=2}^{m} kn_k\right) / \left(\sum_{k=1}^{m} kn_k\right)$ <br> $\hat{\gamma} = \left(\sum_{k=3}^{m} kn_k\right) / \left(\sum_{k=2}^{m} kn_k\right)$ <br> $\hat{\delta} = \left(\sum_{k=4}^{m} kn_k\right) / \left(\sum_{k=3}^{m} kn_k\right)$ |
| Alpha Factor | $\hat{Q}_t = \frac{1}{mN_D}\sum_{k=1}^{m} kn_k$ <br> $\hat{\alpha_k} = n_k / \left(\sum_{k=1}^{m} n_k\right) \quad (k = 1, \ldots, m)$ |

where

$n_1 =$ Number of observed independent failures,
$n_2 =$ Number of observed events involving double CCF.

Calculate the system unreliability for a mission time of 10 hours.

*Solution:*

According to Table 6.3,

$$\hat{\beta} = \frac{2n_2}{n_1 + 2n_2}.$$

Apply this formula to $\beta_{PR}$, $\beta_{PS}$, and $\beta_{VO}$ by using appropriate values for $n_1$ and $n_2$.

$n_{PS} = n_1 + 2n_2 = 12,$
$n_{PR} = n_1 + 2n_2 = 54,$
$n_{VO} = n_1 + 2n_2 = 17.$

Accordingly, using (3.31) and (3.61) for estimating $\lambda$ and $q$, respectively

$$q_{PS} = \frac{12}{500} = 2.4 \times 10^{-2}/D, \qquad \beta_{PS} = \frac{2}{12} = 0.17.$$
$$\lambda_{PR} = \frac{54}{10{,}000} = 5.4 \times 10^{-3}/\text{hr}, \qquad \beta_{PR} = \frac{4}{54} = 0.07.$$
$$q_{VO} = \frac{17}{10{,}000} = 1.7 \times 10^{-3}/D, \qquad \beta_{VO} = \frac{2}{17} = 0.12.$$

Therefore, using $\Pr(C_i)$ in Example 6.1, the estimates of the system failure probabilities are

$$\Pr(C_1) = (1-0.17)^2(2.4 \times 10^{-2})^2 + (1-0.07)^2(5.4 \times 10^{-3} \times 10)^2 = 2.9 \times 10^{-3}.$$
$$\Pr(C_2) = 0.17 \times 2.4 \times 10^{-2} + 0.07 \times 5.4 \times 10^{-3} \times 10 = 7.9 \times 10^{-3}.$$
$$\Pr(C_3) = (1-0.12)^2(1.7 \times 10^{-3})^2 = 2.2 \times 10^{-6}.$$
$$\Pr(C_4) = 0.12 \times 1.7 \times 10^{-3} = 2.0 \times 10^{-4}.$$
$$\Pr(C_5) = (2.4 \times 10^{-2} + 5.4 \times 10^{-3} \times 10)(1.7 \times 10^{-3}) = 1.3 \times 10^{-4}.$$
$$\Pr(C_6) = \Pr(C_s) = 1.3 \times 10^{-4}.$$

Thus, $\Pr(\text{Sys. failure}) = 1.1 \times 10^{-2}$.

---

## 6.2 Uncertainty Analysis

Uncertainty arises primarily due to lack of reliable information, e.g., lack of information about the ways a given system may fail. Uncertainty may also arise due to linguistic imprecision, e.g., the expression "System A is reliable." Furthermore, uncertainty may refer to random variability or the structure of a model. For example, if a Weibull distribution is used to represent the time to failure, the choice of the model involves uncertainty. We may even be uncertain about the way we represent uncertainty. For example, our uncertainty about the variability of $\alpha$ and $\beta$ parameters of the Weibull distribution representing time to failure distribution may be depicted by another distribution, e.g., a lognormal distribution.

The most common practice in measuring uncertainty is the use of the probability concept. In this book, we have only used this measure of uncertainty. As we discussed in Chapter 2, there are different interpretations of probability. This also affects the way uncertainty analysis is performed. In this section, we first briefly discuss uncertainty in models and then present methods of measuring the uncertainty about the parameters of the model. Then we discuss methods of propagating uncertainty in a complex model. For example, in a fault tree model representing a complex system, the uncertainty assigned to each leaf of the tree can be propagated to obtain a distribution of the top event probability.

### *6.2.1 Measurement of Uncertainty*

The simplest way to measure uncertainty is to use sample mean $\bar{x}$ and variance $S^2$, described by (2.73) and (2.74). The coefficient of variation $V_x$ defined as

$$V_x = \frac{S}{\bar{x}}, \tag{6.3}$$

We have discussed earlier in Chapter 2 that estimations of $\bar{x}$ and $S$ are themselves subject to some uncertainty, it is important to describe this uncertainty by confidence intervals of $\bar{x}$ and $S$, e.g., by using (2.83). This brings another level of uncertainty The confidence intervals associated with different types of distributions were discussed in Chapter 3. For a binomial model, the confidence intervals can be obtained from (3.62) and (3.63). Similarly, if the data are insufficient, then the subjectivist definition of probability can be used and probability intervals obtained using (3.69).

### *6.2.2 Parameter Uncertainties*

In Chapter 3, we discussed measures for quantifying uncertainties of parameter values of distribution models for both frequentist and subjectivist (Bayesian) methods. Examples of these parameters are mean $\mu$, failure rate $\lambda$, and probability of failure per demand $p$ of a component. Uncertainty in the parameters is primarily governed by the amount of field data available about failures and repairs of the items. Because of these factors, a parameter does not take a fixed and known value, and has some random variability. In Section 6.2.4, we discuss how the parameter uncertainty is propagated in a system to obtain an overall uncertainty about the system failure.

### *6.2.3 Model Uncertainties*

There are two aspects of modeling uncertainty at the component level or system level. In estimating uncertainty associated with unreliability or unavailability of a basic component, a modeling error can occur as a result of using an incorrect distribution model. Generally, it is very difficult to estimate an uncertainty measure for these cases. However, in a classical (frequentist) approach, the confidence level associated with a goodness-of-fit test can be used as a measure of uncertainty.

For the reliability analysis of a system, one can say that a model describes the behavior of a system as viewed by the analyst. However, the analyst can make mistakes due to a number of constraints, namely, his degree of knowledge and understanding of the system design, its full spectrum of operation and limitations, and his assumptions about the system, as reflected in the reliability model (e.g., a fault tree).

Clearly one can minimize these sources of uncertainty, but one cannot eliminate them. For example, a fault tree based on the analyst's understanding of the success criteria of the system can be incorrect if the success criteria used are in error. For this reason, a more accurate dynamic analysis of the system may be needed to obtain correct success criteria.

Definition and quantification of the uncertainty associated with a model are very complex and cannot easily be associated with a quantitative representation (e.g., probabilistic representation). The readers are referred to Morgan and Henrion (1990) for more discussion on this topic.

### *6.2.4 Propagation Methods*

Consider a general case of a system output $Y$ (e.g., system reliability or unavailability). Based on a model of the system, a general function of uncertain quantities $x_i$ and uncertain parameters $\theta_i$ can describe this system such that

$$Y = f(x_1, x_2, \ldots x_n, \theta_1, \theta_2, \ldots \theta_m). \tag{6.4}$$

For example, $Y$ can be the probability of the top event of a fault tree, in which case $x_i$ will be the probability (unavailability) of each component represented in the fault tree, and $\theta_i$ will be the parameters of the distribution models representing $x_i$.

Clearly the uncertainty related to the accuracy of $Y$ is a model uncertainty and can not be described probabilistically. However, the variability of $Y$ as a result of the variability of the basic parameters $x_i$ and $\theta_i$ can be estimated by the methods of propagation. For simplicity, we describe $Y$ in the following form:

$$Y = g(\gamma_i), \qquad i = 1, 2, \ldots, n. \tag{6.5}$$

We will discuss two distinct sets of propagation methods: classical statistical and Bayesian methods.

### *6.2.5 Classical Statistics Methods*

Obtaining the point estimate and confidence intervals of Y in (6.5) is the primary objective of a classical statistical uncertainty analysis. The point estimate of $Y$ can be easily obtained from

$$\hat{Y} = g[\hat{\gamma}_i], \qquad \text{where } \hat{\gamma}_i \text{ is the estimate of } \gamma_i. \tag{6.6}$$

Here we are interested in knowing how large or how small $Y$ might be, consistent with available data. The degree of consistency is defined by the confidence level. This depicts how uncertain one should be about $\hat{Y}$ because of the nature and extent of the available data that describe $\gamma_i$. Exact statistical confidence intervals for $Y$ are generally not available, so approximation methods are often used. We will describe three such methods: Taylor's series, the bootstrap method, and the System Reduction method. The Taylor's series method is also applicable to probabilistic propagation.

### a) Taylor's Series

This method is one of the oldest for propagating uncertainty. Suppose that from the available data, $\hat{\gamma}_i$ and $S^2(\gamma_i)$ are estimated. The first-order expansion of $g(\gamma_i)$ can approximately estimated from

$$S^2(Y) \approx \sum_{i=1}^{n} \mathrm{Var}(\gamma_i) \left[\frac{\partial Y}{\partial \gamma_i}\right]^2_{\gamma_i=\hat{\gamma}_i} . \tag{6.7}$$

For simple cases of weighted sum, e.g., when we are dealing with the cut sets of a system, then $Y = \sum_{i=1}^{n} a_i x_i$,

$$\hat{Y} = \sum_{i=1}^{n} a_i \hat{x}_i, \tag{6.8}$$

$$S^2(Y) = \sum_{i=1}^{n} a_i^2 S^2(x_i) + 2\sum_{i=1}^{n-1} \sum_{j=i+1}^{n} a_i a_j \mathrm{Cov}(x_i, x_j).$$

Expression (6.8) is exact. If we assume that $x_i$s are independent, then the $\mathrm{Cov}(x_i, x_j)$ is zero and (6.8) is further simplified. Equations (6.7) and (6.8) also apply to probabilistic propagation, in which case $E(x_i)$ and $\mathrm{Var}(x_i)$ replace $\hat{x}_i$ and $S^2(x_i)$, and $E(Y)$ and $\mathrm{Var}(Y)$ are calculated.

### b) Bootstrap Method

The bootstrap method first described by Efron (1979) is a simulation technique in which new samples are generated from the data of an original sample. The method's name, derived from the old saying about pulling yourself up by your own bootstraps, reflects the fact that one available sample gives rise to many others. The result of the bootstrap technique is a distribution from which estimates of upper and lower confidence limits can be directly obtained. The variance of the bootstrap sample and the point estimate may alternately be used to obtain an estimated binomial approximation. The estimated binomial distribution in turn may be used to obtain the confidence limits. In this book, the estimated binomial technique is applied to the bootstrap results to estimate the confidence limits.

Consider the determination of the point estimate and confidence interval of $Y$ in (6.5). Bootstrap samples are generated by randomly generating one value for each of the parameters $\Pr(\gamma_i)$, and then computing a sample for $Y$. $\Pr(\gamma_i)$ can be obtained directly from data. For example, if $F$ failures in $N$ starts of an item are observed, then $\Pr(\gamma_i)$ is assumed to be a binomial distribution, with $\hat{P}_{\gamma_i} = F/N$. Or, if failures occur with time, $N$ replaces $T$.

The point estimate $\hat{Y}$ of $Y$ is obtained by selecting the point estimate for each $\Pr(\gamma_i)$ (i.e., $\hat{P}_{\gamma_i}$) and computing the value of $Y$. Bootstrap samples are generated by randomly generating one value for each of the probabilities $\Pr(\gamma_i)$, and then computing a sample value for $Y$. The sampling process is repeated many times to obtain a well-defined probability distribution for $Y$. The samples generated in this manner are called bootstrap samples. The distribution of the parameter $Q$ obtained from the bootstrap samples can then be treated as if it were a distribution of real samples.

Bootstrap samples may be generated by using the Monte-Carlo simulation approach discussed later in this chapter.

The bootstrap method does not always guarantee a true picture of the statistical accuracy of the sample estimate. For example, if no failures are observed, the distribution becomes degenerate and a probability of 1 is assigned to an outcome of 0. Some approximations may, however, be used to remedy this situation. A conservative assumption is to assign a small value (such as 0.1 or in very rare events, 0.05) to the value of $F$.

The distributions generated through bootstrap samples can be used to determine confidence limits on the sample parameter $Y$. In an alternate method, the variance of the bootstrap sample may be taken as the estimated variance of $Y$. Confidence limits on $Y$ can then be obtained by recasting $\hat{Y}$ and the bootstrap variance estimates to obtain parameters of an equivalent effective binomial distribution. For example, the variance from the bootstrap method can be used along with the point estimate for $Y$ (i.e., $\hat{Y}$ and $S^2(Y)$) to determine the parameters of an estimated binomial distribution (i.e., $P$ and $N$). If the bootstrap point estimate of $Y$ is $\hat{Y}$ and the bootstrap variance is $S^2(Q)$, then

$$N = \hat{Y}(1 - \hat{Y})/S^2(Y), \text{ and}$$
$$P = \hat{Y} \tag{6.9}$$

represent an estimated binomial distribution for $Q$. The equivalent binomial distribution can be used to determine confidence limits of $Y$ by using (3.53).

### c) System Reduction Method

The system reduction method (also referred to as the Maximus method–1980) was developed to estimate confidence limits on the reliability of serial/parallel networks of components, each of which in turn is described by a binomial distribution failure model. In this method, subsystems that consist of only serial components or only parallel components are sequentially defined and combined to yield an equivalent binomial distribution for the system as a whole. The upper and lower confidence limits are then estimated from this equivalent distribution.

This technique yields a single overall $F$ (number of failures) and N (number of trials or number of demands) representative of $Y$ in (6.5). Knowing the overall $F$ and $N$ for $Y$, one can estimate confidence limits by evaluating the confidence limits of the binomial distribution defined by overall binomial distribution described by $F$ and $N$.

Consider a series configuration described by function $Y$ composed of $k$ parameters, each modeled by a binomial distribution. The reduction to an equivalent $F$ and $N$ involves three steps.

1) The point estimate for the overall occurrence probability of $Y$ is determined by

$$\hat{Y} = 1 - \prod_{i=1}^{k}(1 - \hat{P}_{\gamma_i}). \tag{6.10}$$

2) The value of $N$, which is representative of the effective number of demands for $Q$ as a whole, is given by

$$N = \min_i N_i. \tag{6.11}$$

3) The value of $F$, which is representative of the effective number of failures for $Y$ as a whole, is given by

$$F = N \cdot \hat{Y}. \tag{6.12}$$

Consider a parallel system configuration $Y$ composed of $m$ events, such as $m$ cut sets of a fault tree, each modeled by a binomial distribution. Reduction to an equivalent $F$ and $N$ involves a four-step procedure

1) Compute

$$\hat{Y} = \prod_{i=1}^{m} \hat{P}_{\gamma_i} \tag{6.13}$$

2) Compute

$$\hat{Y}' = \prod_{i=1}^{m} (F_i + 1)/(N_i + 1) \tag{6.14}$$

3) $N$ is given by

$$N = (1 - \hat{Y}')/(\hat{Y}' - \hat{Y}). \tag{6.15}$$

4) The equivalent $F$ is computed from $F = N \cdot \hat{Y}$.

It is evident that with this method, 0 failures in $N$ demands or 0 failures in a length of time $T$ can also be easily handled. The bootstrap method is very weak in this respect.

---

*Example 6.3*

Consider a fault tree containing two cut sets described by the expression $T = C_1 + C_2 \cdot C_3$. If the following data are reported for the components, determine a point estimate and 95% confidence interval for $T$ using the system reduction method and the bootstrap method.

| Component | $F$ | $N$ |
|---|---|---|
| $C_1$ | 1 | 1,785 |
| $C_2$ | 8 | 492 |
| $C_3$ | 4 | 371 |

*Solution:*

The second cut set can be considered as a parallel system containing components $C_2$ and $C_3$. Equivalent distribution for each cut set can be considered as series subsystems, and equivalent binomial distribution of $T$ can be obtained for $C_2$ and $C_3$, $CS_2 = C_2C_3$.

$$\hat{P}_{CS_2} = \frac{8}{492} \cdot \frac{4}{371} = 1.75 \times 10^{-4},$$
$$\hat{P}' = \frac{8+1}{492+1} \cdot \frac{4+1}{371+1} = 2.45 \times 10^{-4},$$
$$N_{CS_2} = \frac{1 - 2.45 \times 10^{-4}}{2.45 \times 10^{-4} - 1.75 \times 10^{-4}} = 14,282,$$
$$F_{CS_2} = 1.75 \times 10^{-4} \times 14,282 = 2.5.$$

Now consider $CS_1$ and $CS_2$ as servers. $(CS_1 = C_1$ here.)

$$\hat{P}_T = 1 - \left(1 - \frac{1}{1,785}\right)\left(1 - \frac{2.5}{14,282}\right) = 7.35 \times 10^{-4},$$
$$N_T = \min(N_{CS_1}, N_{CS_2}) = 1,785,$$
$$F_T = 1,785 \times 7.35 \times 10^{-4} = 1.3.$$

Using (3.62) and (3.63), $6.1 \times 10^{-5} \leq P_T \leq 3.4 \times 10^{-3}$.

If we used the bootstrap method with 2,500 trials each would contain a value taken randomly from the binomial distribution of $C_1, C_2$, and $C_3$, and the respective estimate of $T_i$ (i = trial number). After ordering $T_i$s, one can calculate sample mean, sample variance, and confidence limits as follows:

$\hat{P}_T = 7.66 \times 10^{-4}$

$S^2(T) = 2.17 \times 10^{-7}$

According to (6.9), $N_T = 3,525$ and $F_T = 2.7$. This result reasonably agrees with the system reduction technique for $\hat{P}_T$. Using (3.62) and (3.63), yields

$$1.43 \times 10^{-4} \leq P_T \leq 2.10 \times 10^{-3}.$$

The lower bound predicted by the system reduction method is more conservative.

---

Clearly, the system reduction technique is limited to purely parallel series systems, additionally the "rules" discussed above for parallel systems cannot be proven mathematically, although extensive demonstration calculations support their heuristic foundations.

### *6.2.6 Probabilistic Methods*

#### a) Monte-Carlo Simulation

The Monte-Carlo method is a very simple and direct technique for propagating uncertainties. For example, in (6.5), each parameter

$\gamma_i$ is represented by a *pdf* obtained from a Bayesian type analysis. The Monte-Carlo process involves an evaluation of $\Pr(Y)$ for a large number of trials. Each trial is obtained by a random sampling from the distributions assigned to the input variables $\gamma_i$. Reactor Safety Study (1975) describe the details of the process. The procedure is repeated in each trial, and various $Y$ outcomes are sorted to obtain empirical estimates of some desired top event attributes, such as mean, median, and 95th and 5th percentiles.

As the number of trials increases, the precision of the empirical percentile estimates of $Y$ improves. However, Martz et al. (1983) note that the rate of convergence to the true distribution decreases as the number of samples increases. The Monte-Carlo method provides flexibility in the selection of distributions for the parameters $\gamma_i$, and is easy to implement. However, the process should be performed by a computer, and the computer time exponentially increases as the desired precision and thus the number of trials increases.

Latin Hypercube Sampling (LHS) was developed by Iman et al. (1980) to improve the precision of the Monte-Carlo simulation. LHS is a stratified sampling approach in which $n$ different samples are selected from the *pdf* of $\gamma_i$, by dividing the *pdf* into $n$ intervals, each with the same probability. Within each interval, one value of $\gamma_i$'s probability is randomly selected. The $n$ values are then combined with values from another $\gamma_j$ such that the pairwise correlations are near 0. The result is an $n \times m$ matrix, where $n$ is the total number of $\gamma_i$. The $j^{th}$ row of the matrix represents values for all $\gamma_i$'s to be used in the $j^{th}$ run for calculating $Y$. The distribution and quantities of $Y$ are then empirically estimated from the $m$ runs. LHS provides improved accuracy and ensures that the tails of $\gamma_i$ distributions are included. However, there is a possibility of bias in the calculations, particularly when the sample size is small.

## b) Method of Moments

The method of moments described by Morchland and Weber (1972) and Apostolakis and Lee (1977) propagates the uncertainty associated with parameters $\gamma_i$ by generating lower order moments, such as the mean and variance for $Y$, from the lower order moments of the distribution for $\gamma_i$. For this purpose, a Taylor series expansion of $Y$ is used. Its coefficients are related to the moments of the distributions of $\gamma_i$. A detailed treatment of this method is covered in a comparison study of the uncertainty analysis method by Martz et al. (1983). For special cases where $Y = \sum_{i=1}^{n} \gamma_i$, (2.70) is used to obtain $\hat{Y}$. In this case,

$$\mathrm{Var}[Y] = \sum_{i=1}^{n} \mathrm{Var}[\gamma_i] + 2\sum_{i=1}^{n-1} \sum_{j=i+1}^{n} \mathrm{Cov}[\gamma_i, \gamma_j]. \tag{6.16}$$

On the other hand, if $Y = \prod_{i=1}^{n} \gamma_i$, and $\gamma_i$ are independent, then

$$E(Y) = \prod_{i=1}^{n} E(\gamma_i), \quad \text{and}$$

$$\mathrm{Var}(Y) = \left[\sum_{i=1}^{n} \frac{\mathrm{var}(\gamma_i)}{E^2(\gamma_i)}\right] \cdot E^2(Y). \tag{6.17}$$

If $\gamma_i$s are dependent, the equations will be more complex. However, (2.71) and (6.16) can be used, i.e., $ln(Y) = \sum_{i=1}^{n} ln(\gamma_i)$, in which $\ln(\gamma_i)$ replaces $\gamma_i$.

Dezfuli and Modarres (1984) have expanded this approach to efficiently estimate a distribution fit for $Y$ when $\gamma_i$ are highly dependent. The method of moments provides a quick and accurate estimation of lower moments of $Y$ based on the moments of $\gamma_i$, and the process is simple. However, for highly nonlinear expressions of $Y$, the use of only low-order moments can lead to significant inaccuracies, and the use of higher moments is complex.

### c) Discrete Probability Distributions

The method of discrete probability distribution (DPD) introduced by Kaplan (1981) is a technique in which the distribution of $\gamma_i$s is discretized, and a discrete analog of the $Y$ distribution is obtained. Each distribution $\gamma_i$ is divided into $m$ intervals. The probability (area under the pdf) associated with each interval is calculated for each of the $n$ pdfs representing $\gamma_i$s. Using (6.5) the probability of $Y$ can be calculated $n^m$ times. Thus, a discrete probability distribution can be constructed for $Y$.

The DPD method is theoretically valid. However, the technique can become quite complex with respect to computer storage if too many parameters $\gamma_i$s are presented in (6.5).

### *6.2.7 Uncertainty Importance*

Uncertainty importance is a measure of the contribution of uncertainty of each $\gamma_i$ in (6.5) to the uncertainty of $Y$. This measure gives insight into which $\gamma_i$s are mainly causing the uncertainty of $Y$. Thus, collection of more data and further analysis are necessary to reduce this uncertainty. The uncertainty importance $UI_i$ for $\gamma_i$ is calculated as

$$UI_i = [Var(\gamma_i)]^{1/2}\, E\left(\frac{\partial Y}{\partial \gamma_i}\right). \tag{6.18}$$

Other uncertainty importance models are described by Bier (1983), Nakashima and Kazuharu (1982), and Wheeler and Spulak (1985).

### *6.2.8 Graphic Representation of Uncertainty*

The results of an uncertainty analysis should be presented in a clear manner that aids analysts in developing appropriate qualitative insights. Generally, we will discuss three different ways of presenting probability distributions: plotting the pdf or the cpdf, or displaying selected fractiles, as in a Tukey (1979) box plot. Fig. 6.1 shows examples. The density function shows the relative probabilities of different values of the parameters. One can easily see the areas or ranges where high densities (occurrences) of the r.v. occur (e.g., the modes). One can easily judge symmetry and skewness and the general shape of the distribution (e.g., bell-shaped vs. J-shaped). The cpdf is best for displaying fractiles, including the median and confidence intervals. It is easily used for both continuous and discrete distributions.

The standard Tukey box shows a horizontal line from the 10th to 90th percentiles, a box between the lower percentiles (e.g., from the 25th to 75th percentiles), and a vertical line at the median, and points at the minimum and maximum observed values. This method clearly shows the important quantities of the r.v.

In cases where statistical uncertainty bounds are estimated, a Tukey box can be used to describe the confidence intervals. Consider a case where the distribution of a variable $Y$ is calculated and described by a pdf. The pdf of time to failure is represented by an exponential distribution and the value of $\lambda$ for this exponential distribution is represented by a lognormal distribution. Then, $f(Y|\lambda)$ for various values of $\lambda$ can be plotted and families of curves can be developed, as shown in Fig. 6.2.

In general, a third method can be shown by actually displaying the densities of $\lambda$ in a multidimensional form. For example, Fig. 6.3 presents such a case for a two-dimensional distribution. In this figure $f(t|\lambda)$ is shown for various values of $\lambda$ areas, where densities of $\lambda$ are relatively larger and proportionally more curves are presented.

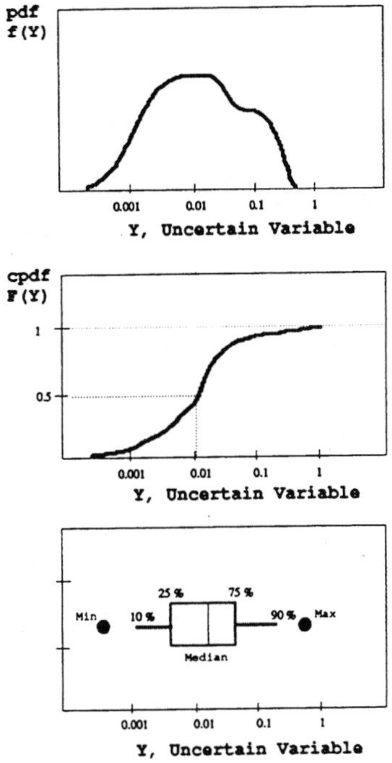


*Fig. 6.1 Three Conventional Methods of Displaying Distribution*

## 6.3 Human Reliability

It has long been recognized that human error has a substantial impact on the reliability of complex systems. Accidents at Three Mile Island and Chernobyl clearly show how human error can defeat engineered safeguards and play a dominant role in the progression of accidents. 70% of aviation accidents are caused by human malfunctions, similar figures apply to the shipping and process industry. The Reactor safety Study (1975) revealed that more than 60% of the potential accidents in the nuclear industry are related to human errors. In general, the human contribution to overall system performance is at least as important as that of hardware reliability.

To obtain a precise and accurate measure of system reliability, human error must be taken into account. Analysis of system designs, procedures, and postaccident reports shows that human error can be an immediate accident initiator or can play a dominant role in the progress of undesired events. Without incorporating human error

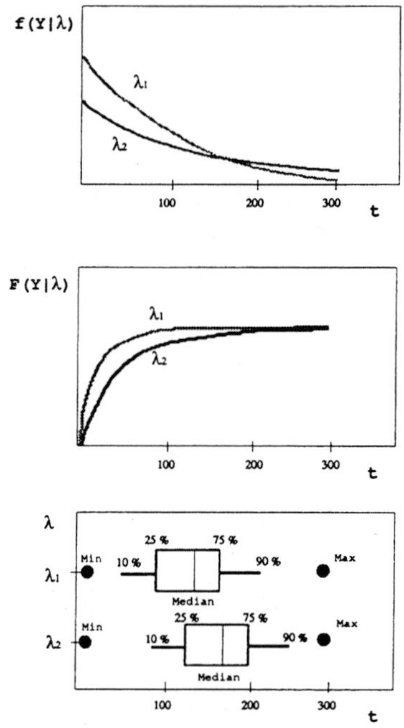


*Fig. 6.2 Representation of Uncertainties*

probabilities, the results are incomplete and often underestimated.

To estimate human error probabilities (and thus human reliability), one needs to understand human behavior. However, human behavior is very difficult to model. Literature shows that there is not a strong consensus on the best way to capture all human actions and quantify human error probabilities. The assumptions, mechanisms, and approaches used by any one specific human model cannot be applied to all human activities. Current human models need further advancement, particularly in capturing and quantifying intentional human errors. Limitations and difficulties in current human reliability analysis (HRA) include the following:

a) Human behavior is a complex subject that cannot be described as a simple component or system. Human performance can be affected by social, environmental, psychological, and physical factors that are difficult to quantify.
b) Human actions cannot be considered to have binary success and failure states, as in hardware failure. Furthermore, the

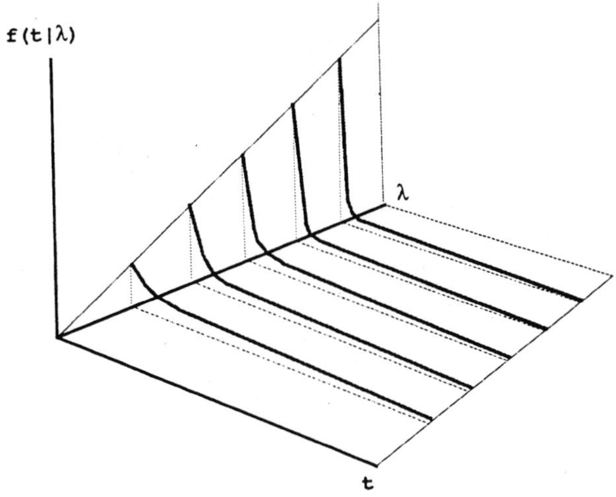


*Fig. 6.3 Two Dimensional Uncertainty Representation*

full range of human interactions have not been fully analyzed by HRA methods.

c) The most difficult problem with HRA is the lack of data on human behavior.

Human error may occur in any phase of the design, manufacturing, construction, and operation of a complex system. Design, manufacturing, and construction errors are also the cause of many types of errors during system operation. The most notable errors are dependent failures whose occurrence can cause loss of system redundancy. These may be discovered in manufacturing and construction, or during system operation. Normally, quality assurance programs are designed and implemented to minimize the occurrence of these types of human error.

In this book, we are concerned with human reliability during system operation, where human operators are expected to maintain, supervise, and control complex systems. In the remainder of this section, human models are reviewed, and important models are described in some detail. Emphasis is on the basic idea, advantages, and disadvantages of each model, and applicability to different situations. Then, we describe the important area of data analysis in HRA. After the links between models and data are reviewed, the problems of human data sources and data acquisition are addressed.

*6.3.1 Human Reliability Analysis Process*

A comprehensive method of evaluating human reliability is the method called Systematic Human Action Reliability Procedure (SHARP) by Hannaman and Spurgin (1984). The SHARP defines seven steps to perform HRA. Each step consists of inputs, activities, rules and outputs. The inputs are derived from prior steps, reliability studies, and other information sources, such as procedures and accident reports. The rules guide the activities which are needed to achieve the objectives of each step. The output is the product of the activities performed by analysts.

The goals for each step are as follows:

1) Definition: Ensure that all different types of human interactions are considered.
2) Screening: Select the human interactions that are significant to system reliability.
3) Qualitative Analysis: Develop a detailed description of important human actions.
4) Representation: Select and apply techniques to model human errors in system logic structures, e.g., fault trees, event trees, MLD, or reliability block diagram.
5) Impact Integration: Explore the impact of significant human actions identified in the preceding step on the system reliability model.
6) Quantification: Apply appropriate data to suitable human models to calculate probabilities for various interactions under consideration.
7) Documentation: Include all necessary information for the assessment to be understandable, reproducible, and traceable.

The relationships among these steps are shown in Fig. 6.4. These steps in human reliability consideration are described in more detail below.

Step 1: Definition

The objective of Step 1 is to ensure that key human interactions are included in the human reliability assessment. Any human actions with a potentially significant impact on system reliability must be identified at this step to guarantee the completeness of the analysis.

Human activities can generally be classified as follows:

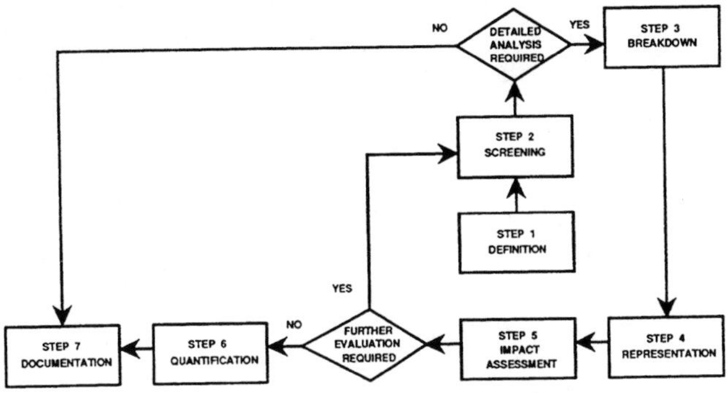


*Fig. 6.4 Systematic Human Action Reliability Procedure, Hannaman and Spurgin (1984)*

Type 1: Before any challenge to a system, an operator can affect availability, reliability, and safety by restoring safeguard functions during testing and maintenance.

Type 2: By committing an error, an operator can initiate a challenge to the system causing the system to deviate from its normal operating envelope.

Type 3: By following procedures during the course of a challenge, an operator can operate redundant systems (or subsystems) and recover the systems to their normal operating envelope.

Type 4: By executing incorrect recovery plans, an operator can aggravate the situation or fail to terminate the challenge to the systems.

Type 5: By improvising, an operator can restore initially failed equipment to terminate a challenge.

As recommended by the SHARP, HRA should use the above classification and investigate the system to reveal possible human interactions. Analysts can use the above-mentioned characteristics for different types of activities. For example, Type 1 interactions generally involve components, whereas Type 3 and Type 4 interactions are mainly operating actions that can be considered at system level. Type 5 interactions are recovery actions that may affect both systems and components. Type 2 interactions can generally be avoided by confirming that human-induced errors are included as contributors to the probability of all possible challenges to the system. The output from this step can be used to revise and enrich system reliability models, such as event trees and fault trees, to fully account for human interactions. This output will be used as the input to the next step.

Step 2: Screening

The objective of screening is to reduce the number of human interactions identified in Step 1 to those that might potentially challenge the safety of the system. This step provides the analysts with a chance to concentrate their efforts on the key human interactions. This is generally done in a qualitative manner. The process is judgmental.

Step 3: Qualitative Analysis

To incorporate human errors into equipment failure modes, analysts need more information about each key human interaction identified in the previous steps to help in representing and quantifying these human actions. The two goals of qualitative analysis are to:

1) Postulate what operators are likely to think and do, and what kind of actions they might take in a given situation, and
2) Postulate how an operator's performance may modify or trigger a challenge to the system.

This process of qualitative analysis may be broken down into four key stages.

1) Information gathering
2) Prediction of operator performance and possible human error modes
3) Validation of predictions
4) Representation of output in a form appropriate for the required function

In summary, the qualitative analysis step requires a thorough understanding of what performance-shaping factors (e.g., task characteristics, experience level, environmental stress, and social-technical factors affect human performance. Based on this information, analyst can predict the range of plausible human action. The psychological model proposed by Rasmussesn (1987) is a useful way of conceptualizing the nature of human cognitive activities. The full spectrum of possible human actions following a misdiagnosis is typically very hard to recognize. Computer simulations of performance described by Woods et al. (1988) and Amendola et al. (1987) offer the potential to assist human reliability analysts in predicting the probability of human errors.

Step 4: Representation

To combine the HRA results with the system analysis models of Chapter 4, human error modes need to be transformed into appropriate representations. Representations are selected to indicate how human actions can affect the operation of a system.

Three basic representations have been used to delineate human interactions: the operator action tree (OAT) described by Wreathall (1981), the confusion matrix described by Potash et al. (1981), and the HRA event trees described by Swain and Guttman (1983). Fig. 6.5 shows an example of OAT. The HRA tree is discussed in Section 6.3.2.

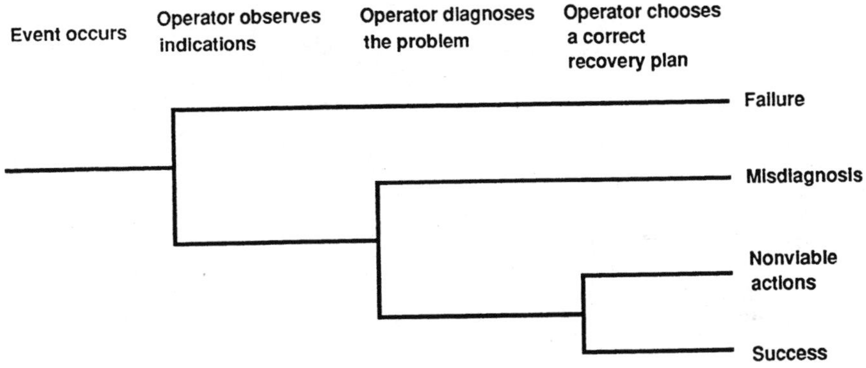


*Fig. 6.5 Operator Action Tree*

Step 5: Impact Integration

Some human actions can introduce new impacts on the system response. This step provides an opportunity to evaluate the impact of the newly identified human actions on the system.

The human interactions represented in Step 4 are examined for their impact on challenges to the system, system reliability, and dependent failures. Screening techniques are applied to assess the importance of the impacts. Important human interactions are found, reviewed, and grouped into suitable categories. If the reexamination of human interactions identifies new human-induced challenges or behavior, the system analysis models (e.g., MLD, fault tree) are reconstructed to incorporate the results.

Step 6: Quantification

The purpose of this step is to assess the probabilities of success and failure for each human activity identified in the previous steps. In this step, analysts apply the most appropriate data or models to produce the final quantitative reliability analysis. Selection of the models should be based on the characteristics of each human interaction.

Guidance for choosing the appropriate data or models to be adopted is provided below.

1) For procedural tasks, the data from Swain and Guttman (1983) or equivalent can be applied.
2) For diagnostic tasks under time constrains, time-reliability curves from Hall et al. and Fragola (1982) or the human cognitive reliability (HCR) model from Hanamman et al. (1984) can be used.
3) For situations where suitable data are not available, expert opinion approaches, such as paired comparison by Hunns and Daniels (1980) and the success likelihood index method by Embry et al. (1984) can be used.
4) For situations where multiple tasks are involved, the dependence rules discussed by Swain and Guttman (1983) can be used to assess the quantitative impact.

Step 7: Documentation

The objective of Step 7 is to produce a traceable description of the process used to develop the quantitative assessments of human interactions. The assumptions, data sources, selected model and criteria for eliminating unimportant human interactions should be carefully documented. The human impact on the system should be stated clearly.

### *6.3.2 HRA Models*

The HRA models can be classified into the following categories. Representative models in each are also summarized.

1) Simulation Models
   a) Maintenance Personnel Performance Simulation (MAPPS)
   b) Cognitive Environment Simulation (CES)
2) Expert Judgment Methods
   a) Paired Comparison
   b) Direct Numerical Estimation (Absolute Probability Judgment)

c) Success Likelihood Index Methodology (SLIM)

3) Analytical Methods
   a) Technique for Human Error Rate Prediction (THERP)
   b) Human Cognitive Reliability Correlation (HRC)
   c) Time Reliability Correlation (TRC)

We will briefly discuss each of these models. Human error is a very complex subject. There is no single model that captures all important human errors and predicts their probabilities. Poucet (1988) reports the results of a comparison of the HRA models. He concludes that the methods could yield substantially different results, and presents their suggested use in different contexts. Table 6.4 compares the methods described above.

### Simulation Models

These models primarily rely on computer methods based on models that mimics human behavior under different conditions.

Maintenance Personnel Performance Simulation (MAPPS)

MAPPS, developed by Siegel et al. (1984), is a computerized simulation model that provides human reliability estimation for testing and maintaining tasks. To perform the simulation, analysts must first find out the necessary tasks and substasks that individuals must perform. Environmental, motivational, task, and organizational variables that influence personnel performance reliability are input into the program. Using the Monte-Carlo simulation technique, the model can output the probability of success, time to completion, idle time, human load, and level of stress. The effects of a particular parameter or subtask performance can be investigated by changing the parameter and repeating the simulation.

The simulation output of task success is based on the difference between the ability of maintenance personnel and the difficulty of the subtask. The model used is

$$\Pr(\text{success}) = \exp(x)/(1 + \exp(x)), \tag{6.19}$$

where $x$ = the difference between personnel ability and task difficulty.

Cognitive Environment Simulation (CES)

Woods (1988) has developed a model based on techniques from artificial intelligence (AI). The model is designed to simulate a limited resources problem solver in a dynamic, uncertain, and complex situation. The main focus is on the formation of intentions, situations and factors leading to intentional failures, forms of intentional failures, and the consequence of intentional failures.

*Table 6.4 Comparison of the Human Reliability Models, Hannaman and Worledge(1988)*

| | DESIRABLE FEATURES | Calculates probability of action | Structures error state | States assumptions explicitly | Provides basis of model | Applied in PRA nuclear plant | Includes time dependence | Has been validated or compared | 1. Design features include | MM interface | Communication link | Job performance aids | 2. Mgt characteristics include | Training | Crew structure | Allocation of priorities | Personnel policy | 3. Personal factors included | Skills | Stress response | Motivation | Knowledge (plant, etc.) | OPERATIONS—Included | Decision making | Procedure following | Recovery action | Maintenance/test | Continuous control/supervision |
|---|---|---|---|---|---|---|---|---|---|---|---|---|---|---|---|---|---|---|---|---|---|---|---|---|---|---|---|---|
| *Model reviewed* | *Attributes* | | | | | | | | *Shaping factors* | | | | | | | | | | | | | | | | | | | |
| THERP diagnosis model | | Y | | Y | E | Y | Y | | | | | | | Y | | | | | | Y | | | | Y | | Y | | |
| Operator action tree | | Y | Y | Y | | Y | | | | | | | | | | | | | | | | | | | | | | |
| Time–reliability correlation | | Y | | | D | Y | Y | P | | | | | | | | | | | | | | | | Y | Y | Y | | |
| Success likelihood index method | | Y | Y | Y | | Y | | | | Y | | | | | | | | | | | | | | | | | | |
| Siegel and Wolf models | | | | Y | E | | | P | | | | Y | | | Y | | Y | | Y | | Y | | | | | | Y | |

Key: Y, yes P, partially; I, implicitly; E, expert judgment;
D, data; blank, not addressed or addressed by judgment.

Similar to the MAPPS model, the CES model is a simulation approach that mimics the human decision process during an emergency condition. But CES is a deterministic approach, which means the program will always obtain the same results if the input is unchanged. The first step in CES is to identify the conditions leading to human intentional failures. CES provides numerous performance-adjusting factors to allow the analysts to test different working conditions. For example, analysts may change the number of people interacting with the system (e.g., the number of operators), the depth or breadth of working knowledge, or the human-machine interface. Human error prone points can be identified by running the CES for different conditions. The human failure probability is evaluated by knowing, a priori, the likelihood of occurrence of these error prone points.

In general, CES is not a human rate quantification model. It is primarily a tool to analyze the interaction between problem-solving resources and task demands.

## Expert Judgment Methods

The primary reason for using expert judgment in HRA is that there often exist little or no relevant or useful human error data. Expert judgment is discussed in more detail in Section 6.4. There are two requirements for selected experts.

- They must have substantial expertise.
- They must be able to accurately translate this expertise into probabilities.

### Direct Numerical Estimation

For the direct numerical estimation method described by Stillwell et al. (1982), experts are asked to directly estimate the human error probabilities and the associated upper/lower bounds for each task. A consistency analysis might be taken to check for agreement among these judgments. Then, individual estimations are aggregated by either arithmetic or geometric average.

### Paired Comparison

Paired comparison, described by Hunns and Daniel (1980), is a scaling technique based on the idea that judges are better at making simple comparative judgments than making absolute judgments. An interval scaling is used to indicate the relative likelihood of occurrence of each task. Saaty (1980) describes this general approach in the context of a decision analysis technique. The method is equally applicable to HRA.

Success Likelihood Index Methodology (SLIM)

The success likelihood index methodology (SLIM) developed by Embry et al. (1984) is a structural method that uses expert opinion to estimate human error rates. The underlying assumption of SLIM is that the success likelihood of tasks for a given situation depends on the combination of effects from a small set of performance-shaping factors (PSFs) relevant to a group of tasks under consideration.

In this procedure, judges are asked to assess the relative importance (weight) of each PSF with regard to its impact on the tasks of interest. An independent assessment is made to the level or the value of the PSFs in each task situation. After identifying and agreeing on the small set of PSFs respective weights and ratings for each PSF are multiplied. These products are then summed to produce the success likelihood index (SLI), varying from 0 to 100 after normalization. This value indicates the judge's belief regarding the positive or negative effects of PSFs on task success.

The SLIM approach assumes that the functional relationship between success probability and SLI is logarithmic, i.e.,

$$\log\left[\Pr(\text{Operator Success})\right] = a(\text{SLI}) + b, \tag{6.20}$$

where $a$ and $b$ are empirically derived constants. To calibrate a and b, at least two human tasks of known reliability must be used in (6.20), from which constant a and b are calculated.

This technique has been implemented as an interactive computer program. The first module, called Multi-Attribute Utility Decomposition (MAUD), analyzes a set of tasks to define their relative likelihood of success given the influence of PSFs. The second module, Systematic Approach to the Reliability Assessment of Humans (SARAH), is then used to calibrate these relative success likelihoods to generate absolute human error probability. The SLIM technique has a good theoretical basis in decision theory. Once the initial data base has been established with the SARAH module, evaluations can be performed rapidly. This method does not require extensive decomposition of a task to an elemental level. For situations where no data are available, this approach enables HRA analysts to reasonably estimate human reliability. However, this method makes extensive use of expert judgment, which requires a team of experts to participate in the evaluation process. The resources required to set up the SLIM-MAUD data base are generally greater than other techniques.

## Analytical Methods

These methods generally use a model based on some key parameters that form the value of human reliabilities.

Technique For Human Error Rate Prediction (THERP)

The oldest and most widely used HRA technique is the THERP analysis developed by Swain and Guttman (1983) and reported in the form of a handbook. The THERP approach uses conventional system reliability analysis modified to account for possible human error. Instead of generating equipment system states, THERP produces possible human task activities and the corresponding human error possibilities. THERP is carried out in the five steps described below.

1) Define system failures of interest

From the information collected by examining system operation and analyzing system safety, analysts identify possible human interaction points and task characteristics, and their impact on the systems. Then, screening is performed to determine critical actions that require detailed analysis.

2) List and analyze related human actions

The next step is to develop a detailed task analysis and human error analysis. The task analysis delineates the necessary task steps and the required human performance. The analyst then determines the errors that could possibly occur. The following human error categories are defined by THERP:

- Error of omission (omit a step or the entire task)
- Error of commission
  - Selection error (select the wrong control, choose the wrong procedures)
  - Sequence error (actions carried out in the wrong order)
  - Time error (actions carried out too early/too late)
  - Qualitative error (action is done too little/too much)

At this stage, opportunities for human recovery actions (recovery from an abnormal event or failure) should be identified. Without considering recovery possibilities, overall human reliability might be dramatically underestimated.

The basic tool used to model tasks and task sequences is the HRA event tree, According to the time sequence or procedure order, the tree is built to represent possible alternative human actions. Therefore, if appropriate error probabilities of each subtask are known and the tree adequately depicts all human action sequences, the overall reliability of this task can be calculated. An example of an HRA event tree is shown in Fig. 6.6.

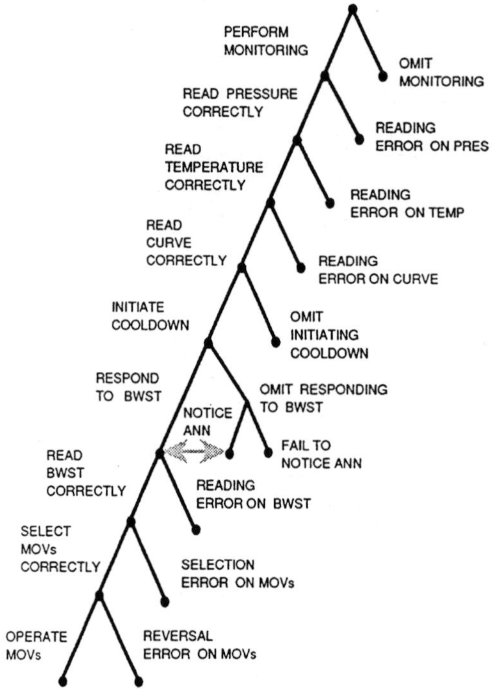


*Fig. 6.6 HRA Event Tree Depicting Some Operator Actions During a Small-break Loss of Coolant in Nuclear Plants, Hannaman and Spurgin (1984)*

3) Estimate relevant error probabilities

As explained in the previous section, human error probabilities (HEPs) are required for the failure branches in the HRA event tree. Chapter 20 of Swain and Guttman (1983) THERP contains a data source containing the following information.

- Data tables containing nominal human error probabilities
- Performance models explaining how to account for PSFs to modify the nominal error data
- A simple dependence model for converting independent failure probabilities into conditional failure probabilities

In addition to the data source of THERP, analysts may use other data sources, such as the data from recorded incidents, trials from simulations, and subjective judgment data, if necessary.

4) Estimate effects of error on system failure events

In a system reliability framework, the human error tasks are incorporated into the system model, such as a fault tree. Hence, the probabilities of undesired events can be evaluated and the contribution of human errors to system reliability or availability can be estimated.

5) Recommend changes to system design and recalculate system reliability

A sensitivity analysis can be performed to identify dominant contributors to system unreliability. System performance can then be improved by reducing the sources of human error or redesigning the safeguard systems.

THERP's approach is very similar to the equipment reliability methods described in Chapter 4. The integration of human reliability analysis and equipment reliability analysis is a straightforward using the THERP process. Therefore, it is easily understood by system analysts. Compared with the data for other models, the data for THERP are much more complete and easier to use. The handbook contains guidance for modifying the listed data for different environments. The dependencies among subtasks are formally modeled, although subjective. Conditional probabilities are used to account for this kind of task dependence.

Very detailed THERP analysis can require a large amount of effort. In practice, by reducing the details of the THERP analysis to an appropriate level, the amount of work can be minimized. THERP is not appropriate for evaluating errors involving high-level decisions or diagnostic tasks. In addition, THERP does not model underlying psychological causes of errors. Since it is not an ergonomic tool, this method cannot produce explicit recommendation for design improvement.

Human Cognitive Reliability (HCR) Correlation

during the development SHARP, a need was identified to find a model to quantify the reliability of control room personnel responses to abnormal system operations. The HCR correlation, described by Hannaman et al. (1984), is essentially a normalized time-reliability correlation (described below) whose shape is determined by the available time, stress, human-machine interface, etc. Normalization is needed to reduce the number of curves required for a variety of situations. It was found that a set of three curves (skill-, rule-, and knowledge-based ideas, developed by Rasmussen,1982) could represent all kinds of human decision behaviors. The application of HCR is straightforward. The HCR correlation curves can be developed for different situations from the results of simulator experiments. Therefore, the validity can be verified continuously. This approach also has the capability of accounting for cognitive and environmental PSFs.

Some of the disadvantages of the HCR correlation are.

- The applicability of the HCR to all kinds of human activities is not verified.
- The relationships of PSFs and nonresponse probabilities are not well addressed.
- This approach does not explicitly address the details of human thinking processes. Thus, information about intentional failures cannot be obtained.

Time-Reliability Correlation (TRC)

Hall et al. (1982) concentrate on the diagnosis and decision errors of nuclear power plant opertors after the initiation of an accident. They criticize the behavioral approach used by THERP and suggest that a more holistic approach be taken to analyze decision errors.

The major assumption of TRC is that the time available for diagnosis of a system fault is the dominant factor in determining the probability of failure. In other words, the longer people take to think, the more unlikely they are to make mistakes. The available time for decision and diagnosis is delimited by the operator's first awareness of an abnormal situation and the initiation of the selected response. Because no data were available when the TRC was developed, an interim relationship was obtained by consulting psychologists and system analysts. Recent reports confirm that the available time is an important factor in correctly performing cognitive tasks. A typical TRC is shown in Fig. 6.7.

Daugherty and Fragola (1988) is a good reference for TRC as well as other HRA methods. TRC is very easy and fast to use. However, TRC is still a premature approach. The exact relationship between time and reliability requires more experimental and actual observations. This approach overlooks other important PSFs, such as experience level, task complexity, etc. TRC focuses only on limited aspects of human performance in emergency conditions. The

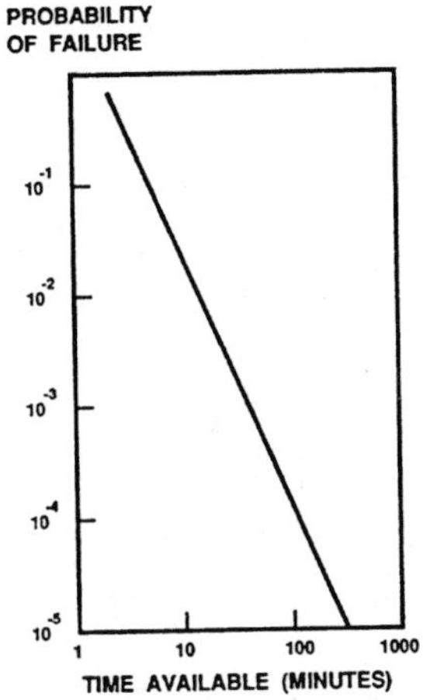


*Fig. 6.7 Time-Reliability Correlation for Operators*

time available is the only variable in this model. Therefore, the estimation of the effect of this factor should be very accurate. However, TRC does not provide guidelines or information on how to reduce human error contributions.

### *6.3.3 Human Reliability Data*

There is general agreement that a major problem for HRA is the scarcity of data on human performance that can be used to estimate human error rates and performance time. To estimate human error probabilities, one needs data on the relative frequency of the number of errors and/or the ratio of "near-misses" to total number of attempts. Ideally, this information can be obtained from observing a large number of tasks performed in a given application. However, this is impractical for several reasons. First, error probabilities for many tasks, especially for rare emergency conditions, are very small. Therefore, it is very difficult to observe enough data within a reasonable amount of time to get statistically meaningful results. Second, possible penalties assessed against people who make errors in e.g., a nuclear power plant or in aircraft cockpit, discourages free reporting of all errors. Third, the costs of collecting and analyzing data could be unacceptably large. Moreover, estimation of performance times presents difficulties since data taken from different situations might not be applicable.

Data can be used to support HRA quantification in a variety of ways, e.g., to confirm expert judgment, develop human reliability data, or support development of an HRA model. Currently, available data sources can be divided into the following categories: 1) actual data, 2) simulator data, 3) interpretive information, and 4) expert judgment.

Psychological scaling techniques, such as paired comparisons, direct estimation, SLIM, and other structured expert judgment methods, are typically used to extrapolate error probabilities. In many instances, scarcity of relevant hard data makes expert judgment a very useful data source. This topic is discussed further in the next section.

## 6.4 Use of Expert Opinion for Estimating Reliability Parameters

The use of expert opinions is often desired in reliability analysis, and in many cases is unavoidable. One reason for using experts is the lack of a statistically significant amount of empirical data necessary to estimate new parameters. Another reason for using experts is to assess the likelihood of a one-time event, such as the chance of rain tomorrow.

However, the need for expert judgment that requires extensive knowledge and experience in the subject field is not limited to one-time events. For example, suppose we are interested in using a new and highly capable microcircuit device currently under development by a manufacturer and expected to be available for use soon. The situation requires an immediate decision on whether or not to design an electronic box around this new microcircuit device. Reliability is a critical decision criterion for the use of this device. Although reliability data on the new device are not available, reliability data on other types of devices employing similar technology are accessible. Therefore, reliability assessment of the new device requires both knowledge of and expertise in similar technology, and can be achieved through the use of expert opinion.

Recent specific examples of expert use are the Reactor Safety Study (1975); IEEE-Standard 500 (1984); and "Severe Accident Risk: An Assessment for Five U.S. Nuclear Power Plants" (1990), where expert opinion was used to estimate the probability of components failure and other rare events. The Electric Power Research Institute study (1986) has relied on expert opinion to assess seismic hazard rates. Other applications include weather forecasting. For example, Clemens et al. (1990) discusses the use of expert opinion by meteorologists. Another example is the use of expert opinion in assessing human error rates discussed by Swain and Guttman (1983).

The use of expert opinion in decision making is a two-step process: eliciation and analysis of expert opinion. The method of elicitation may take the form of individual interviews, interactive group sessions, or the Delphi approach discussed by Dalkey and Helmer (1963). The relative effectiveness of different elicitation methods has been addressed extensively in the literature. Techniques for improving the accuracy of expert estimates include calibration, improvement in questionnaire design, motivation techniques, and other methods, although clearly no technique can be applied to all situations. The analysis portion of expert use involves combining expert opinions to produce an aggregate estimate that can be used by reliability analysts. Again, various aggregation techniques for pooling expert opinions exists, but of particular interest are those adopting the form of mathematical models. The usefulness of each model depends on both the reasonableness of the assumptions (implicit and explicit) carried by the model as it mimics the real world situation, and the ease of implementation from the user's perspective. The term "expert" generally refers to any source of information that provides an estimate and includes human experts, measuring instruments, and models.

Once the need for expert opinion is determined and the opinion is elicited, the next step is to establish the method of opinion analysis and application. This is a decision task for the analysts, who may simply decide that the single best estimate of the value of interest is the estimate provided by the arithmetic average of all estimates, or an aggregate from a nonlinear pooling method, or some other opinons. Two methods of aggregating expert opinion are discussed in more detail, the geometric averaging technique and the Bayesian technique.

### *6.4.1 Geometric Averaging Technique*

Suppose $n$ experts are asked to make an estimate of the failure rate of an item. The estimates can be pooled using the geometric averaging technique. For example, if $\lambda_i$ is the estimate of the $i^{th}$ expert, then an estimate of the failure rates is obtained from

$$\hat{\lambda} = \left[\prod_{i=1}^{n} \lambda_i\right]^{1/n} \tag{6.21}$$

This was the primary method of estimating failure rates in IEEE-Standard 500 (1984). The IEEE-Standard 500 contains rate data for electronic, electrical, and sensing components. The reported values were synthesized primarily from the opinions of some 200 experts (using a form of the Delphi procedure). Each expert reported "low," "recommended," and "high" values for each failure rate under normal conditions, and a "maximum" value that would be applicable under all conditions (including abnormal conditions). The estimates were pooled using (6.21). For example, for maximum values,

$$\hat{\lambda}\text{max} = \left[\prod_{i=1}^{n} \lambda\text{max}, i\right]^{1/n}.$$

As discussed by Mosleh and Apostolakis (1983), the use of geometric averaging implies that 1) all the experts are equally competent, 2) the experts do not have any systematic biases, 3) experts are independent, and 4) the preceding three assumptions are valid regardless of which value the experts are estimating, e.g., high, low, or recommended.

The estimates can be represented in the form of a distribution. Apostolakis et al. (1980) suggests the use of a lognormal distribution for this purpose. In this approach, the "recommended" value is taken as the median of the distribution, and the error factor (EF) is defined as

$$\text{EF} = \left(\frac{\hat{\lambda}\text{max}}{\hat{\lambda}_{\text{low}}}\right)^{1/2}, \tag{6.22}$$

where

$$\sigma_t = \frac{\ln \text{EF}}{1.645}.$$

### *6.4.2 Bayesian Approach*

As discussed by Mosleh and Apostolakis (1983), the challenge of basing estimates on the expert opinion is to maintain coherence throughout the process of formulating a single best estimate based on the experts' actual estimates and their credibilitites. Coherence is a notion of internal consistency within a person's state of belief. In the subjectivist school of thought, a probability is defined as a measure of personal uncertainty. This definition assumes that a coherent person will provide his or her probabilistic judgments in compliance with the axioms of probability theory.

An analyst often desires a modeling tool that can aid him or her in formulating a single best estimate from expert opinion(s) in a coherent manner. Informal methods such as simple averaging will not guarantee this coherence. Bayes' theorem, however, provides a framework to model expert belief, and ensures coherence of the analysts in arriving at a new degree of belief in light of expert opinion. According to the general form of the model given by Mosleh and Apostolakis, the state-of-knowledge distribution of a failure rate $\lambda$, after receiving an expert estimate $\hat{\lambda}$, can be obtained by using Bayes' theorem in the following form:

$$\pi(\lambda|\hat{\lambda}) = k^{-1} L(\hat{\lambda}|\lambda)\pi_0(\lambda), \tag{6.23}$$

where

$\pi_0(\lambda)$ = Prior distribution of $\lambda$,
$\pi(\lambda|\hat{\lambda})$ = Posterior distribution of $\lambda$,
$L(\hat{\lambda}|\lambda)$ = Likelihood of receiving the estimate $\hat{\lambda}$ given the true failure rate $\lambda$,
$k$ = Normalizing factor.

One of the models suggested for the likelihood of observing $\hat{\lambda}$ given $\lambda$, is a lognormal distribution in the following form:

$$L(\hat{\lambda}|\lambda) = \frac{1}{\sqrt{2\pi}\sigma\hat{\lambda}} \exp\left[-\frac{1}{2}\left(\frac{\ln(\hat{\lambda}) - \ln(\lambda) - \ln(b)}{\sigma}\right)^2\right], \tag{6.24}$$

where $b$ is a bias factor ($b = 1$ when no bias is assumed) and $\sigma$ is the logarithmic standard deviation of $\hat{\lambda}$, given $\lambda$. When the analyst believes no bias exists among the experts, he or she can set $b = 1$. The quantity $\sigma$, therefore, represents the degree of accuracy of the experts' estimate as viewcal by the analyst. The work by Kim (1991), which includes a Bayesian model for a relative ranking of experts, is an extension of the works by Mosleh and Apostolakis.

### *6.4.3 Statistical Evidence on the Accuracy of Expert Estimates*

Among the attempts to verify the accuracy of expert estimates, two types of expert estimates are studied, assessment of single values and assessment of distributions.

Notable among the studies on the accuracy of expert assessments of a single estimate is Snaith's study (1981). In this study, observed and predicted reliability parameters for some 130 pieces of different equipment and systems used in nuclear power plants were evaluated. The predicted values included both direct assessments by experts and the results of analysis. The objective was to determine correlations between the predicted and observed values. Fig. 6.8 shows the ratio ($R = \lambda/\hat{\lambda}$) of observed to predicted values plotted against their cumulative frequency. As shown, the majority of the points lie within the dashed boundary lines. Predicted values are within a factor of 2 from the observed values, and 93% are within a factor of 4. The figure also shows that $R = 1$ is the median value, indicating that there is no systematic bias in either direction. Finally, the linear nature of the curve shows that $R$ tends to be lognormally distributed, at least within the central region. This study clearly supports the use and accuracy of expert estimation.

Among the studies of expert estimation are the works by cognitive psychologists. For example, Lichtenstein et al. (1977) described the results of testing the adequacy of probability assessments and concluded that "the overwhelming evidence from research on uncertain quantities is that people's probability distributions tend to be biased." Commenting on judgmental biases in risk perception, Slovic et al. (1980) stated: "A typical task in estimating uncertain quantities like failure rates is to set upper and lower bounds such that there is a 98% chance that the true value lies between them. Experiments with diverse groups of people making different kinds of judgments have shown that, rather than 2% of true values falling outside the 98% confidence bounds, 20% to 50% do so. Thus, people think that they can estimate such values with much greater precision than is actually the case."

Based on the above conclusion Apostolakis (1982) has suggested the use of the 20th and 80th percentiles of lognormal distributions instead of the 5th and 95th when using (6.22), to avoid a bias toward low values, overconfidence of experts, or both. When using the Bayesian estimation method in (6.23) and (6.24), the bias can be accounted for by using larger values of $\sigma$ and $b$ in (6.24).

## 6.5 Measures of Importance

During the design of a system, the choice of the components and their arrangement may render some to be more critical than others. For example, placing a component in series within a system causes it to have a much higher importance for system reliability than the same component would have if it were in parallel within the system. In this section, we describe three methods of measuring the importance of components: Birnbaum, Criticality, and Vesely-Fussell measures of importance.

### *6.5.1 Birnbaum Measure of Importance*

Birnbaum (1969) introduced the concept of **importance**. In the Birnbaum measure, the importance improves as the reliability

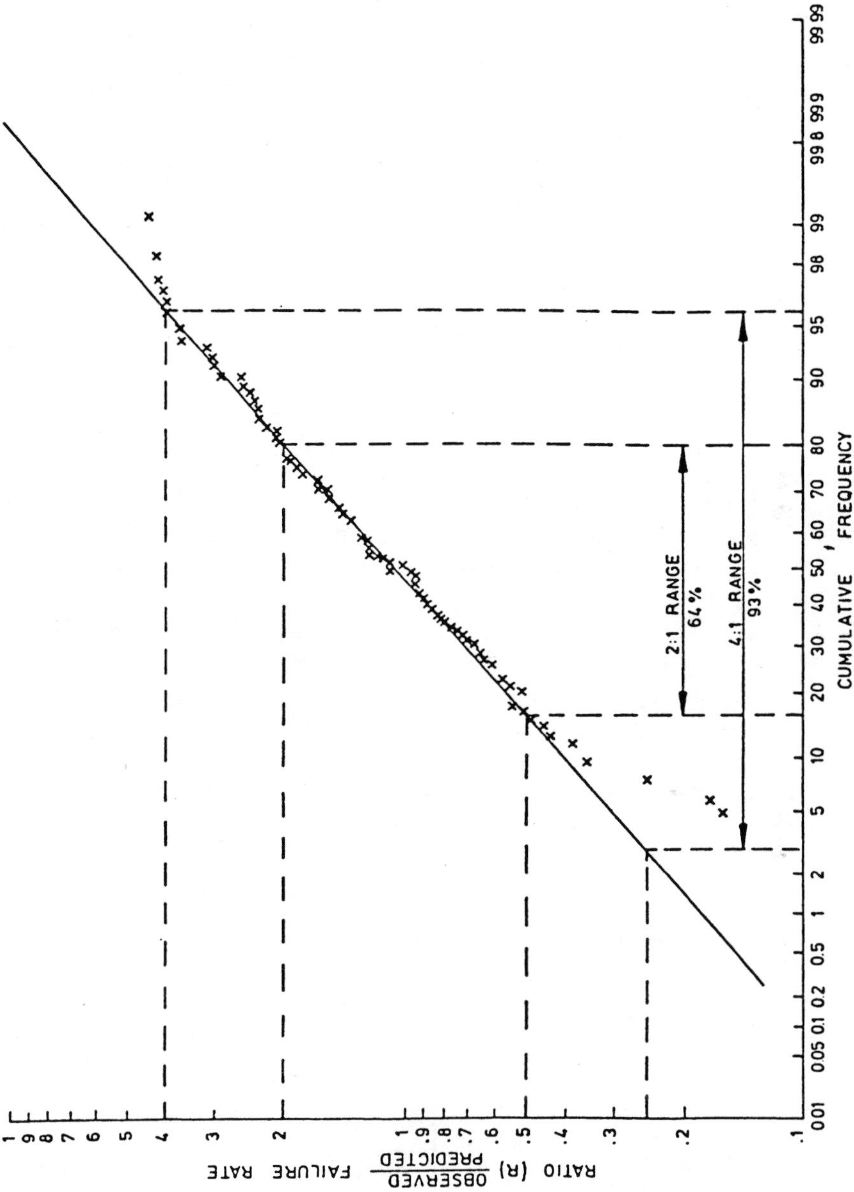


*Fig. 6.8 Frequency Distribution of the Failure Rate Ratio, Snaith (1981).*

of component $i$ improves. Accordingly, Birnbaum's measure of component importance for success space as described by Sharirli (1985) is

$$I_i^B = \frac{\partial R_s[r(t)]}{\partial r_i(t)}, \tag{6.25}$$

where

$I_i^B$ = Birnbaum importance of component $i$,
$R_s[r_i(t)]$ = Reliability of the system as a function of the reliability of individual component $r(t)$,
$r_i(t)$ = Reliability of component $i$.

Another form of (6.25) is

$$I_i^B(t) = R_s[r_i(t) = 1, r(t)] - R_s[r_i(t) = 0, r(t)], \tag{6.26}$$

where $R_s[r_i(t) = 1, r(t)]$ and $R_s[r_i(t) = 0, r(t)]$ represent the reliability of the system with the reliability of component $i$ set to 1 and 0, respectively. The properties associated with $I_i^B$ are as follows

1) $0 \leq I_i^B \leq 1$.
2) When $R_s[r(t)]$ is a linear function of $r(t)$, and if all components $i$ are independent, then $I_i^B$ does not depend on $r_i(t)$.

Equation (6.25) and (6.26) are often used in conjunction with the unreliability function $F_s[q_i(t)]$, where $F_s[q_i(t)]$ is the unreliability or unavailability function of the system as a function of component unreliability or unavailability $q(t)$. In this case, (6.26) is replaced by

$$I_i^B(t) = \frac{\partial F_s[q(t)]}{\partial q_i(t)} = F_s\left[q_i(t) = 1, q(t)\right] - F_s\left[q_i(t) = 0, q(t)\right]. \tag{6.27}$$

---

*Example 6.4*

Consider the system shown below. Determine the importance of each component at $t = 720$ hours. Assume an exponential time to failure.

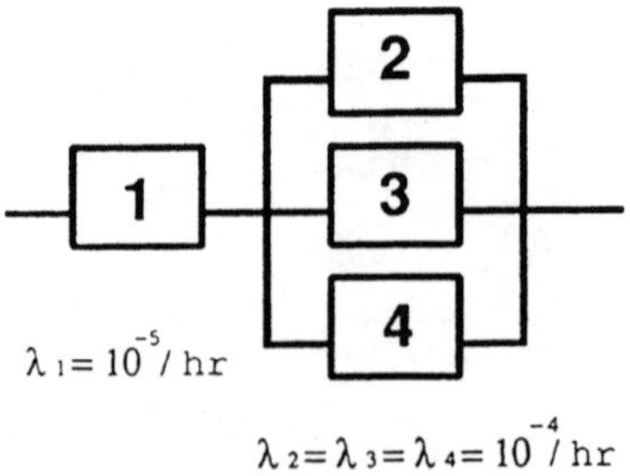

*Solution:*

$$r_1(t = 720) = 0.993, \quad r_2(t = 720) = r_3(t = 720) = r_4(t = 720) = 0.93.$$

The reliability function of the system is

$$R_s(r(t)) = r_1(t) \cdot [1 - (1 - r_2(t))(1 - r_3(t))(1 - r_4(t))].$$

Using (6.26),

$$I_1^B(t) = R_s[r_1(t) = 1, r(t)] - R_s[r_1(t) = 0, r(t)],$$

$$I_1^B(t) = [1 - (1 - r_2(t))(1 - r_3(t))(1 - r_4(t)], \quad therefore$$

$$I_1^B(t = 720) \approx 1,$$

$$I_2^B(t) = r_1(t)[(1 - r_3(t))(1 - r_4(t))],$$

$$I_2^B(t = 720) \approx 0,$$

similarly,

$$I_3^B(t = 720) = I_4^B(t = 720) \approx 0.$$

It can be concluded that the rate of improvement in Component 1 has far more importance (impact) on system reliability than Components 2, 3, and 4. However, if the reliability of the parallel units decreases by an order of magnitude, clearly the importance of Components 2, 3, and 4 increases (e.g., for $\lambda_2 = \lambda_3 = \lambda_4 = 10^{-3}/hr$, $I_2^B = I_3^B = I_4^B = 0.26$, and $I_1^B = 0.87$). Similarly, if an identical unit is in parallel with Component 1, the importance changes.

---

*6.5.2 Criticality Importance*

Birnbaum's importance for Component $i$ is independent of the reliability of Component $i$ itself. Therefore, $I_i^B$ is not a function of $r_i(t)$. It is clear that it would be more difficult to improve the more reliable components than to improve the less reliable ones. From this, the criticality importance of Component i is defined as

$$I_i^{CR}(t) = \frac{\partial R_s[r(t)]}{\partial r_i(t)} \cdot \frac{r_i(t)}{R_s[r(t)]}, \quad \text{or} \tag{6.28}$$

$$I_i^{CR}(t) = I_i^B \cdot \frac{r_i(t)}{R_s[r(t)]}.$$

From (6.28), it is clear that the Birnbaum importance is corrected for reliability of the individual components relative to the reliability of the whole system. Therefore, if the Birnbaum importance of a component is high, but the reliability of the component is low with respect to the reliability of the system, then criticality importance assigns a low importance to this component. Similarly, (6.28) can be represented by the unreliability or unavailability function

$$I_i^{CR}(t) = I_i^B \cdot \frac{q_i(t)}{F_s[q(t)]}. \tag{6.29}$$

*6.5.3 Vesely-Fussell Importance*

In cases where Component $i$ contributes to system reliability, but is not necessarily critical, the Vesely-Fussell importance measure can be used. This measure, is introduced by W.E. Vesely and later applied by Fussell (1975), is in the form of

$$I_i^{VF}(t) = \frac{R_i[r(t)]}{R_s[r(t)]}, \tag{6.30}$$

where $R_i[r(t)]$ is the contribution of Component $i$ to the reliability of the system. Similarly, using unreliability or unavailability functions,

$$I_i^{VF}(t) = \frac{F_i[q(t)]}{F_s[q(t)]}, \tag{6.31}$$

where $F_i[q(t)]$ denotes the probability that Component $i$ is contributing to system failure.

The Vesely-Fussell importance measure has been applied to cut sets to determine the importance of individual cut sets. For example, consider importance $I_k$ of the $k^{th}$ cut set representing a system failure. In that case, (6.31) replaces

$$I_k^{VF}(t) = \frac{Q_k(t)}{Q_s(t)}, \tag{6.32}$$

where

$Q_k(t) =$ Probability as a function of time that minimal cut set $k$ occurs,

$Q_s(t) =$ Total probability as a function of time that the system fails (due to all cut sets).

Generally, the minimal cut sets with the largest values of $I_k$ are the most important ones. Consequently, system improvements should initially be directed toward the minimal cut sets with the largest importance values.

If the probability of all minimal cut sets are known, then the following approximate expression can be used to find the importance of individual components.

$$I_i^{VF} \approx \frac{\sum_{j=1}^{m} Q_j(t)}{Q_s(t)}, \tag{6.33}$$

where

$Q_j(t) =$ Probability as a function of time that the $j^{th}$ minimal cut set which contains component $i$ is failed, and

$m =$ Number of minimal cut sets that contain component $i$

Expression (6.33) is an approximation; the situation of two minimal cut sets containing component $i$ failing at the same time is neglected since its probability is very small. Generally, the importance measures described are used in connection with the unreliability or unavailability functions.

---

*Example 6.5*

Consider the water-pumping system below. Determine the Birnbaum, Criticality, and Vesely-Fussell importance measures of the valve (V), pump-1 (P-1) and pump-2 (P-2) using both reliability and unreliability versions of the importance measures.

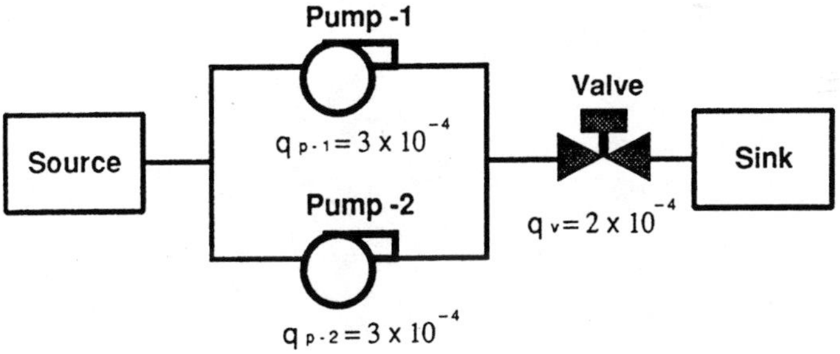


*Solution:*

The reliability function is

$$R_s[r(t)] = r_v \cdot [r_{P-1} + r_{P-2} - r_{P-1} \cdot r_{P-2}].$$

Using rare event approximation, the unreliability function is

$$F_s[q(t)] = q_{P-1} \cdot q_{P-2} + q_v.$$

1) Birnbaum's importance:

$$I_v^B = r_{P-1} + r_{P-2} - r_{P-1} \cdot r_{P-2} \approx 1,$$
$$I_{P-1}^B = r_v - r_v \cdot r_{P-2} \approx 0,$$
$$I_{P-2}^B = r_v - r_v \cdot r_{P-1} \approx 0.$$

Using the unreliability function,

$$I_v^B = 1,$$
$$I_{P-1}^B = q_{P-2} \approx 0,$$
$$I_{P-2}^B = q_{P-1} \approx 0.$$

2) Criticality Importance:

$$I_v^{CR} = 1 \times \frac{0.9998}{0.9998} = 1,$$
$$I_{P-1}^{CR} = I_{P-2}^{CR} = 0.0003 \times \frac{3 \times 10^{-4}}{2 \times 10^{-4}} \approx 0.$$

3) Vesely-Fussell Importance:

$$R_v[r(t)] = R_S[r(t)] \approx 1,$$
$$R_{P-1}[r(t)] = r_v \cdot r_{P-1} - r_v \cdot r_{P-1} \cdot r_{P-2} \approx 1$$
$$R_{P-2}[r(t)] = r_v \cdot r_{P-2} - r_v \cdot r_{P-1} \cdot r_{P-2} \approx 1,$$
$$I_v^{VF} = \frac{0.9998}{0.9998} = 1,$$
$$I_{P-1}^{VF} = I_{P-2}^{VF} = \frac{0.9995}{0.9998} \approx 1.$$

Using the unreliability function,

$$F_v[r(t)] = q_v, \;\; F_{P-1}[q(t)] = q_{P-1}, \;\; F_{P-2}[q(t)] = q_{P-2}.$$

Then,

$$I_v^{VF} \approx \frac{0.0002}{0.0002} = 1,$$
$$I_{P-1}^{VF} = I_{P-2}^{VF} = \frac{9 \times 10^{-8}}{0.0002} \approx 0.$$

### *6.5.4 Practical Aspects of Importance Measures*

There are two principal factors that determine the importance of a component in a system: 1) the structure of the system, and 2) the reliability or unreliability of the components. Depending on the measure selected, one of the above may be pertinent. Also, depending on whether we use reliability or unreliability, some of these measures behave differently. (In Example 6.5, this is seen in $I^V_{p-1}$and $P^{VF}_{p-2}$.) Where their importance in success space is almost 1 and in the failed space is 0.

The Birnbaum measure of importance completely depends on the structure of the system (e.g., whether the system is dominated by a parallel or series configuration). Therefore, it should only be used to determine the degree of redundancy and appropriateness of the system's logic.

The criticality importance is related to that of Birnbaum's. However, it is also affected by the reliability/unreliability of the components and the system. This measure allows for the evaluation of the importance of a component in light of its potential to improve system reliability. The effect of improvements on one component may result in changes in the importance of other components of the system.

The Vesely-Fussell measure of importance has been widely used in practice, mostly for measuring importance in the failure space using unreliability/unavailability functions). The measure is more influenced by the actual reliability/unreliability of the components and the systems as well as the logical structure of the system. Because of its simplicity, this measure has been widely used.

Generally, the importance of components should be used during design or evaluation of systems to determine which components or subsystems are important to the reliability of the system. Those with high importance could prove to be candidates for further improvements. In an operational context, items with high importance should be watched by the operators, since they are critical for the continuous operation of the system.

A number of other measures of importance have been introduced, as well as computer program importance calculations. The readers are referred to Lambert (1975), Sharirli (1985), and NUREG/CR-4550 (1990).

## 6.6 Failure Analysis

Statistical, probabilistic, or deterministic methods are used to analyze item failures. While all three methods or combinations of them can be used, in this section we rely primarily on the statistical methods for the analysis of failures. However, for evaluating the results of the analysis, mainly deterministic techniques are used. Probabilistic (Bayesian) and deterministic techniques are equally applicable. However,since Bayesian techniques may require expert or prior knowledge about equipment failures, they should be used only when observed failure data are sparse.

The statistical methods described in this section are based on the classical inference methods discussed in Chapters 3 and 5. That is, the history of failure or event occurrences is first studied to determine whether or not a statistically significant trend can be detected. If not, the traditional maximum likelihood parameter estimating method is used to determine the failure characteristic of the item, e.g., to determine the failure rate or demand failure probability of an item.

If the trend analysis method discussed in Chapter 5 shows a significant trend in the data, it is important to determine the nature and degree to which the failure characteristic of the item is changed. Classical statistics methods are used to determine the failure characteristics of equipment if no trends are exhibited.

When the failure characteristics of an item with or without trend are determined, one needs to evaluate them to determine whether or not they show any change in the capability of the item. Both statistical and nonstatistical techniques can be used to detect changes.

When significant changes are detected, it is very important to search for possible reasons for such changes. This may require an analysis to determine the root causes of the detected changes or observed failure events. No standard practice exists for determining root-cause failures. Engineers often use ad hoc techniques for this purpose.

Fig. 6.9 shows the overall approach employed in this section. The basis for the statistical methods used in this document are explained in the remainder of this section.

### *6.6.1 Detecting Trends in Observed Failure Events*

The use of statistical estimators for equipment failure characteristics is justified only after it has been proven that the failure occurrence is reasonably constant, i.e., there is no evidence of an increasing or decreasing trend.

In Chapter 5, we described the Centroid method to test for the possibility of a trend. In (5.1), since $U$ is a sensitive measure, one could use the following practical criteria to ensure detection of

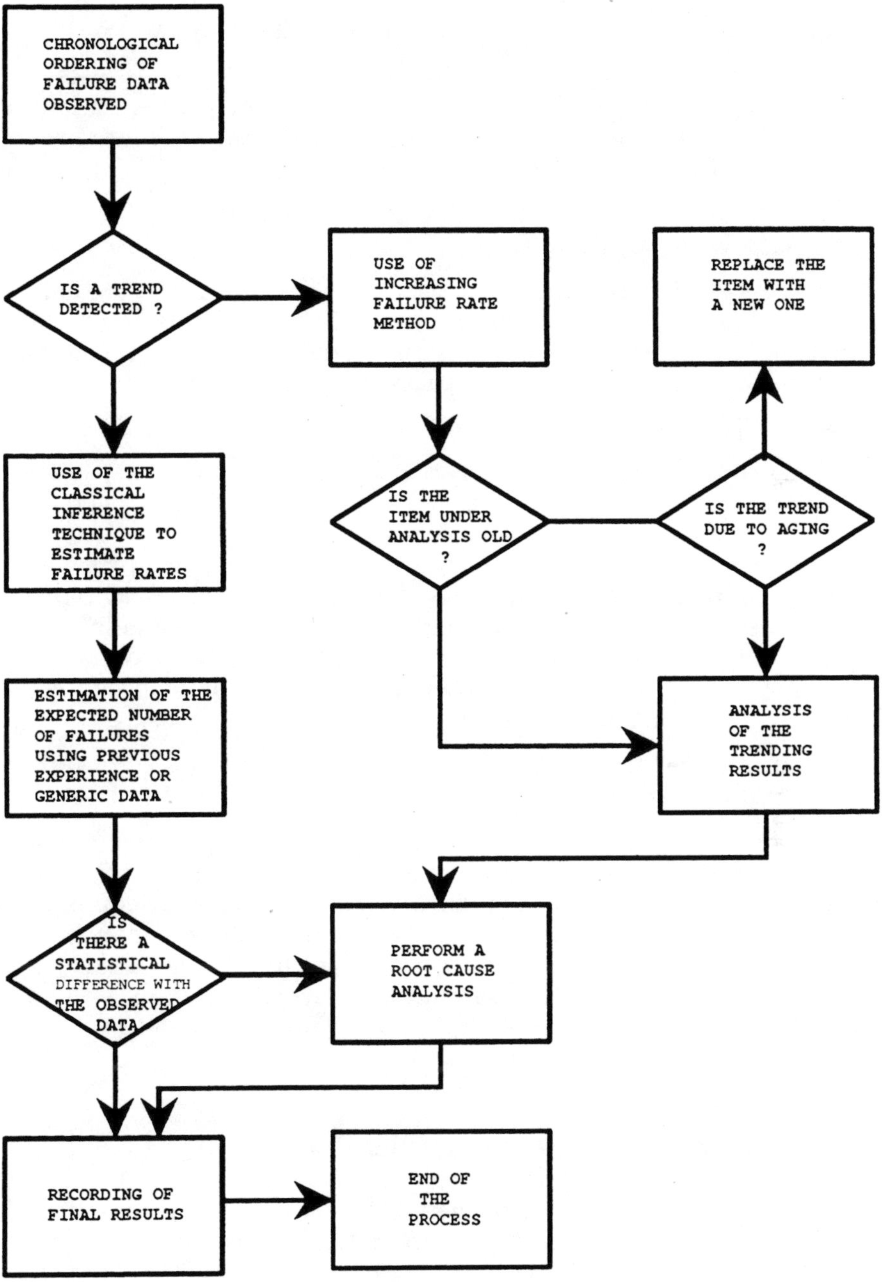


*Fig. 6.9 Failure Analysis Process*

trends, especially when the amount of data are limited:

1. When $U > 0.5$ or $U < -0.5$, assume a reasonable trend exists.
2. Otherwise, depending on the age and the item's recent failure history, assume a mostly constant failure rate (or failure probability) or a mild trend exists.

### *6.6.2 Failure Rate and Failure Probability Estimation for Data With No Trend*

Section 3.5 dealt with statistical methods for estimating failure rate and failure probability parameters of components when there is no trend in failures. The objective is to find a **point estimate** and a **confidence interval** for the parameters of interest.

#### Parameter Estimation When Failures Occur by Time

When failures of equipment occur by time (r failures in T hours), the exponential distribution is most commonly used. Therefore, when the time to failure of a component is believed to occur at a constant rate (i.e., no trend), the exponential model is reasonable and the parameter estimation should proceed. In this case, $\lambda$ parameter must be estimated. The point estimator is for the failure rate parameter ($\lambda$) of the exponential distribution obtained from $\hat{\lambda} = r/t$. Depending on the method of observing data, the confidence interval of $\lambda$ can be obtained from one of the expressions in Table 3.2.

#### Parameter Estimation When Failures Occur on Demand (Binomial Model)

When the data are in the form of $X$ failures in $N$ trials (or demands), no time relationship exists and the binomial distribution best represents the data. This situation often occurs for equipment in the standby mode, e.g., a redundant pump that is demanded for operation $N$ times in a fixed period of time. In a binomial distribution, the only parameter of interest is $p$. An estimate of $p$ and its confidence interval can be obtained from (3.61) through (3.63).

### *6.6.3 Failure Rate and Failure Probability Estimation for Data With Trend*

The existence of a trend in the data indicates that the interarrivals of failures are not statistically similar, and thus (5.2) should be used. Chapter 5 describes the method of estimating the rate of failure occurrence $\lambda(t)$ using (5.3) through (5.7).

### 6.6.4 *Evaluation of Statistical Data*

After the data are analyzed, it is important to determine whether or not any significant changes between the past data and more recent data can be detected. If such changes are detected, it is important to formulate a procedure for dealing with them.

## Evaluation of Data with No Trend

Two methods of evaluation are considered, statistical and nonstatistical. One effective statistical technique is the chi-square method. The nonstatistical technique only considers degrees of change in the failure characteristics of an item (e.g., in the form of a percent difference from a generic value or prior experience). Proper action is suggested based on a predefined criterion.

As mentioned earlier, the chi-square method can be adapted to the type of problems considered here. The chi-square method was described in Chapter 2. In failure analysis, the chi-square test can be used to determine whether or not the observed failure data are statistically different from generic data, or from past history of the same or a similar item. For example, consider Fig. 6.10. If the expected number of failures, based on generic failure data or previously calculated values (e.g., using statistical analysis), are determined and compared with the observed failures, one can statistically measure the difference.

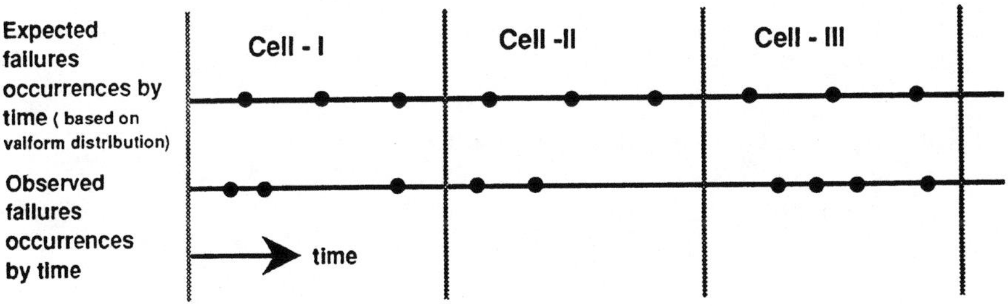


*Fig. 6.10 Comparison of Expected and Observed Failure Occurrences by Item*

It is easy to divide the time line (or in a demand type item, the number of demands) into equal time demand intervals (e.g., three intervals as in Fig. 6.10) and compare them to see whether or not the observed and expected failures in each interval are statistically different.

For example, in Fig. 6.10, the following chi-square value can be calculated:

$$W = (3-3)^{2/3} + (3-2)^{2/3} + (3-4)^{2/3} = 2/3.$$

This shows that there is a slight difference between the observed and expected data, but depending on the desired level of confidence, this may or may not be acceptable.

The nonstatistical technique uses only a percent difference between the estimated failure rate $\hat{\lambda}$ and the generic failure rate $\lambda_g$. For example, by using

$$e = \left| \frac{\hat{\lambda} - \lambda_g}{\lambda_g} \right| \times 100, \tag{6.34}$$

if the difference is large (more than 100), one can assume the data are different and further root-cause analysis is required.

## Evaluation of Data With Trend

Generally, there is no set rule for this purpose. One approach is to use the doubling failure concept. If two consecutive intervals of $(t_1, t_2)$ and $(t_2, t_3)$ are such that $t_2 - t_1 = t_3 - t_2$, and the expected number of failures in each interval ($N_1$ and $N_2$ respectively) are such that $N_2/N_1 = 2$, then it is easy to prove, using (5.2) and (5.3), that $\beta = 1.58$. Accordingly, for $N_2/N_1 = 5$, $\beta = 2.58$. These can be used as guidelines for determining the severity of the trend. For example, one can assume the following:

- If $1 \leq \beta \leq 1.58$, the trend is mildly increasing. Suggest a root-cause analysis and implement a careful monitoring system.
- If $1.58 < \beta \leq 2.58$, the trend is major. Suggest replacement or root-cause analysis.
- If $\beta > 2.58$, the trend is significant. Cease operation of the item and determine the root cause of the trend.

### *6.6.5 Root-Cause Analysis*

Root causes are the most basic causes that can be reasonably identified by experts and can be corrected so as to minimize their recurrence. The process of identifying root causes is generally performed by a group of experts (investigators), sometimes with the help of computers. Modarres et al. (1989) explains the application of expert systems in root-cause analysis. The goal of the experts is to identify the basic causes. The more specific they can be about the reasons an incident occurred, the easier it is to arrive at a recommendation that will prevent recurrence of the failure events. However, investigation of root causes should not be carried to the extreme. The analysis should get the most out of the time spent, and only identify root causes for which a reasonable corrective action exists. Therefore, very complex and specific mechanisms of failure do not need to be identified, especially when corrective actions can be determined at a higher level of abstraction. The recommended corrective actions should be specific and should directly address the root causes identified during the analysis.

Root-cause analysis involves three steps:

1) Determining events and causal factors
2) Coding and documentating root causes
3) Generating recommendations

Charting the event and causal factors provides a road map for experts to organize and analyze the information that they gather, identify their findings, and highlight gaps in knowledge as the investigation progresses. For example, a sequence diagram similar to that in Fig. 6.11 is developed, showing the events leading up to and following an occurrence as well as the conditions and their causes surrounding the failure event. The process is performed inductively and in progressively more detail.

Fig. 6.11(a) shows the causal relations leading to a "failure event," including the conditions, events, and causal factors.

Following this step, the causal factors and events should be documented. One method suggested by the Root-Cause Analysis Handbook (1991) uses a root cause tree involving six levels. From the event and causal factors chart, these levels are described and documented. Fig. 6.11(b) shows an example of the levels used and Fig. 6.11(c) shows a example of a reporting made based on this classification.

The final and most important step in this process is to generate of recommendations. This process is based on the experience of the experts. However, as a general guideline, the following items should be considered in the recommended corrective actions:

1) At least one corrective action should be identified for each root cause .
2) The corrective action should directly and unambiguously address the root cause.

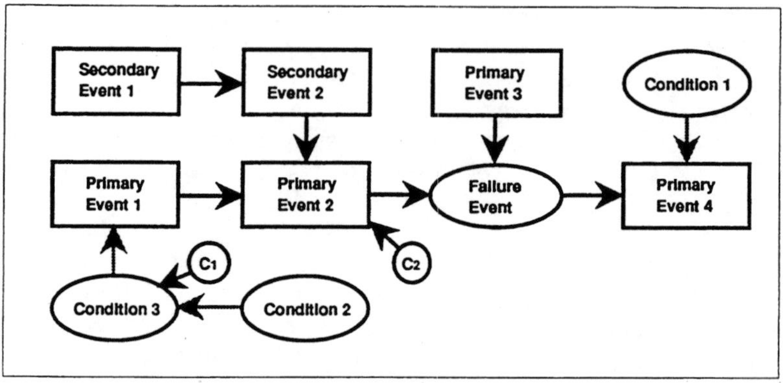


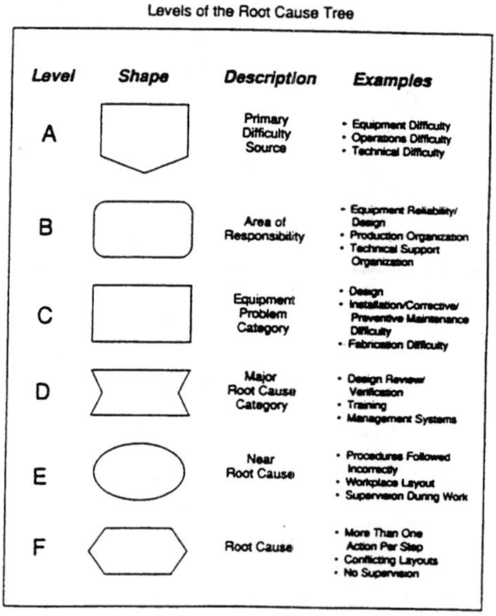
Levels of the Root Cause Tree
Level
Shape
Description
Examples
A
Primary Difficulty Source
• Equipment Difficulty
• Operations Difficulty
• Technical Difficulty
B
Area of Responsibility
• Equipment Reliability/ Design
• Production Organization
• Technical Support Organization
C
Equipment Problem Category
• Design
• Installation/Corrective/ Preventive Maintenance Difficulty
• Fabrication Difficulty
D
Major Root Cause Category
• Design Review/ Verification
• Training
• Management Systems
E
Near Root Cause
• Procedures Followed Incorrectly
• Workplace Layout
• Supervision During Work
F
Root Cause
• More Than One Action Per Step
• Conflicting Layouts
• No Supervision

| Causal Factor | Path Through Root Cause Tree | Recommendations |
|---|---|---|
| **Operator had not previously been to Motor Control Center**<br><br>BACKGROUND:<br><br>The Operator who went to verify the position of the Motor-Generator (M-G) switchgear had not been required to use this particular switchgear in the past. If he had been shown the switchgear as part of his training, it is unlikely that he would have forgotten its location. | **. Operations Difficulty**<br>**. Production Organization**<br>**. Training**<br><br>The Operator had never been trained on location of equipment ( including switchgear ) in Motor Control Center | **. Include a tour of the Motor Control Centers as part of on-the-job training. Provide specific instructions on how to use drawings for verifying the positions of important pieces of equipment. ( Production Training Department )** |

*Fig. 6.11 Events and Causal Factors Chart*

3) The corrective action should not have secondary degrading effects.
4) The consequences of the recommended (or not recommended) corrective actions should be identifiable.
5) The cost associated with implementation of the corrective action should be estimated.
6) The need for special resources and training for implementation of the action should be identified.
7) The effect on the frequency of item failure should be estimated.
8) The impact the corrective action is expected to have on other items or on workers should be addressed.
9) The effect of the corrective action should be easily measurable.
10) Other possible corrective actions that are more resource intensive but more effective should be listed.

The root-cause analysis is a major field of study. For further reading in this subject, see Chu (1989), Ferry (1988), Kendrick (1987), and Kendrick (1990).

## 6.7 Reliability Growth Models

Usually during the development of a system, the design changes as development progresses. The aim of these changes is to improve reliability based on known design deficiencies. As the life cycle of the system progresses from the development stage to manufacturing, construction, production, and operation, one would expect the implementation of these changes to reduce the incidence of failure. This elimination of design weaknesses and improvement in system reliability is known as reliability growth.

The concepts of reliability growth are discussed by a number of authors. Balaban (1978) presents the mathematical models of reliability growth; Keller (1987) reviews the mathematical modeling of the aging and repair process; and O'Connor (1991) discusses general methods for sequential testing, reliability demonstration, and growth monitoring. Generally, the reliability growth models are applicable to the following situations:

1) Predicting long-term reliability
2) Monitoring the effectiveness of design changes
3) Planning replacement strategy or maintenance (e.g., for age-related degradation in a system or plant)
4) Performing life-cycle costing

## Duane Method

This reliability growth model involves developing an empirical relationship based on observating improvement in the MTBF in certain aircraft items. Duane observed that the cumulative MTBF plotted against time on a log-log scale is a straight line. The slope of this line can be used as a measure of the MTBF or reliability growth. Clearly other reliability characteristics such as failure rate and cumulative number of failures, can be equally expressed as a function of time. Accordingly, if the expression $\phi(t) = 1/\lambda(t)$ is the cumulative MTBF, then, according to the Duane model,

$$\ln \lambda(t) = b + a \ln t, \tag{6.35}$$

where a and b are constant.

$$\ln \phi(t) = \ln \phi_0 + m(\ln t - \ln t_0),$$

where $m$ is a constant, and $\phi_0$ and $t_0$ are the cumulative MTBF and the time at the beginning of the monitoring period, respectively. Then

$$\ln H(t) = -\beta \ln \alpha + \beta \ln t, \tag{6.36}$$

where $H(t)$ is the cumulative number of failures between 0 and $t$, and $\alpha$ and $\beta$ are constants. Thus, assuming an exponential time-to-failure model, $\lambda(t)$ and $H(t)$ are obtained from

$$1/\lambda(t) = t/H(t) = \frac{\text{Total operating time (experience)}}{\text{Total number of failures observed}}. \tag{6.37}$$

The nonlogarithmic form of (6.36) can be expressed by

$$H(t) = (t/\alpha)^{\beta}. \tag{6.38}$$

Since $H(t)$ represents cumulative failures, the differentiation of (6.38) with respect to $t$ yields the hazard rate.

$$\frac{dH(t)}{dt} = h(t) = \frac{\beta}{\alpha}\left(\frac{t}{\alpha}\right)^{\beta-1}. \tag{6.39}$$

Expression (6.39) is the hazard rate function of a Weibull distribution. Accordingly, $\beta < 1$ represents reliability growth; $\beta > 1$ represents reliability degradation. The values of $\beta$ and $\alpha$ can be estimated either by plotting failure data on Weibull paper or using maximum likelihood estimators, using (3.59) and (3.60) or (5.5) and (5.6). Implicit in the model for repairable systems is the assumption that after a failure, the system is returned to an as-good-as-new state. Accordingly, if $\phi(t) = 1/\lambda(t)$,

$$\ln \phi(t) = \beta \ln \alpha - (\beta - 1)\ln(t). \tag{6.40}$$

A comparison between (6.35) and (6.40) yields

$$\beta = 1 - m, \quad \text{and}$$

$$\alpha = t_0 \left(\frac{\phi_0}{t_0}\right)^{1/\beta}. \tag{6.41}$$

The Duane method can also be applied to estimate the time (e.g., test time) before an item attains a target MTBF. This time can be estimated as a function of $\alpha$. Therefore, when MTBF is fixed using (6.40), it is possible to develop guidelines for values of $\beta$ based on past experience. O'Connor (1991) discusses the following guidelines:

$m =$ 0.4–0.6. The program's top priority is the elimination of failure modes. The program uses accelerated tests and suggests immediate analysis and effective corrective action for all failures.

$m =$ 0.3–0.4. The program gives priority to reliability improvement. The program uses normal environmental tests and well-managed analysis. Corrective action is taken for important failure modes.

$m =$ 0.2–0.3. The program gives routine attention to reliability improvement. The program does not use applied environmental tests. Corrective action is taken for important failure modes.

$m =$ 0.0–0.2. The program gives no priority to reliability improvement. Failure data are not analyzed. Corrective action is taken for important failure modes, but with low priority.

---

*Example 6.6*

The table below provides failure data for a new device. Determine the Duane model. Calculate the failure rate at the end of the data observation.

| Failure Number | 1 | 2 | 3 | 4 | 5 | 6 | 7 | 8 | 9 | 10 |
|---|---|---|---|---|---|---|---|---|---|---|
| Time to Failure ($t_i$) | 540 | 1121 | 1910 | 2472 | 1810 | 2201 | 2402 | 2521 | 2442 | 2661 |

*Solution:*

From (5.5) and (5.6),

$$t_0 = \sum_{i=1}^{n} t_i = 20,080$$

$$\hat{\beta} = \frac{10}{\ln \frac{20,080}{540} + \ln \frac{20,080}{1121} + \ldots} = 0.42,$$

$$\hat{\alpha} = \frac{20,080}{10^{1/0.42}} = 83.70.$$

Accordingly, the slope of the Duane model is

$$m = 1 - \beta = 1 - 0.42 = 0.58.$$

This corresponds to a program dedicated to the reduction of failures. From (6.39),

$$\lambda(t) = \frac{0.42}{83.7}\left(\frac{t}{83.7}\right)^{-0.58},$$

by assuming that the present time is the time at which the last failure is observed then for

$$t = 20,080, \quad \lambda(20,080) = 2.1 \times 10^{-4}/\text{hr}.$$

---

For cases where failure observations are made at certain intervals and the number of failures observed in each interval is known, the treatment is more complex. For more information in this regard, see Keller (1987).

## 6.8 Stress-Strength Analysis

Generally, failures can be viewed as overstress. Conversely, when applied stress exceeds strength, a failure may occur. The probability that an item will work (will not fail) is equal to the probability that the applied stress is less than the item's strength. Then,

$$R = \Pr(S > s), \tag{6.42}$$

where

$R =$ Reliability of the item,
$s =$ Applied stress, and
$S =$ Item's strength.

Examples of stress failures include the following:

1) Misalignment of a journal bearing, lack of lubricants, or incorrect lubricants generate an internal load that causes the bearing to fail.
2) The voltage applied to a transistor gate is too high, causing a high temperature that melts the transistor's semiconductor material.
3) Cavitation causes pump failure, which in turn causes a violent vibration that ultimately breaks the rotor.
4) Lack of heat removal from a feed pump in a power plant results in heating of the pump seals, causing the seals to break.

Engineers should try to ensure that the strength of an item exceeds the stress placed on it for all possible stress situations. Traditionally, in the deterministic design process, safety factors are used to cover the spectrum of possible applied stresses. This is generally a good engineering principle, but failures occur despite these safety factors. On the other hand, safety factors that are too stringent results in overdesign, high cost, and sometimes ineffectiveness.

If the range of major stresses is known or can be estimated, a probabilistic approach can be used to address the problem. This approach eliminates overdesign and high cost and failures caused by stresses not considered early in the design. If the distribution of $S$ and $s$ can be estimated as $F(S)$ and $g(s)$, then

$$
\begin{aligned}
R &= \int_0^\infty F(S) dg(s) \qquad (6.43)\\
&= \int_0^\infty f(S)[\int_S^\infty f(s) ds] dS.
\end{aligned}
$$

Fig. 6.12 shows typical $F(S)$ and $g(s)$ distributions.

The **Safety Margin (SM)** is defined as

$$
\mathrm{SM} = \frac{E(S) - E(s)}{[Var(S) + Var(s)]^{1/2}}. \qquad (6.44)
$$

The SM shows the relative difference between the mean values for stress and for strength. The larger the SM, the more reliable the item will be. Use of (6.44) is a more objective way of measuring the safety of items. It also allows for calculation of reliability and probability of failure as compared with the traditional deterministic approach

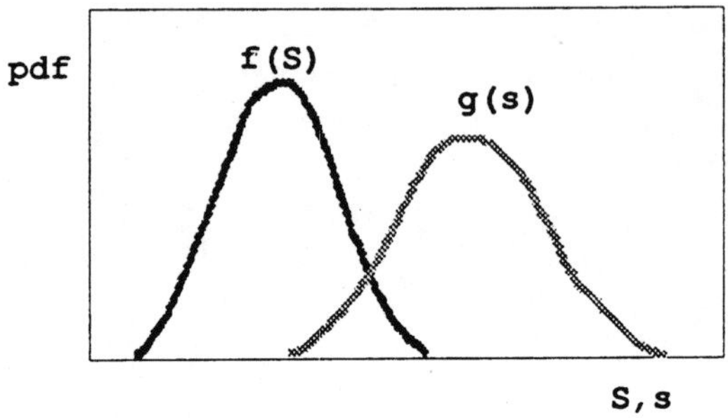


*Fig. 6.12 Stress-Strength Relationship*

using safety factors. However, good data on the variability of stress and strength are often not easily available. In these cases, engineering judgment can be prudently used to obtain the distribution including engineering uncertainty. The section on expert judgment explains methods for doing this in more detail.

The distribution of stress is highly influenced by the way the item is used and the internal and external operating environments. The design determines the strength distribution, and the degree of quality control in manufacturing primarily influences the strength variation.

It is easy to show that for a normally distributed $S$ and $s$,

$$R = \Phi(SM), \tag{6.45}$$

where $\Phi(SM)$ is the probability obtained from Table A.1 with $z = SM$.

---

*Example 6.7*

Consider the stress and strength of a beam in a structure represented by the following normal distributions:

$\mu_S = 420kg/cm^2$ and $\sigma_S = 32kg/cm^2$.
$\mu_s = 310kg/cm^2$ and $\sigma_s = 72kg/cm^2$.

What is the reliability of this structure?

*Solution:*

SM = $\frac{420-310}{[32^2+72^2]^{1/2}} = 1.4$,
with $z = 1.4$ and using Table A.1,
$R = \phi(1.4) = 0.91$.

---

If both the stress and strength distributions are exponential with parameters $\lambda_s$ and $\lambda_S$, receptively, the reliability can be expressed simply by

$$R = \frac{\lambda_s}{\lambda_s + \lambda_S}. \tag{6.46}$$

For additional information about stress-strength methods in reliability analysis, the readers are referred to O'Connor (1991) and Kapur and Lamberson (1977).

## 6.9 Reliability-Centered Maintenance

The reliability-centered maintenance (RCM) methodology is a systematic approach directed towards defining and developing applicable and effective preventive maintenance (PM) strategies. The RCM approach shifts away from a primary emphasis on PM tasks that attempt to restore equipment to an ideal condition. Rather, if focuses on those tasks necessary to maintain system function. RCM has been originally developed in the late 1960's by the commercial air transport industry and later implemented by the Department of Defense and more recently by the nuclear industry. The need for this new approach to PM activities developed as the size and complexity of aircrafts increased, thus causing significant increases in PM costs. RCM has proved to be both successful and beneficial to the airline and more recently to the nuclear industry.

The RCM methodology involves a systematic and logical step-by-step consideration of:

1) The function of a system or component
2) The methods of failure of that function
3) The importance of the function and its failure
4) A priority-based consideration that identifies those PM activities that both reduce failure potential and are cost-effective

The key steps of this process include:

- Definition of system boundaries. Boundaries must be clearly identified and clear explanation of the level of detail for the analysis be presented.
- Determination of the functions of a system, its subsystems, or components. Each component within the system or subsystem may have one or more functions. These should by explained inputs and outputs of functions across system boundaries must also be identified.
- Determination of functional failures. A functional failure occurs when a system or subsystem fails to provide its required function.

- Determination of dominant failure modes. One of the logical system analysis methods (e.g., fault tree or MLD) along with FMEA should be used to identify the modes that are the leading (high probability) causes for functional failures.
- Determination of corrective actions. Applicable and effective course of action for each failure mode should be identified. This action may be to implement a PM task, accept the likelihood of failure, or initiate redesign.
- Integration of the results. The results of the PM task along with other specifics of implementation are integrated into the Maintenance Plan.

From the above steps, it is clear that RCM methodology can be divided into two basic phases. First, the system and its boundaries are defined and then the system is decomposed to subsystems and components, and their functions are identified along with those failures that are likely to cause loss of the functions. Second, each of the functional failures is examined to determine the associated failure mode and to determine whether or not there are effective PM strategies (or task) that eliminate or minimize occurrence of the failure mode identified.

For those failure modes for which an effective PM task is specified, further definition is necessary. Each task should be labeled as either time-directed, condition-monitoring, or failure-finding. Time-directed tasks are generally applicable when the probability of failure increases with the time, that is the failure mode has a positive trend as discussed in Chapter 5. Time can be measured in several different ways, including actual run time or the number of startups (demands) or shutdowns of the component (with the given failure mode). Condition-monitoring tasks are generally applicable when one can efficiently correlate functional failures to detectable and measurable parameters of the system. For example, vibration of a pump can be measured to predict alignment problems. Failure-finding tasks are not preventive, but are intended to discover failures that are otherwise hidden. If no effective PM task can be identified for a hidden failure, a scheduled functional failure-finding task may be devised.

## BIBLIOGRAPHY

1. Amendola, A. U. Bersini, P.C. Cacciabue and G. Mancini (1987). Modeling Operators in Accident Conditions: Advances and Perspectives on Cognitive Model, *Int. J. Man-Machine Studies*, 27: 599.

2. Apostolakis, G. (1982). "Data Analysis in Risk Assessment," *Nuclear Engineering and Design*, 71: 375-381 (1982).
3. Apostolakis, G., S. Kaplan, B.J. Garrick and R.J. Duphily (1980). "Data Specialization for Plant-Specific risk Studies," *Nuclear Engineering and Design*, 56: 321-329.
4. Apostolakis, G. and V.T. Lee (1977). "Methods for the Estimation of Confidence Bounds for the Top Event Unavailability of Fault Trees," *Nuclear Engineering and Design*, Vol. 41, pp. 411-419.
5. Atwood, C.L. (1983). "Common Cause Failure Rates for Pumps," NUREG/CR-2098, US Nuclear Regulatory Commission, Washington, D.C..
6. Balaban, H.S. (1978). "Reliability Analysis for Complex Repairable Systems," Reliability and Biometry, SIAM.
7. Bier, V.M. (1983). "A Measure of Uncertainty Importance for Components in Fault Trees," Transactions of the 1983 Winter Meeting of the Am. Nucl. Soc., San Fransisco.
8. Birnbaum, Z.W. (1969). "On the Importance of Different Components in a Multicomponent System" in *Multivariate Analysis-II* (P.R. Krishnaiah, ed.), Academic Press, New York.
9. Chu (1989). "Root Cause Guidebook: Investigation and resolution of Power Plant Problems," Failure Prevention, Inc., San Clemente, CA.
10. Clemens, R.J. and R.L. Winkler (1990). "Unanimity and Compromise Among Probability Forecasters," *Mgmt. Science*, 36: 767-779.
11. Dalkey, N. and O. Helmer (1963). "An experimental Application of the Delphi Method to the Use of Experts," *Mgmt. Science*, 9: 458-467.
12. Dezfuli, H. and M. Modarres (1984). "Uncertainty Analysis of Reactor Safety Systems with Statistically Correlated Failure Data," *Reliability Engineering Journal*, Vol. 11, 1, pp. 47-64.
13. Dougherty E.M. and J.R. Fragola (1988). *Human reliability Analysis : A System Engineering Approach with Nuclear Power Plant Applications*, John wiley and Son, New York.
14. Efron, B.A. (1979). "Bootstrap Methods: Another Look at the Jacknife," *Annals of Statistics*, Vol. 19, pp. 1-26.
15. Embry, D.E., P.C. Humphreys, E.A. Rosa, B. Kirwan and K. Rea (1984). "SLIM-MAUD: An Approach to Assessing Human Error Probabilities Using Stuctured Expert Judgment," U.S. Nuclear Regulatory Commission, NUREG/CR-3518, Washington DC.

16. Ferry (1988). *Modern Accident Investigation and Analysis*, 2nd ed., John Wiley and Son, New York.
17. Fleming K.N., A. Mosleh and R.K. Deremer (1986). "A systematic Procedure for the Incorporation of Common Cause Event, Into Risk and Reliability Models," *Nuclear Engineering and Design*, 58, 415-424.
18. Fleming, K.N. (1975). "A Reliability Model for Common Mode Failures in Redundant Safety Systems," Proceeding of the Sixth Annual Pittsburgh Conference on Modeling and Simulations, Instrument Society of America, Pittsburgh, PA.
19. Fussell (1975). "How to Hand Calculate System Reliability and Safety Characteristics," *IEEE Transaction on Reliability*, Vol. R-24, No. 3.
20. Hall, R.E., J. Wreathall and J.R. Fragola (1982). "Post Event Human Decision Errors: Operator Action/Time reliability Correlation", U.S. Nuclear Regulatory Commission, NUREG/CR-3010, Washington, DC.
21. Hannaman, G.W. and A.J. Spurgin (1984). "Systematic Human Action Reliability Procedure (SHARP), Electric Power Research Institute," NP-3583, Palo Alto, CA.
22. Hannaman, G.W., A.J. Spurgin, Y.D. Lukic (1984). "Human Cognitive Reliability Model for PRA Analysis", NUS Corporation, NUS-4531, San Diego, CA.
23. Hunns, D.M. and B.K. Daniels (1980). "The Method of Paired Comparisons", *3rd European reliability Data Bank Seminar*, University of Bradford, National Center of System Reliability, United Kingdom.
24. IEEE Standard-500 (1984). "IEEE guide to the Collection and Presentation of Electrical, Electronic and Sensing Component Reliability Data for Nuclear Powered Generation Stations," Institute of Electrical and Electronic Engineers, Piscataway, NJ.
25. Iman, R.L., J.M. Davenport and D.K. Zeigler (1980). "Latin Hypercube Sampling (Program User's Guide)," SAND79-1473, Sandia National Laboratories, Albuquerque, NM.
26. Kaplan, S. (1981). "On the Method of Discrete Probability Distributions in Risk and Reliability Calculation–Application to Seismic Risk Asssessment," *Risk Analysis Journal*, 1, pp. 189-196.
27. Kapur, K.C. and L.R. Lamberson (1977). *Reliability in Engineering Design*, John Wiley and Son, New York.
28. Keller, A.Z. (1987). *Reliability Aging and growth Modeling in Reliability Modeling and Applications*, A.G. Colombo and A.Z.

Keller (eds.), ECSC, EEC, EAEC, Brussels and Luxembourg.
29. Kendrick (1990). *Systematic Safety Training*, Marcel Dekker, New York.
30. Kendrick (1987). *Investigating Accidents with STEP*, Marcel Dekker, New York.
31. Kim, J.H. (1991). "A Bayesian Model for Aggregating Expert Opinions," Ph.D. Dissertation, University of Maryland, Department of Materials and Nuclear Engineering, College Park, MD.
32. Lambert, H.E. (1975). "Measures of Importance of events and Cut Sets in Fault trees" in *Reliability and Fault Tree Analysis*, R. Barlow, J. Fussell and N. Singpurwalla (eds.), SIAM, Philadelphia, PA.
33. Lichtenstein, S.B., B. Fischoff and L.D. Phillips (1977). "Calibration of Probabilities: The State of the Art," *Decision Making and Change in Human Affairs*, J. Jungerman and G.deZeeuw (ed.), D. Reidel, Dordrecht, Holland.
34. Martz, H.F. (1983). A Comparison of Methods for Uncertainty Analysis of Nuclear Plant Safety System Fault Tree Models, U.S. Nuclear regulatory Commission and Los Alamos National Laboratory, NUREG/CR-3263, Los Alamos, NM.
35. Maximus Inc. (1980). Handbook for the Calculation of Lower Statistical Confidence Bounds on System Reliability.
36. Modarres, M., L. Chen and M. Danner (1989). "A knowledge-Based Approach to Root-Cause Failure Analysis," Proceeding of the Expert Systems Applications for the Electric Power Industry Conference, Orlando, FL.
37. Morchland, J.D. and G.G. Weber (1972). "A Moments Method for the Calculation of Confidence Interval for the Failure Probability of a System," Proceeding of the 1972 Annual Reliability and Maintainability Symposium, pp. 505-572.
38. Morgan, M.G. and M. Henrion (1990). *Uncertainty: A Guide to Dealing with Uncertainty in Quantitative Risk and Policy Analysis*, Cambridge Press, Cambridge, U.K.
39. Mosleh, A. (1991). "Common Cause Failures; An Analysis Methodology and Examples," *Reliability Engineering and System Safety,* 34, 249-292.
40. Mosleh, A. and N.O. Siu (1987) "A Multi-parameter, Event-based Common-cause Failure Model," Proc. of the Ninth International Conference on Structural Mechanics in Reactor Technology, Lausanne, Switzerland.
41. Mosleh, A. et al. (1988). "Procedure for Treating Common Cause Failures in Safety and Reliability Studies," U.S. Nu-

clear Regulatory Commission, NUREG/CR-4780, Vol. I and II, Washington, DC.

42. Mosleh, A. and G. Apostolakis (1983). "Combining Various Types of Data in Estimating Failure Rates," Transaction of the 1983 Winter Meeting of the American Nuclear Society, San Fransisco, CA.
43. Nakashima, K. and Y. Kazuharu (1982). " Variance - Importance of System Components," *IEEE Transaction on Reliability*, vol. R-31, No.31.
44. NUREG/CR-4550 (1990). Analysis of Core Damage Freqeuncy from Internal Events, U.S. Nuclear Regulatory Commission, Vol. 1, Washington, DC.
45. O'Connor (1991). *Practical Reliability Engineering*, 3rd ed., John Wiley and Son, New York.
46. Potash, L., M. Stewart, P.E. Diets, C.M. Lewis and E.M. Dougherty (1981). "Experience in Integrating the Operator Contribution in the PRA of Actual Operating Plants," Proceedings of American Nuclear Society, Topical Meeting on Probabilistic Risk Assessment, Port Chester, New York.
47. Poucet, A. (1988). "Survey of Methods Used to Assess Human reliability in the Human Factors Reliability Benchmark Exercise," *Reliability Engineering and System Safety*, 22, pp. 257-268.
48. Poucet, A. (1988). "State of the Art in PSA Reliability Modeling as Resulting from the International Benchmark Exercise Project," NUCSAFE 88 Conference, Avignon, France.
49. Rasmussen, J. (1987). "Cognitive Control and Human Error Mechanisms," Chapter 6 in J. Rasmussen, K. Duncan and J. LePlate (ed.), *New Technology and Human Error*, John Wiley and Son, New York.
50. Rasmussen, J. (1982). "Skills, Rules and Knowledge: Signals, Signs and Symbols and Their Distinctions in Human Performance Models," *IEEE Transactions on Systems, Man and Cybernetics*, Vol. SMC-3, (3), pp. 257-268.
51. Reactor Safety Study: An Assessment of Accidents in U.S. Commercial Nuclear Power Plants (1975). U.S. Regulatory Commission, WASH-1400, Washington, DC.
52. Root-Cause Analysis Handbook (1991). Facility Safety Evaluation Section, Safety Department, Westinghouse Savannah River Company, Aiken, SC.
53. Saaty, T.L. (1980). *The Analytic Hierarchy Process*, McGraw Hill, New York .

54. Severe Accident Risk: An Assessment for Five U.S. Nuclear Power Plants, U.S. Nuclear Regulatory Commission, NUREG-1150, Washington, DC (1990)
55. Sharirli (1985). "Methodology for System Analysis Using Fault Trees, Success Trees and Importance Evaluations," Ph.D. dissertation, University of Maryland, Department of Chemical and Nuclear Engineering, College Park, MD.
56. Siegel, A.I., N.D. Bartter, J.J. Wolf, H.E. Knee and P.M. Haas (1984). "Maintenance Personnel Performance Simulation (MAPPS) Model," U.S. Nuclear regulatory Commission, NUREG/CR-3626, Vol. I and II, Washington, DC.
57. Slovic, P., B. Fischhoff and S. Lichtenstein (1980). "Facts Versus Fears: Understanding Perceived Risk," *Societal Risk Assessment*, R.C. Schwing and W.A. Albers, Jr. (ed.), Plenum, New York.
58. Snaith, E.R. (1981). "The Correlation Between the Predicted and Observed Reliabilities of Components, Equipment and Systems," National Center of Systems Reliability, U.K. Atomic Energy Authority, NCSR-R18.
59. Stillwell, W., D.A. Seaver and J.P. Schwartz (1982). "Expert Estimation of Human Error Problems in Nuclear Power Plant Operations: A Review of Probability Assessment and Scaling," U.S. Nuclear Regulatory Commission, NUREG/CR-2255, Washington, DC.
60. Swain, A.D., and H.E. Guttman (1983). "Handbook of Human Reliability Analysis with Emphasis on Nuclear Power Applications," U.S. Nuclear regulatory Commission, NUREG/CR-1278, Washington, DC.
61. Tukey (1979). "Protection Against Depletion of Stratospheric Ozone by Chlorofluorocarbons," Report by the Committee on Impacts of Stratospheric Change and the Committee on Alternative for the Reduction of Chlorofluorocarbon Emission, National Research Council, Washington, DC.
62. Wheeler, T.A., R.G. Spulak (1985). " The Importance of Data and Related Uncertainties in Probabilistic Risk Assessments," Amer. Nucl. Soc. PSA Topical Meeting, San Fransisco.
63. Woods, D.D., E.M. Roth and H. Pole (1988). "Modeling Human Intention Formation for Human Reliability Assessment," *Reliability Engineering and System Safety*, 22: 169-200.
64. Wreathall, J. (1981). "Operator Action Tree Method," IEEE Standards Workshops on Human Factors and Nuclear Safety, Myrtle Beach, SC.

## Exercises

6.1 Consider two resistors in parallel configuration. The mean and standard deviation for the resistance of each are as follows:

$$\mu_{R1}=25\Omega \quad \sigma_{R1} = 0.1\mu_{R1}$$
$$\mu_{R2}=50\Omega \quad \sigma_{R2} = 0.1\mu_{R2}$$

Using one of the statistical uncertainty techniques, obtain:
a) mean and standard deviation of the equivalent resistor,
b) in what ways the uncertainty associated with the equivalent resistance is different from the individual resistor? Discuss the results.

6.2 The results of a bootstrap evaluation gives: $\mu = 1 \times 10^{-4}$, and $\sigma = 1 \times 10^{-3}$. Evaluate the pseudo failures F, in N trials for an equivalent binomial distribution. Estimate the 95% confidence limits of $\mu$.

6.3 Repeat Exercise 4.6 and assume that a common cause failure between the valves and the pumps exist. Using the generic data in Table C.1, calculate the probability that the top event occurs. Use a $\beta$-factor method with $\beta = 0.1$ for valves and pumps. Discuss if the selection of $\beta = 0.1$ is sensitive to the end result.

6.4 Repeat Exercise 6.3 and calculate the Birnbaum and Vesely-Fussell importance measures for all events modeled in the fault tree (including the common cause failures).

6.5 The following data are given for a prototype of a system which undergoes design changes. A total of 10 failures have been observed since the beginning of the design. Determine the Duane reliability growth model that represents these data. Discuss the results.

| Failure Number | 1 | 2 | 3 | 4 | 5 | 6 | 7 | 8 | 9 | 10 |
|---|---|---|---|---|---|---|---|---|---|---|
| Cumulative Test Time (hr) | 12 | 75 | 102 | 141 | 315 | 330 | 342 | 589 | 890 | 1,007 |

# 7
# Risk Analysis

Risk analysis is a technique for identifying, characterizing, quantifying, and evaluating hazards. It is widely used by private and government agencies to support regulatory and resource allocation decisions. Risk analysis (also called risk assessment) consists of two distinct phases: a qualitative step of identifying, characterizing, and ranking hazards; and a quantitative step of risk evaluation, which includes estimating the likelihood (e.g., frequencies) and consequences of hazard occurrence. After risk has been quantified, appropriate risk-management options can be devised and considered; risk-benefit or cost-benefit analysis may be performed; and risk-management policies may be formulated and implemented. The main goals of risk management are to minimize the occurrence of accidents by reducing the likelihood of their occurrence (e.g., minimize hazard occurrence); reduce the impacts of uncontrollable accidents (e.g., prepare and adopt emergency responses); and transfer risk (e.g., via insurance coverage). The estimation of likelihood or frequency of hazard occurrence depends greatly on the reliability of the system's components, the system as a whole, and human-system interactions. These topics have been extensively addressed in previous chapters of this book. In this chapter we discuss how the reliability evaluation methods addressed in the preceding chapters are used, collectively, in a risk-analysis process. We have also discussed some relevant topics which are not discussed in the previous chapters (e.g., risk perception).

## 7.1 Risk Perception and Acceptability

### *7.1.1 Risk Perception*

Perceptions of risk often differ from objective measures and may distort or politicize risk-management decisions. Subjective judgment, beliefs, and societal bias against events with low probability but high consequences may influence the understanding of the results of a risk analysis. Public polls indicate that societal perception of risk for certain unfamiliar or incorrectly publicized activities is far out of proportion to the actual damage or risk measure. For example, according to Litai (1980), the risk of motor and aviation accidents is perceived to be less than its actual value by a factor of 10 to 100 by the public, but the risk of nuclear power and food coloring is overestimated by a factor of greater than 10,000. Risk conversion and compensating factors must often be applied to determine risk tolerance thresholds accurately, to account for public bias against risks that are unfamiliar (by a factor of 10), catastrophic (by a factor of 30), involuntary (by a factor of 100), or uncontrollable (by a factor of 5 to 10), or have immediate consequences (by a factor of 30). For example, people perceive a voluntary action to be less risky by a factor of 100 than an identical involuntary action. Although the exact values of above conversion factors are debatable, they generally show the direction and the degree of bias in people's perception.

Different risk standards often apply in the workplace, where risk exposure is voluntary and exposed workers are indemnified. Stricter standards apply to public risk exposure, which is involuntary. The general guide to risk standards is that occupational risk should be small compared with natural sources of risk. Some industrial and voluntary risks may be further decreased by strict enforcement or adequate implementation of known risk-avoidance measures (e.g., wearing seat belts, not drinking alcohol, or not smoking). Therefore, some of these risks are controllable by the individual (who can choose whether to fly, to work, to drive or to smoke), while others are not (e.g., chemical dumps, severe floods, and earthquakes).

### *7.1.2 Risk Acceptability*

Risk acceptance is a complex subject and is often subject of controversial debate. However, using the results of risk assessment in a relative manner is a common method of ranking risk-exposure levels. For example, consider Tables 7.1. In this table societal risk of individual death due to the leading causes are ranked. An assessed risk from any controllable activity should be required to be lower than the risk of these causes, so as to be defined acceptable. These de facto levels of socially tolerated (acceptable) levels of risk exposure can define acceptable risk thresholds of risk. Although regulators often strive to assess absolute levels of risk, the relative ranking of risks is a better risk-management strategy for allocating resources toward regulatory controls. Cost-benefit analysis is often required as an adjunct to formulating risk-control strategies to socially acceptable levels.

Another form of risk ranking is to use odds or probability of hazard exposure per unit of time. For example, Table 7.2 is a typical ranking for some societal causes. It should be noticed that for an objective ranking the risk exposure should be the same group. For example, risk of breast cancer is different for different age group, and largely applies to women.

As the third and perhaps a more objective method of risk comparison, sometimes risk exposure is normalized both to the population exposed and to the duration of the exposure and is used for comparison purpose. To compare the risk associated with each cause, consistent units are used (such as number of fatalities or dollar loss per year, per 100,000 population, per event, per person-year of exposure). Table 7.3 shows a risk comparison based on amount of exposure that yields the same risk value.

The typical guideline for establishing risk-acceptance criteria for involuntary risks to the public has been that fatality rates from the activity of interest should never exceed average individual fatality rates from natural causes (about 0.07 per 100,000 population, from all natural causes) and should be further reduced by risk-control measures to the extent feasible and practical. For example, the U.S. Nuclear Regulatory Commission (1986) has recently suggested quantitative safety goals which implicitly define acceptable risk in nuclear power plants. These safety goals state that the risk (prompt fatality or cancer fatality) from state that the risk (prompt fatality or cancer fatality) from nuclear power plants should not exceed 0.1% of the sum of prompt fatality or cancer fatality risk to which all other risks that individual U.S. residents and the public as a whole are generally exposed. Also it requires that reactors be designed such that the overall mean frequency of a large radioactive release to the environment from a reactor accident be less than $1 \times 10^{-6}$ per year of reactor operation.

The societal benefits and the cost trade-offs for risk reduction are widely used guides to set and justify risk acceptability limits. By comparing the risks and benefits associated with certain activities, fair, balanced and consistent limits for risk acceptability can be set and institutional controls on risk can be established. Rowe (1977) describes methods of risk-benefit and cost trade-off for risk analysis.

## 7.2 Determination of Risk Values

There are two major parts in risk analysis:

- Determination of the likelihood, ( e.g. Prob. $P_i$ or frequency of occurrence, $F_i$), of an undesirable event, $E_i$.

*Table 7.1 Major Causes of Death in the United States in 1985, Lewis ( 1990 )*

| No. | Cause | Number |
|---|---|---|
| 1 | Cardiovascular diseases | 978,000 |
| | ( Heart disease ) | ( 771,000 ) |
| 2 | Malignancies | 462,000 |
| | ( Lung, respiratory, etc. ) | ( 127,000 ) |
| 3 | Accidents | 93,500 |
| | ( Motor vehicle ) | ( 45,900 ) |
| 4 | Pulmonary diseases, Chronic | 75,000 |
| 5 | Pneumonia | 68,000 |
| 6 | Diabetes | 37,000 |
| 7 | Suicide | 29,500 |
| 8 | Liver diseases | 27,000 |
| 9 | Homicide ( including police ) | 20,000 |
| 10 | Other | 296,000 |
| | Total | 2,086,000 |

*Table 7.2 Risk of Dying from Selected Causes, Paulos ( 1991 )*

| Cause | Odds |
|---|---|
| Breast Cancer ( at age 60 ) | 1 in 500 |
| Breast Cancer ( at age 40 ) | 1 in 1,000 |
| Car crash | 1 in 5,300 |
| Drowning | 1 in 20,000 |
| Choking | 1 in 68,000 |
| Bicycle crash | 1 in 75,000 |

*Table 7.3 Risk Exposures That Increase Chance of Death by 1 in 1,000,000 per year, Wilson ( 1979 )*

| Nature of Risk Exposure | Cause of Death |
|---|---|
| Smoking 1.4 Cigarettes | Cancer, heart disease |
| Spending 1 hour in a coal mine | Black lung disease |
| Spending 3 hours in a coal mine | Accident |
| Living 2 days in New York or Boston | Air pollution |
| Traveling 10 miles by bicycle | Accident |
| Traveling 300 miles by car | Accident |
| Traveling 10000 miles by jet | Accident |
| Having chest X-ray taken in a good hospital | Cancer caused by radiation |
| Living 50 years within 5 miles of a nuclear plant | Cancer caused by plant |

Sometimes the likelihood estimates are generated from a detailed analysis of past experience and available historical data; sometimes they are judgmental estimates based on an expert's view of the situation, or simply a best guess. This assessment of event likelihood can be useful, but the confidence in such estimates depends on the quality and quantity of the data and the methods used to determine event likelihood.

- Evaluation of the consequence, $C_i$, of this hazardous event. The choice of the type of consequence may affect the acceptability threshold and the tolerance level for risk.

Risk analysis generally consists of the following three steps, sometimes called the "Risk Triplet" which is represented by (1.4).

1) Selection of a specific hazardous reference event $E_i$ or scenarios (sequence or chain of events) for quantitative analysis (hazard identification)
2) Estimation of the likelihood of events, $P_i$
3) Estimation of the consequences of these events, $C_i$

In most risk assessments the likelihood of event $E_i$ is expressed in terms of the probability of that event. Alternatively, a frequency per year or per event (in units of time) may be used. Consequence $C_i$, is a measure of the impacts of event $E_i$. This can be in the form of mission loss, payload damage, damage to property, number of injuries, number of fatalities, dollar loss, etc.

The results of the risk estimation are then used to interpret the various contributors to risk, which are compared, ranked, and placed in perspective. This process consists of:

1) Calculating and graphically displaying a **risk profile** based on individual failure event risks, similar to the process presented in Fig. 7.1. This method will be discussed in more detail in this section.
2) Calculating a total expected risk value $R$ from

$$R = \sum_i P_i \times C_i. \tag{7.1}$$

Naturally, all the calculations described involve some uncertainties, approximations, and assumptions. Therefore, uncertainties must be considered explicitly, as discussed is in Section 6.2. Using expected losses and the risk profile, one can evaluate the amount of investment that is reasonable to control risks, alternative risk-management decisions to avoid risk (i.e., decrease the probability) and alternative actions to mitigate consequences. Therefore, the following two additional planning steps are usually included in risk analysis:

1) Identification of cost-effective risk management alternatives
2) Adoption and implementation of risk-management methods

The risk estimation results are often shown in a general form similar to Table 7.4. There are two useful ways to interpret such results: determining expected risk values, $R_i$, and constructing risk profiles. Both methods are used in quantitative risk analysis.

*Table 7.4 General Form of Output from the Analytic Phase of Risk Analysis*

| Undesirable Event | Likelihood | Consequences | Risk Level |
|---|---|---|---|
| $E_1$ | $P_1$ | $C_1$ | $R_1 = P_1C_1$ |
| $E_2$ | $P_2$ | $C_2$ | $R_2 = P_2C_2$ |
| $E_3$ | $P_3$ | $C_3$ | $R_3 = P_3C_3$ |
| . | . | . | . |
| . | . | . | . |
| $E_n$ | $P_n$ | $C_n$ | $R_n = P_nC_n$ |

Expected values are most useful when the consequences $C_i$ are measured in financial terms or other directly measurable units. The expected risk value $R_i$ (or expected loss) associated with event $E_i$ is the product of its probability $P_i$ and consequence values, as described by (7.1). Thus, if the event occurs with a frequency of 0.01 per year, and if the associated loss is $1 million, then the expected loss (or risk value) is: $R_i = 0.01 \times \$1,000,000 = \$10,000$. Conversely, if the frequency of event occurrence is 1 per year, but the loss is $10,000, the risk value is still $R_i = 1 \times \$10,000 = \$10,000$. Thus, the risk value for these two situations is the same, i.e., both events are equally risky.

Since this is the expected annual loss, the total expected loss over 20 years (assuming a constant dollar value) would be $200,000.

This assumes the parameters do not vary significantly with time, and ignores the low probability of multiple losses over the period. Expression (7.1) can be used to obtain the total expected loss per year for a whole set of possible events. This expected risk value assumes that all events ($E_i$) contributing to risk exposure have equal weight. Occasionally, for risk decisions, value factors (weighting factors) are assigned to each event contributing to risk. The relative values of the terms associated with the different hazardous events give a useful measure of their relative importance, and the total risk value can be interpreted as the average or "expected" level of loss over a period of time.

As discussed earlier another method for interpreting the results is construction of a risk profile. With this method, the probability values are plotted against the consequence values. Fig.7.1 shows two methods for doing this. Fig. 7.1(a) shows the use of logarithmic scales, which are usually used because one can cover a wide range of values. The error brackets denote uncertainties in the probability estimate (vertical) and the consequences (horizontal). This approach provides a means of easily illustrating events with high probability, high consequence, or high uncertainty. It is useful when discrete probabilities and consequences are known. Fig. 7.1(b) shows the construction of the complementary cumulative probability risk profile (sometimes known as a Farmer's curves (1960)). In this case, the logarithm of the probability that the total consequence C exceeds $C_i$ is plotted against the logarithm of $C_i$. The most notable application of this method was in the landmark Rector Safety Study (1975). With this method, the low probability/high consequence risk values and high probability/low consequence risk values can be easily seen. That is, the extreme values of the estimated risk can be easily displayed.

## 7.3 Formalization of Risk Assessment

The hazardous events $E_i$ discussed in the previous section can occur as a result of a chain of some basic events. In combination, these events are called a "scenario." The risk-assessment process is therefore primarily one of scenario development, with the risk contribution from each possible scenario that leads to the outcome or event of interest. This concept described in terms of the triplet represented by (1.4). Because the risk-assessment process focuses on scenarios that lead to hazardous events, the general methodology becomes one that allows the identification of all possible scenarios, calculation of their individual probability, and a consistent descrip-

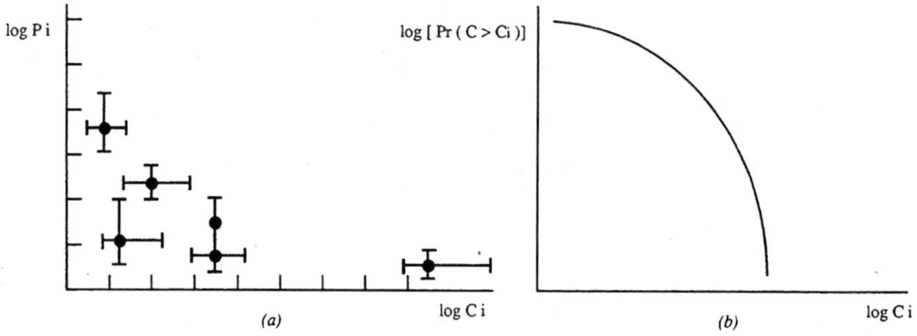


*Fig. 7.1 Construction of a Risk Profile*

tion of the consequences that result from each. Scenario development requires a set of descriptions of how a barrier confining a hazard is threatened, how the barrier fails, and the effects on the subject when it is exposed to the uncontained hazard. This means that one needs to formally address the items described below.

## Identification of hazards

A survey of the process under analysis should be performed to identify the hazards of concern. These hazards can be categorized as follows:

- Chemical hazard (e.g., toxic chemicals released from a chemical process)
- Thermal hazard (e.g., high-energy explosion from a chemical reactor)
- Mechanical hazard (e.g., kinetic or potential energy from a moving object)
- Electrical hazard (e.g., potential difference, electrical and magnetic fields, electrical shock)
- Ionizing radiation (e.g., radiation released from a nuclear plant)

- Nonionizing radiation (e.g., radiation from a microwave oven)

Presumably, each of these hazards will be part of the process and will use normal process boundaries as the containment. This means that, provided there is no disturbance in the process, the barrier that contains the hazard will be unchallenged. However, in a risk scenario one postulates the challenges to such barriers and tries to estimate the probability of these challenges.

## Identification of barriers

Each of the identified hazards must be examined to determine all the physical barriers that contain it or can intervene to prevent or minimize exposure to the hazard. These barriers may physically surround the hazard (e.g., walls, pipes, valves, fuel clad, structures); they may use distance to minimize exposure to the hazard (e.g., minimize exposure, to radioactive materials); or they may provide direct shielding of the subject from the hazard (e.g., protective clothing, bunkers.).

## Identification of challenges to the barriers

Identification of each of the individual barriers is followed by a concise definition of the requirements for maintaining each one. This can be done by developing an analytical model that has a hierarchical character. One can also simply identify what is needed to maintain the integrity of each barrier. These are due to the degradation of strength of the barrier and high stress in the barrier. We discussed these aspect in Section 6.8.

- Barrier strength degrades because of:
  - reduced thickness (due to geometrical change, erosion, corrosion etc.),
  - change in material properties (e.g., toughness, yield strength). This may be affected by the local environment, e.g., temperature).
- Stress on the barrier increases by:
  - internal forces or pressure,
  - penetration or distortion by external objects or forces.

The above causes of degradation are often the result of one or more of the following conditions:

- Malfunction of process equipment (e.g., the emergency cooling system in a nuclear plant)
- Problems with man-machine interface
- Poor design or maintenance
- Adverse natural phenomena
- Adverse human-made environment.

### Estimation of hazard exposure

The next step in the risk-assessment procedure is to define those scenarios in which the barriers may be breached, and then make the best estimate possible of the probability or frequency for each sequence. Those scenarios that pose similar levels of hazard under similar conditions of hazard dispersal are grouped together, and the probabilities or frequencies of the various groups are determined.

### Consequences Evaluation

The range of effects produced by exposure to the hazard may encompass harm to people, damage to equipment, and contamination of land or facilities. These effects are evaluated from knowledge of the toxic behavior of the particular material(s) and the specific outcomes of the scenarios considered. In the case of the dispersal of toxic materials, the size of the release is combined with the potential dispersion mechanisms to calculate the outcome.

From the generic nature of risk analysis, there appears to be a common approach to understanding the ways in which hazard exposure occurs. This understanding is key in the development of logical scenario models that can then be solved. Quantitative and qualitative solutions can provide estimates of barrier adequacy and methods of effective enhancement. This formalization provides a basis from which we can describe a commonly used practice in risk analysis called **probabilistic risk assessment** (PRA). This technique, pioneered by the nuclear industry, is the basis of a large number of formal risk assessments today. We describe this approach in Section 7.4 and provide an example in Section 7.5.

## 7.4 Steps in Conducting a Probabilistic Risk Assessment

The following subsections provide a discussion of the various elements of a PRA as we walk our way through the steps that must be performed. We also describe the methods that are useful for this analysis as described in previous chapters of the book. Fig. 7.2 shows the general PRA process.

### 7.4.1 Methodology Definition

Preparing for a PRA begins with a review of the objectives of the risk analysis. An inventory of possible techniques for the desired analysis should be developed. The available techniques range from required computer codes to facility experts and analytical experts. This, in essence, provides a road map for the analysis. The methods described in the preceding chapters of this book discussed most of the techniques currently used for PRA.

The resources required for each analytical option should be evaluated, and the most cost-effective option selected. The basis for the selection should be documented briefly, and the selection process reviewed to ensure that the objectives of the analysis will be adequately met.

### 7.4.2 Familiarization and Information Assembly

A general knowledge of the physical layout of the system or process (e.g., facility, plant, design), administrative controls, maintenance and test procedures, and protective systems whose function maintains safety is necessary to begin the PRA. All systems, locations, and activities expected to play a roll in the initiation, propagation, or arrest of an upset or hazardous condition must be understood in sufficient detail to construct the models necessary to capture all possible scenarios. A detailed inspection of the process must be performed in the areas expected to be of interest and importance to the analysis.

The following items should be considered in this step:

1) Major safety and emergency systems (or methods) should be identified.
2) Physical interactions among all major systems should be identified and explicitly described. The result should be summarized in a dependency matrix.
3) Past major failures and abnormal events that have been observed in the process should be noted and studied. Such information would help ensure inclusion of important applicable scenarios.
4) Consistent documentation is key to ensuring the quality of the PRA. Therefore, a good filing system must be created at the outset, and maintained throughout the study.

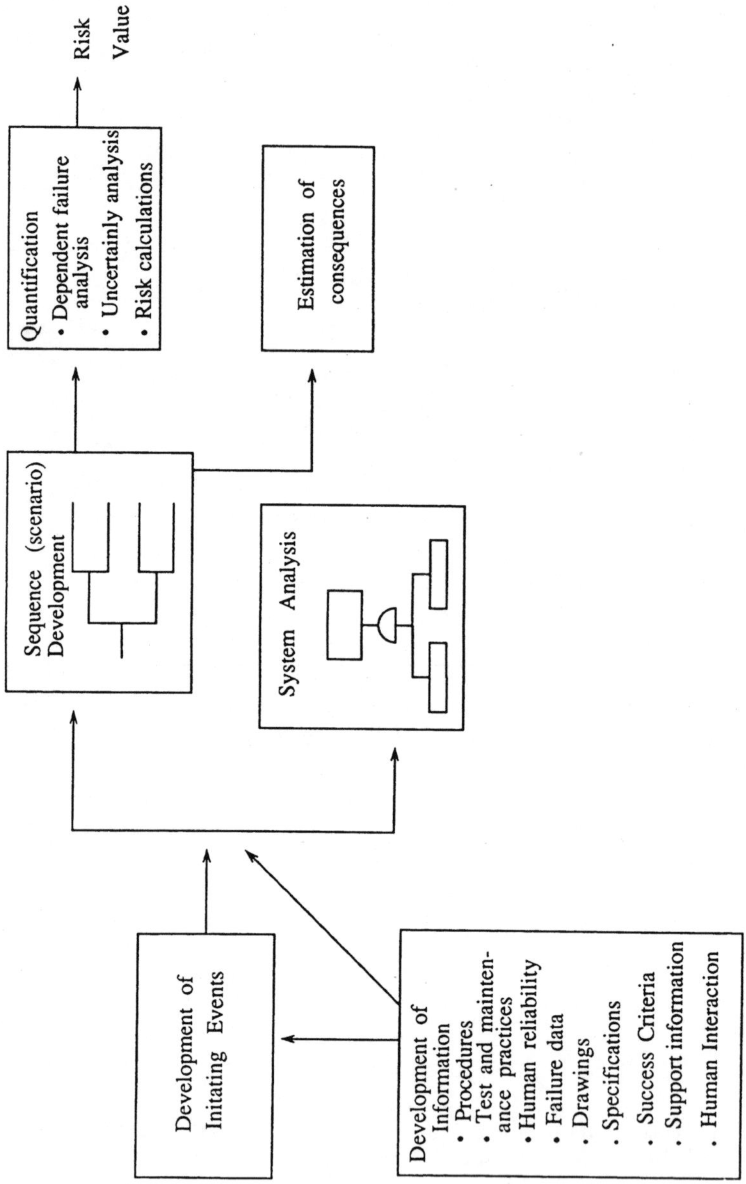


*Fig. 7.2 The Process of Probabilistic Risk Analysis*

With the help of process designers, operators, or owners, one should determine the ground rules for the analysis, the scope of the analysis, and the configuration to be analyzed. One should also determine the faults and conditions to be included or excluded, the operating modes of concern, the "freeze date"design, and the hardware configuration on the design freeze date. The freeze date is an arbitrary date after which no additional changes in the process design and configuration will be modeled. Therefore, the results of the PRA are only applicable to the process at the freeze date.

### *7.4.3 Identification of Initiating Events*

This task involves identifying those events (abnormal events) that could, if not correctly responded to, result in hazard exposure. The first step involves identifying sources of hazard and barriers around these hazards. The next step involves identifying events that can lead to a direct threat to the integrity of the barriers.

A system or process may have one or more operational modes which produce its output. In each operational mode, specific functions are performed that result in the output. Each function is directly related to one or more systems that perform the necessary functional actions. These systems, in turn, are composed of more basic units (e.g., components) that accomplish the objective of the system. As long as a system is operating within its design parameters, there is little chance of challenging the system boundaries in such a way that hazards will escape those boundaries. These operational modes are called normal operation modes.

During normal operation mode loss of certain functions or systems will cause the process to enter an off-normal condition. Once in this condition, there are two possibilities. First, the state of the process could be such that no other function is required to maintain the process in a safe condition. (**safe** refers to a mode where the chance of exposing hazards beyond the process boundaries is incredible.) The second possibility is a state wherein other functions or systems are required to prevent exposing hazards beyond the system boundaries. For this second possibility, the loss of a functional or loss of a system is an initiating event. Since such an event is related to the operating process equipment, it is called, an **operational initiating event**.

Operational initiating events can also apply to shutdown and start-up modes of the process. The terminology remains the same since, for a shutdown or start-up procedure, certain equipment must be functioning. For example, an operational initiating event found during the PRA of a test reactor was Low Primary Coolant System Flow. Flow is required to transfer heat produced in the reactor to heat exchangers and ultimately to the cooling towers and the air. If this coolant flow function is reduced to the point where insufficient heat is transferred, core damage could result. Therefore, another protective system must operate to remove the heat produced by the reactor. By definition, then, Low Primary Coolant System Flow is an operational initiating event.

One method for determining the operational initiating events begins with first drawing a functional diagram of the process (similar to the MLD method described in Chapter 4). From the functional diagram, a hierarchical relationship is produced, with the process objective being successful completion of the desired process. Each function can then be decomposed into its systems, and components can be combined in a logical manner to represent success of that function. (Fig. 7.3 illustrates this hierarchical decomposition.) Potential initiating events are the failures of particular functions, systems, or components, the occurrence of which causes the process to fail. These potential initiating events are grouped such that members of a group require similar process system and safety system responses to cope with the initiators. These groupings are the operational initiator categories.

An alternative to the use of functional hierarchy for identifying initiating events is the use of FMEA, discussed in Chapter 4. The difference between these two methods is noticeable, namely, the functional hierarchy method is deductive and systematic, whereas FMEA is inductive. The use of FMEA for identifying initiating events consists of identifying failure events (modes of failure) whose effect is a threat to hazard barriers. In both of the above methods, one can always supplement the set of initiating events with generic initiating events (if known). For example see NUREG/CR-4550 (1990) for these initiating events for nuclear reactors.

To simplify the process, it is necessary, after identifying all initiating events, to combine those initiating events that pose the same threat to hazard barriers and require the same mitigating functions of the process to prevent hazard exposure. The following inductive procedures should be followed when grouping initiating events:

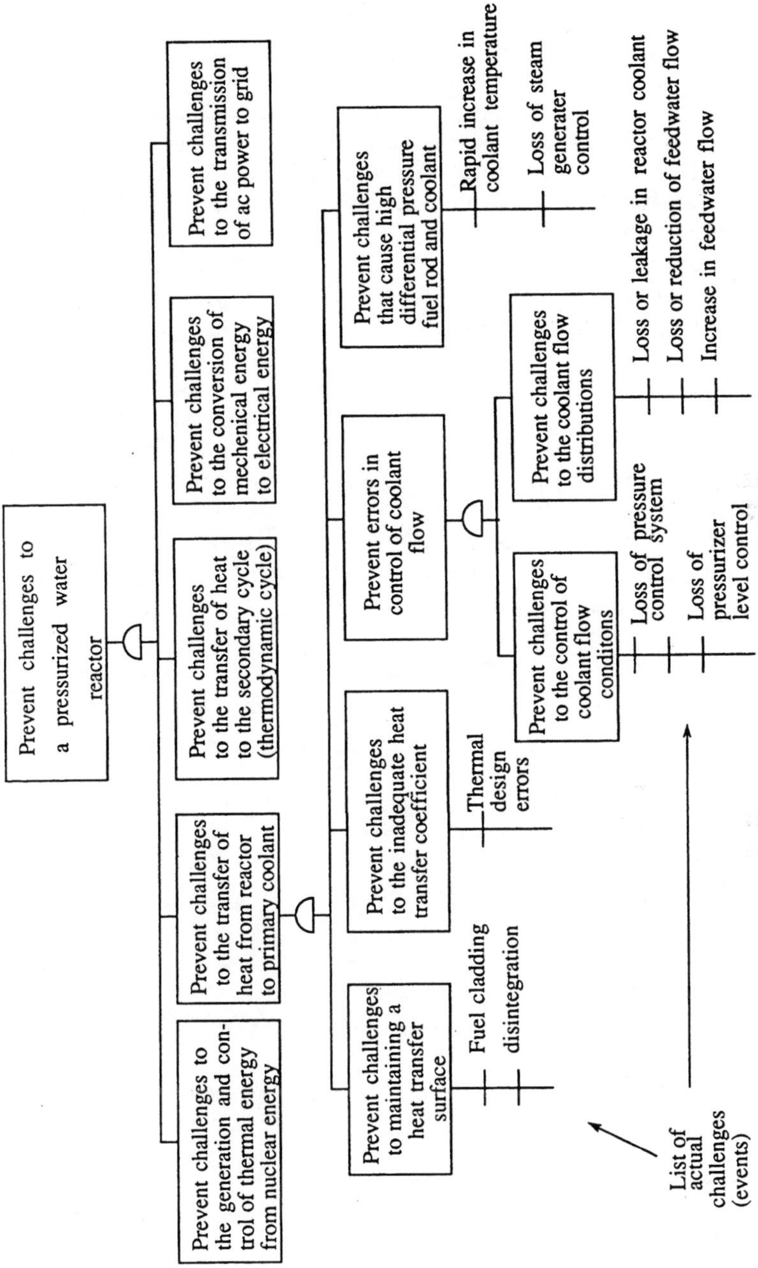


*Fig. 7.3* *Partial Goal Tree to Determine Challenges to a Pressurized Water Reactor*

1) Combine the initiating events that directly break all hazard barriers.
2) Combine the initiating events that break the same hazard barriers (not necessarily all the barriers).
3) Combine the initiating events that require the same set of mitigating human or automatic actions following their occurrence.
4) Combine the initiating events that simultaneously disable the normal process as well as some of the available mitigating human or automatic actions.

Events that cause off-normal operation of the process and require other systems to operate to maintain process materials within their desired boundaries, but are not directly related to a process system or component, are **nonoperational initiating events**. Nonoperational initiating events are identified with the same methods used to identify operating events. However, the events of interest are those that are primarily external to the process. These are discussed in more detail in Sections 7.4.6 and 7.4.7.

The following procedures should be followed in this step of the PRA:

1) Select a method for identifying specific operational and nonopertional initiating events. Two representative methods are functional hierarchy and FMEA. If a generic list of initiating events is available, it can be used as a supplement.
2) Using the method selected, identify a set of initiating events.
3) Group the initiating events such that those having the same effect on the process and requiring the same mitigating functions to prevent hazard exposure are grouped together.

### *7.4.4 Sequence or Scenario Development*

The goal of scenario development is to derive a complete set of scenarios that encompasses all of the potential propagation paths that can lead to loss of confinement of the hazard following the occurrence of an initiating event. To describe the cause and effect relationship between initiators and the event progression, it is necessary to identify those functions (e.g., safety functions) that must be maintained to prevent loss of hazard barriers. The scenarios that describe the functional response of the process to the initiating events are frequently displayed by event-trees.

As discussed in Chapter 4, event trees order and depict (in approximately chronological manner) the success or failure of key mitigating actions (e.g., human actions or mitigative hardware that automatically responds) that are required to respond following an initiating event. In PRA, two types of event trees can be developed: **functional** and **systemic**. The functional event tree uses mitigating functions as its heading. The main purpose of the functional tree is to better understand the scenario of events at a high level following the occurrence of an initiating event. The functional tree also guides the PRA analyst in the development of a more detailed systemic event tree. The systemic event tree, reflects the mitigative scenarios of specific events (specific human actions or mitigative system operations or failures) that lead to a hazardous outcome. That is, the functional event tree can be further decomposed to show specific hardware or human actions that perform the functions described in the functional event tree. Therefore, a systemic event tree fully delineates the process or system response to an initiating event and serves as the main tool for further analysis in the PRA.

The following procedures should be followed in this step of the PRA:

1) Identify the mitigating functions for each initiating event (or group of events).
2) Identify the corresponding human actions, systems or hardware operations associated with each function, along with their necessary conditions for success.
3) Develop a functional event tree for each initiating event (or group of events).
4) Develop a systemic event tree for each initiating event, delineating the success conditions, initiating event progression phenomena, and end effect of each scenario.

### *7.4.5 System Analysis*

Event trees commonly involve branch points at which a given system (or event) either works (or happens) or does not work (or does not happen). Sometimes, failure of these systems (or events) is rare and there may not be an adequate record of observed failure events to provide a dependable data base of failure rates. In such cases, other system analysis methods described in Chapter 4 may be used, depending on the accuracy desired. The most common method used in PRA to calculate the probability of system failure is fault tree analysis. This analysis involves developing a system model in which the system is broken down into basic components or modules for which adequate data exist. In Chapter 4, we discussed how a fault tree can represent the event headings of an event tree.

Different event-tree modeling approaches imply variations in the complexity of the system models that may be required. If only main functions or systems are included as event-tree headings, the fault trees become more complex and must accommodate all dependencies among front-line and support functions (or systems) within the fault tree. If support functions (or systems) are explicitly included as event-tree headings, more complex event trees and less complex fault trees will result.

The following procedures should be followed as part of developing the fault tree:

1) Develop a fault tree for each event in the event tree heading.
2) Explicitly model dependencies of a system on other systems and intercomponent dependencies (e.g., common cause failure as described in Section 6.1).
3) Include all potential causes of failure, such as hardware, software, test and maintenance, and human error, in the fault tree.

### *7.4.6 Internal Events External to the Process*

Events that originate within a complex system are called internal events. Events that adversely affect the process and occur outside of the process boundaries, but within the facility, are defined as internal events external to the process. Typical internal events external to the process are internal fires, internal floods, and high-energy events within the complex system. The effects of these events should be modeled with event trees to show all possible scenarios.

### *7.4.7 External Events*

The clear counterpoint to the type of initiating event discussed in Section 7.4.6 is an initiating event that originates outside of the complex system, called an external event. Examples of external events are fires and floods that originate outside of the system, seismic events, transportation events, volcanic events, and high-wind events. Again, this classification can be used in grouping the event-tree scenarios.

### *7.4.8 Dependent Failure Considerations*

To attain the very low levels of risk, the systems and hardware that comprise the barriers to hazard exposure must have very high levels of reliability. This high reliability is typically achieved through the use of redundant and/or diverse hardware, which provides multiple success paths. The problem then becomes one of ensuring the independence of the paths, since there is always some degree of coupling between their failure mechanisms, either through the operating environment (events external to the hardware) or through functional and spatial dependencies. In Section 6.1, we elaborated on the nature and mathematics of these dependencies. Treatment of dependencies should be carefully included in both event-tree and fault-tree development and analysis in PRA. As the reliability of individual systems and subsystems increases due to redundancy, the contribution from dependent failures becomes more important; at some point, dependent failures may dominate the overall reliability. Including the effects of dependent failures in the reliability models is difficult and requires that sophisticated, fully integrated models be developed and solved to find those failure combinations that lead to mission failure. The treatment of dependent failures is not a single step performed during the PRA; it must be considered throughout the analysis (e.g., in event trees, fault trees, and human actions).

The following procedures should be followed in the dependent analysis:

1) Identify the items that are similar and could cause dependent or common cause failures. For example, similar pumps, motor-operated valves, air-operated valves, diesel generators, and batteries are major components in process plants, and are considered important sources of common cause failures.
2) Items that are potentially susceptible to common cause failure should be explicitly incorporated into the fault trees and event trees where applicable.
3) Functional dependencies should be identified and explicitly modeled in the fault trees and event trees.

*7.4.9 Failure Data Analysis*

A critical building block in assessing the reliability and availability of items in complex systems is the failure data on the performance of items. In particular, the best resources for predicting future availability of equipment are past experiences or tests. Component reliability data are inputs to system reliability studies, and the validity of the results depends highly on the quality of the input information. It must be recognized, however, that historical data have predictive value only to the extent that the conditions under which the data were generated remain applicable. Determining of the various component failure data consists essentially of the following steps: collecting generic data, assessing generic data, statistically evaluating process-specific data, and specializing the failure probability distributions using process-specific data. Three types of events identified during the accident-sequence definition and system modeling must be quantified for the event trees and fault trees to estimate the frequency of occurrence of sequences: initiating events, component failures and human errors.

The quantification of initiating events and components failure probabilities involves two separate activities. First, the reliability model for each event must be established; then the parameters of the model must be estimated. The necessary data include component failure rates, repair times, test frequencies, test downtimes, common-cause probabilities, and uncertainty characterizations. In Chapter 3 we discussed available methods for analyzing data to obtain the probability of failure or the probability of occurrence of equipment failure. In Chapter 5 we discussed analysis of data relevant to repairable systems. Finally, in Chapter 6 we discussed analysis of data for dependent failures and human reliability. The establishment of the data base to be used will generally involve the collection of some equipment or process-specific data or the use of generic data bases.

The following procedures should be followed as part of the data analysis task:

1) Determine generic values of failure rate and demand failure probability for each component identified in the fault-tree analysis. This can be obtained either from process-specific experiences or from generic sources of data (see Chapter 3.)
2) Determine test, repair, and maintenance outages primarily from experience, if available. Otherwise use generic sources.

3) Determine the frequency of initiating events and other component failure events from experience, expert judgment, or generic sources. (see Chapter 3 and 6.)
4) Determine the common cause failure probability for similar items, primarily from generic values. However, when significant specific data are available, they can be used (see Chapter 6.)

### *7.4.10 Quantification*

Fault-tree/event-tree sequences are quantified to determine the frequencies of scenarios and associated uncertainties in the calculation. The approach depends somewhat on the manner in which system dependencies have been handled. We will describe the more complex situation in which the fault trees are not independent, i.e., there are dependencies (e.g., through support systems).

Normally, the quantification will use a Boolean reduction process to arrive at a Boolean representation for each sequence. Starting with fault-tree models for the various systems or event headings in the event trees, and using probability estimates for each of the events in the fault trees, the probability of each event-tree heading is obtained (if the heading is independent of other headings). The fault trees for support systems (e.g., cooling, power.) are merged where needed with the front-line systems and converted into Boolean equation representations. The equations are solved for the minimal cut sets for each of the front-line systems (those identified as headings on the event trees). The minimal cut sets for the front-line systems are then appropriately combined to determine the cut sets for the event-tree sequences. The process is described in Chapter 4.

If all possible cut sets are retained during this process, an unmanageably large collection of terms will almost certainly result. Therefore, the collection of cut sets is truncated (i.e., insignificant members are discarded based on the number of terms in a cut set or on the probability of the cut set.) This is usually a practical necessity because of the overwhelming number of cut sets that can result from the combination of a large number of failures, even though the probability of any of these combinations may be vanishingly small. The truncation process does not disturb the effort to determine the dominant scenarios since we are discarding scenarios that are very often unlikely.

A valid concern is sometimes voiced that even though the individual discarded cut sets may be at least several orders of magnitude less probable than the average of those retained, the large number of them might represent a significant part of the risk. The actual risk might thus be considerably larger than the PRA results indicate. Detailed examination of a few PRA studies of nuclear power plants shows that truncation did not have a significant effect on the total risk result in those particular cases. The process of quantification is generally straightforward, and the methods used in its process are described in Section 4.6. More objective truncation methods are discussed by Dezfuli and Modarres (1985).

The following procedures should be followed as part of the quantification process:

1) Merge corresponding fault trees associated with each failure or success event in the event tree sequences (i.e., combine them in a Boolean form). Develop a reduced Boolean function for each sequence.
2) Calculate the total frequency of each sequence, using the frequency of initiating events, the probability of hardware failure, test and maintenance frequency (outage), common cause failure probability, and human error probability.
3) Use the minimal cut sets of each sequence for the quantification process. If needed, simplify the process by truncating based on the cut sets or probability.
4) Calculate the total frequency of each sequence.

## 7.5 A Simple Example of Risk Analysis

Consider the fire protection system shown in Fig. 7.4. This system is designed to extinguish all possible fires in a plant with toxic chemicals. Two physically independent water extinguishing nozzles are designed such that each is capable of controlling all types of fires in the plant. Extinguishing nozzle 1 is the primary method of injection. Upon receiving a signal from the detector/alarm/actuator device, pump-1 starts automatically, drawing water from the reservoir tank and injecting it into the fire area in the plant. If this pump injection path is not actuated, plant operators can start a second injection path manually. If the second path is not available, the operators will call for help from the local fire department, although the detector also sends a signal directly to the fire department. However, due to the delay in the arrival of the local fire department, the magnitude of damage would be higher than it would be if the local fire extinguishing nozzles were available to extinguish the fire. Under all conditions, if the normal off-site power is not available due to the fire or other reasons, a local generator would provide electric power to the pumps. The power to the detector/alarm/actuator system is provided through the batteries, which is constantly charged by the off-site power. Even if the ac power is not available, the dc power provided through the battery is expected to be available at all times. The manual valves on the two sides of pump-1 and pump-2 are normally open, and only remain closed when they are being repaired. The entire fire system and generator are located outside of the reactor compartment, and are therefore not affected by an internal fire. The risk-analysis process for this situation consists of the steps explained below.

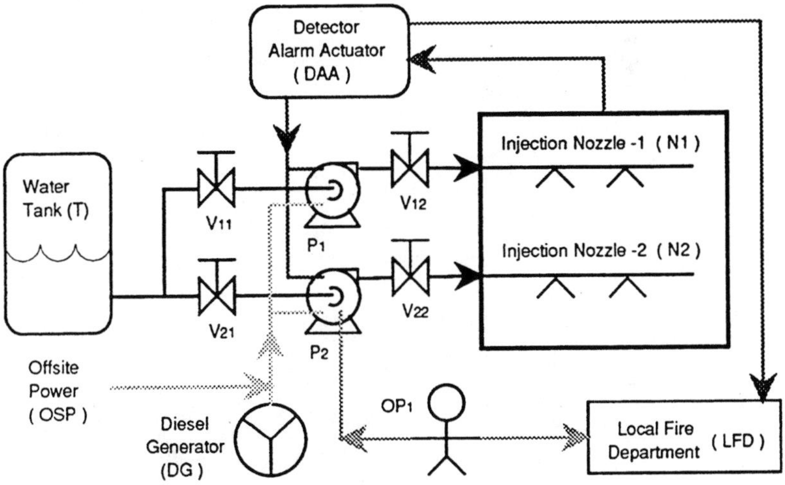


*Fig. 7.4 A Fire Protection System*

1) Identification of Initiating Events

In this step, all events that lead to or promote a fire in the reactor compartment must be identified. These should include equipment malfunctions, human errors, and process conditions. The frequency of each event should be estimated. Assuming that all events would lead to the same magnitude of fire, the ultimate initiating event is fire, the frequency of which is the summation of the frequency of the individual fire-causing events. Assume for this example that the frequency of fire is estimated at $1 \times 10^{-6}$/year. Since fire is the only challenge to the plant in this example, we end up with only one initiating event. However, in more complex situations, a large set of initiating events can be identified, each posing a different challenge to the plant.

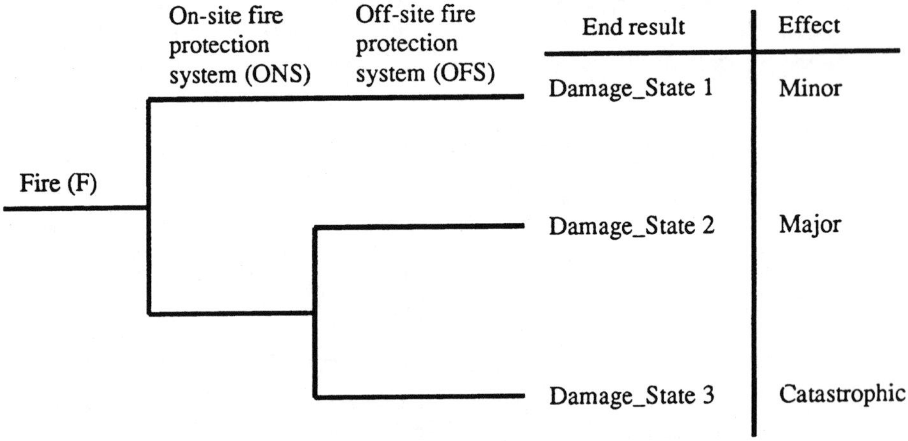


*Fig. 7.5 Scenario of Events Following a Fire Using the Event-Tree Method*

2) Scenario Development

In this step, we should explain the cause and effect relationship between the fire and the progression of events following the fire. We will use the event-tree method to depict this relationship. Generally, this is done inductively, and the level of detail considered in the event tree is somewhat dependent on the analyst. Two protective measures have been considered in the event tree shown in Fig. 7.5: on-site protective measures (on-site pumps, tanks, etc.), and off-site protective fire department measures. The selection of these measures is based on the fact that availability or unavailability of the on-site or off-site protective measures would lead to different plant damage states.

3) System Analysis

In this step, we should identify all failures (equipment or human) that lead to failure of the event-tree headings (on-site or off-site protective measures).

For example, Fig. 7.6 shows the fault tree developed for the On-Site Fire Protection System. In this fault tree, all basic events that lead to the failure of the two independent paths are described. Note that MAA, electric power to the pumps, and the water tank are shared by the two paths. Clearly these are considered physical dependencies. This is taken into account in the quantification step of the risk analysis. In this tree, all external event failures and passive failures are neglected.

Fig. 7.7 shows the fault tree for the Off-Site Fire Protection System. This tree is simple since it only includes all failures that does not lead to an on-time response from the local fire department.

It is also possible to use the MLD for system analysis. An example of the MLD for this problem is shown in Fig. 7.8. However, only the fault trees are used for risk analysis, although MLD can also be used.

4) Failure Data Analysis

It is important at this point to calculate the probability of the basic failure events described in the event trees and fault trees. As indicated earlier, this can be done by using either plant-specific data, generic data, or expert judgment. Table 7.5 describes the data used and their source. In this table, it is assumed that at least 10 hours of operation is needed for the fire to be completely extinguished.

5) Quantification

To calculate the frequency of each scenario defined in Fig. 7.5, we must first determine the cut sets of the two fault trees shown in Fig. 7.6 and Fig. 7.7. From this, the cut sets of each scenario are determined, followed by calculation of the probability of each scenario based on the occurrence of one of its cut sets. These steps are described below.

*i*– The cut sets of the On-Site Fire Protection System Failure are obtained using the technique described in Section 4.2. These cut sets are listed in Table 7.6. Only cut set number 22, which is failure of both pumps is subject to a common cause failure. This is shown by adding a new cut set (cut set number 24), which represents this common cause failure.

*ii*– The cut sets of the Off-Site Fire Protection System Failure are similarly obtained and listed in Table 7.7.

*iii*– The cut sets of the three scenarios are obtained using the following Boolean equations representing each scenario:

$$Scenario - 1 = F \bullet \overline{ONS},$$

$$Scenario - 2 = F \bullet ONS \bullet \overline{OFS},$$

$$Scenario - 3 = F \bullet ONS \bullet OFS.$$

The process is described in Section 4.3.2.

*iv*– The frequency of each scenario is obtained using data listed in Table 7.5. These frequencies are shown in Table 7.8.

*v*– The total frequency of each scenario is calculated using the rare event approximation. These are also shown in Table 7.8.

6) Consequences

In the scenario development and quantification tasks, we identified three distinct scenarios of interest, each with different outcomes and frequencies. The consequences associated with each scenario should be specified in terms of both economic and/or human losses. This part of the analysis is one of the most difficult for several reasons.

1) Each scenario poses different hazards and methods of hazard exposure, and require careful monitoring. In this case, the model should include how the fire can spread through the plant, how people can be exposed, evacuation procedures, the availability of protective clothing, etc.
2) The outcome of the scenario can be measured in terms of human losses. It can also be measured in terms of financial losses, i.e., the total cost associated with the scenario. This involves assigning a dollar value to human life or casualties, which is a source of controversy.

Suppose a careful analysis is performed of the spread of fire and fire exposure, with consideration of the above issues, and ultimately results in damages measured only in terms of economic losses. These results are shown in Table 7.9.

The low value of dollars at risk indicates that fire risk is not important in this plant. However, Scenarios 1 and 2 are significantly more important than Scenario 3. Therefore, if the risk were high, one should improve those components or events that are major contributors to Scenario 1 and 2. Scenario 1 is primarily due to common cause failure between pumps P1 and P2, so reducing this failure is a potential source of improvement.

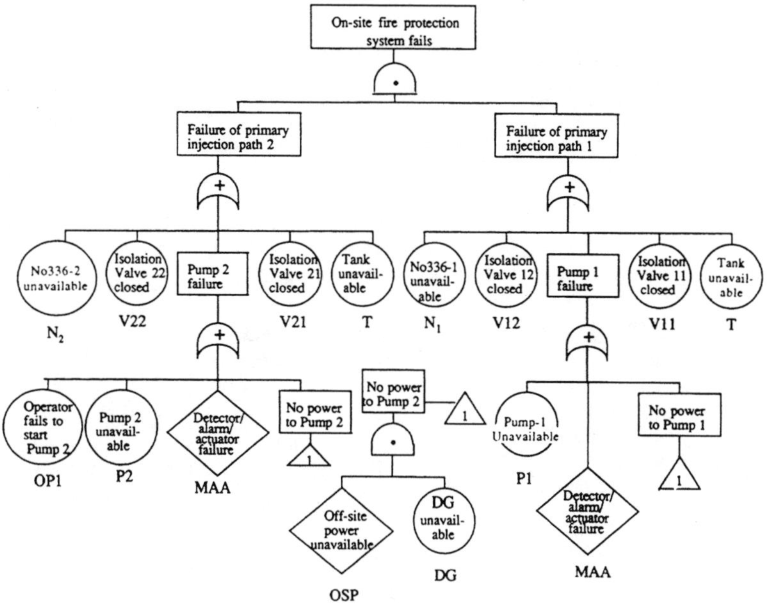


*Fig. 7.6 Fault Tree for On-Site Fire Protection System Failure*

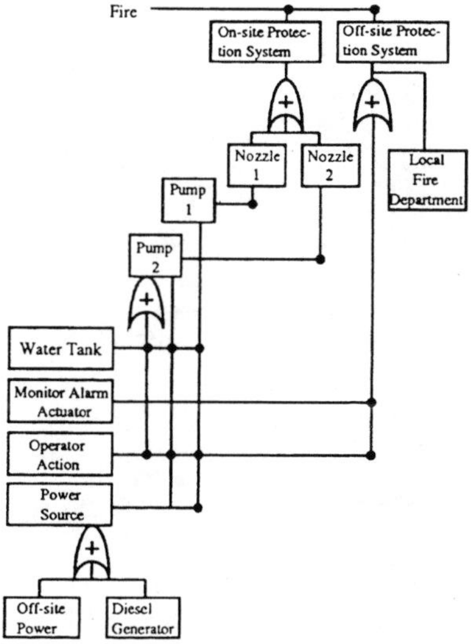


*Fig. 7.7 Fault Tree for Off-Site Fire Protection System Failure*

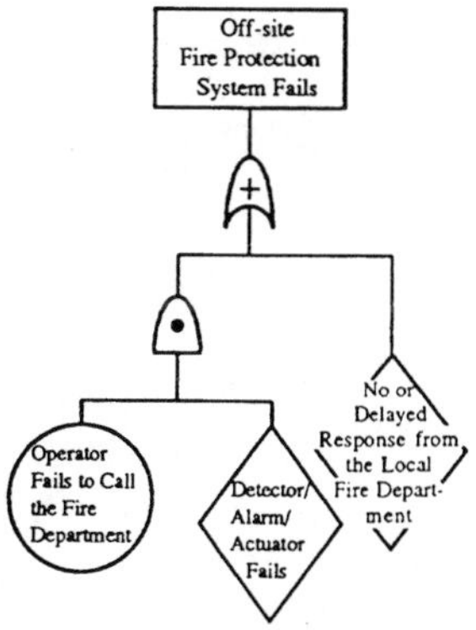


*Fig. 7.8 MLD for the Fire Protection System*

*Table 7.5 Sources of Data and Failure Probabilities*

| Failure Event | Plant-Specific Experience | Generic Data | Probability Used | Comments |
|---|---|---|---|---|
| Fire initiation frequency | No such experience in 10 years of operation. | 5 fires in similar plants. There are 70,000 plant-years of experience. | $F = 5/70{,}000 = 7.1\times10^{-4}$/yr. | Use generic data. |
| Pump 1 and Pump 2 failure | 4 failure of two pumps to start. Monthly tests are performed which takes negligible time. Repair time takes about 10 hours at a frequency of 1 per year. No experience of failure to run. | Failure to run = $1\times10^{-5}$/hr. | $\frac{4}{2\times12\times10} = 1.7\times10^{-2}$ /demand. Unavailability = $1.7\times10^{-2} + \frac{10}{8760} = 1.8\times10^{-2}$. Failure to run $= 1\times10^{-5}$/hr. $P1 = P2 = 1.7\times10^{-2} + 1\times10^{-5}\times10 \approx 1.7\times10^{-2}$. | For failure to start, use plant-specific data. For failure to run, use generic data. If possible use Bayesian updating technique described in Section 3.6. Assume 10 years of experience and 8,760 hours in one year. |
| Common cause failure between Pump 1 and Pump 2 | No such experience. | Using the $\beta$ factor method, $\beta=0.1$ for failure of pumps to start. | Using (6.x) unavailability due to common cause failure: $CCF = 0.1\times1.8\times10^{-2} = 1.8\times10^{-3}$. | Assume no significant common cause failure exists between valves and nozzles. See Section 6.1 for more detail. |
| Failure of isolation valves | 1 failure to leave the valve in open position following a pump test. | Not used. | $V_{11} = V_{12} = V_{21} = V_{22} = \frac{1}{10\times12\times2} = 4.2\times10^{-3}$. | Plant-specific data used. |
| Failure of nozzles | No such experience. | $1\times10^{-5}$/demand | $N_1 = N_2 = 1\times10^{-5}$ /demand. | Generic data used. |
| Diesel generator failure | 3 failures in monthly tests. 40 hours of repair per year. | $3\times10^{-2}$/demand $3\times10^{-3}$/hr. 40 run | $3/(12\times10) = 2.5\times10^{-2}$/demand $3\times10^{-3}$/hr. $DG = 2.5\times10^{-2} + 3\times10^{-3}\times10 = 5.5\times10^{-2}$. | Plant-specific data used for demand failure. Assume 10 years of experience. Generic to run. |
| Loss of off-site power | No experience. | 0.1/yr. | $OSP = 0.1\times\frac{10}{8760} = 1.1\times10^{-4}$/demand. | Assume 104 hours of operation for fire extinguisher and use generic data. |
| Failure of MAA | No experience. | No data available. | $MAA = 1\times10^{-4}$. | This estimate is based on expert judgment. See Section 6.3 for the methods. |

*Table 7.5 Continued*

| Failure Event | Plant-Specific Experience | Generic Data | Probability Used | Comments |
|---|---|---|---|---|
| Failure of operator to start Pump 2 | No such experience. | Using the THERP method for tasks of this kind, $1\times10^{-2}$ is suggested. | $OP1=1\times10^{-2}$ | Use the THERP Handbook data discussed in Section 6.3. |
| Failure of operator to call the fire department | No such experience. | $1\times10^{-3}$ | $OP2=1\times10^{-3}$ | This is based on experience from no response to similar situations. Generic probability is used. |
| No or delayed response from fire department | No such experience. | $1\times10^{-4}$ | $LFD=1\times10^{-4}$ | This is based on response to similar cases from the fire department. Delayed/no arrival is due to accidents, traffic, communication problems, etc. |
| Tank failure | No such experience. | $1\times10^{-5}$ | $T=1\times10^{-5}$ | This is based on data obtained from rupture of the tank or insufficient water content.. |

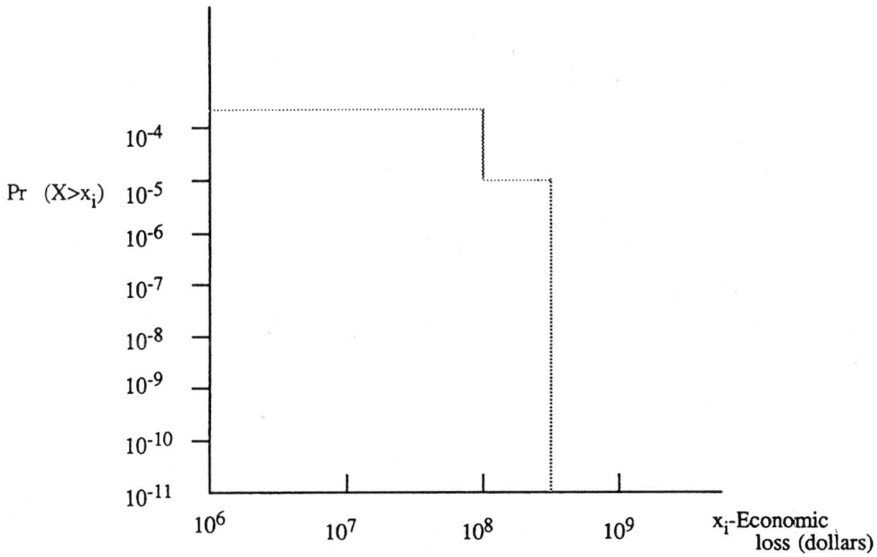


*Fig. 7.9 Risk Profile*

*Table 7.6 Cut Sets of the On-Site Fire Protection System Failure*

| Cut Set No. | Cut Set | Probability/ (% of total) | Cut Set No. | Cut Set | Probability |
|---|---|---|---|---|---|
| 1 | T | $1x10^{-5}$(0.35) | 14 | $V_{21} \cdot P1$ | $7.1x10^{-5}$(2.5) |
| 2 | MAA | $1x10^{-4}$(3.5) | 15 | $V_{21} \cdot V_{11}$ | $1.8x10^{-5}$(0.64) |
| 3 | OSP • DG | $6x10^{-6}$(0.21) | 16 | $OP1 \cdot N_1$ | $1x10^{-7}$(no) |
| 4 | $N_2 \cdot N_1$ | $1x10^{-10}$(0) | 17 | $OP1 \cdot V_{12}$ | $4.2x10^{-5}$(1.5) |
| 5 | $N_2 \cdot V_{12}$ | $4.2x10^{-8}$(~0) | 18 | OP1 • P1 | $1.7x10^{-4}$(6.0) |
| 6 | $N_2 \cdot P1$ | $1.7x10^{-7}$(~0) | 19 | $OP1 \cdot V_{11}$ | $4.2x10^{-5}$(1.5) |
| 7 | $N_2 \cdot V_{11}$ | $4.2x10^{-8}$(~0) | 20 | $P2 \cdot N_1$ | $1.7x10^{-7}$(~0) |
| 8 | $V_{22} \cdot N_1$ | $4.2x10^{-8}$(~0) | 21 | $P2 \cdot V_{12}$ | $7.1x10^{-5}$(2.5) |
| 9 | $V_{22} \cdot V_{12}$ | $1.8x10^{-5}$(0.64) | 22 | P2 • P1 | $2.9x10^{-4}$(0.3) |
| 10 | $V_{22} \cdot P1$ | $7.1x10^{-5}$(2.5) | 23 | $P2 \cdot V_{11}$ | $7.1x10^{-5}$(2.5) |
| 11 | $V_{22} \cdot V_{11}$ | $1.8x10^{-5}$(0.64) | 24 | CCF | $1.8x10^{-3}$(63.8) |
| 12 | $V_{21} \cdot N_1$ | $4.2x10^{-8}$(~0) | | | $Pr(ON) \approx \sum_i C_i$ |
| 13 | $V_{21} \cdot V_{12}$ | $1.8x10^{-5}$(0.35) | | | $=2.8x10^{-3}$ |

*Table 7.7 Cut Sets of the Off-Site Fire Protection System*

| Cut Set No. | Cut Set | Probability |
|---|---|---|
| 1 | LFD | $1 \times 10^{-4}$ |
| 2 | OP2 • MAA | $1 \times 10^{-7}$ |
| | | $Pr(OFF) \simeq 1x10^{-4}$ |

*Table 7.8 Cut Sets of the Scenarios*

| Scenario No. | Cut Sets | Frequency | Comment |
|---|---|---|---|
| 1 | $F \cdot \overline{ON}$ | $7.1\times10^{-4} \cdot (1-2.8\times10^{-2})$ <br> $= 7.1\times10^{-4}$ | Since the probability can be directly evaluated for $\overline{ON}$ without evaluating the need to generate cut sets, only the probability is calculated. |
| 2 | $F \cdot MAA \cdot LFD \cdot OP_2$ | $7.0 \times 10^{-8}$ | 1. Only cut sets from Table 7.6 that have a contribution greater than 1% are shown. <br> 2. Cut set $F \cdot MAA \cdot \overline{LFD} \cdot \overline{MAA}$ is eliminated since $MAA \cdot \overline{MAA} = \phi$. |
| | $F \cdot V_{22} \cdot P_1 \cdot \overline{LFD} \cdot \overline{OP_2}$ | $5.0 \times 10^{-8}$ | |
| | $F \cdot V_{22} \cdot P_1 \cdot \overline{LFD} \cdot \overline{MAA}$ | $5.0 \times 10^{-8}$ | |
| | $F \cdot V_{21} \cdot P_1 \cdot \overline{LFD} \cdot \overline{OP_2}$ | $5.0 \times 10^{-8}$ | |
| | $F \cdot V_{22} \cdot P_1 \cdot \overline{LFD} \cdot \overline{MAA}$ | $5.0 \times 10^{-8}$ | |
| | $F \cdot OP_1 \cdot V_{12} \cdot \overline{LFD} \cdot \overline{OP_2}$ | $2.9 \times 10^{-9}$ | |
| | $F \cdot OP_1 \cdot V_{12} \cdot \overline{LFD} \cdot \overline{MAA}$ | $2.9 \times 10^{-9}$ | |
| | $F \cdot OP_1 \cdot P_1 \cdot \overline{LFD} \cdot \overline{OP_2}$ | $1.1 \times 10^{-7}$ | |
| | $F \cdot OP_1 \cdot P_1 \cdot \overline{LFD} \cdot \overline{MAA}$ | $1.1 \times 10^{-7}$ | |
| | $F \cdot OP_1 \cdot V_{11} \cdot \overline{LFD} \cdot \overline{OP_2}$ | $2.9 \times 10^{-9}$ | |
| | $F \cdot OP_1 \cdot V_{11} \cdot \overline{LFD} \cdot \overline{MAA}$ | $2.9 \times 10^{-9}$ | |
| | $F \cdot P_2 \cdot V_{12} \cdot \overline{LFD} \cdot \overline{OP_2}$ | $5.0 \times 10^{-8}$ | |
| | $F \cdot P_2 \cdot V_{12} \cdot \overline{LFD} \cdot \overline{MAA}$ | $5.0 \times 10^{-8}$ | |
| | $F \cdot P_2 \cdot P_1 \cdot \overline{LFD} \cdot \overline{OP_2}$ | $2.0 \times 10^{-7}$ | |
| | $F \cdot P_2 \cdot P_1 \cdot \overline{LFD} \cdot \overline{MAA}$ | $2.0 \times 10^{-7}$ | |
| | $F \cdot P_2 \cdot V_{11} \cdot \overline{LFD} \cdot \overline{OP_2}$ | $5.0 \times 10^{-8}$ | |
| | $F \cdot P_2 \cdot V_{11} \cdot \overline{LFD} \cdot \overline{MAA}$ | $5.0 \times 10^{-8}$ | |
| | $F \cdot CCP \cdot \overline{LFD} \cdot \overline{OP_2}$ | $1.3 \times 10^{-6}$ | |
| | $F \cdot CCP \cdot \overline{LFD} \cdot \overline{MAA}$ | $1.3 \times 10^{-6}$ | |
| | | $\sum_i = 1.3 \times 10^{-6}$ | |
| 3 | $F \cdot MAA \cdot LFD$ | $7.1 \times 10^{-12}$ | 1. Only cut sets from Tables 7.6 and 7.7 that have the highest contribution to the scenario are shown. |
| | $F \cdot V_{22} \cdot P_1 \cdot LFD$ | $5.0 \times 10^{-12}$ | |
| | $F \cdot V_{21} \cdot P_1 \cdot LFD$ | $5.0 \times 10^{-12}$ | |
| | $F \cdot OP_1 \cdot V_{12} \cdot LFD$ | $2.9 \times 10^{-12}$ | |
| | $F \cdot OP_1 \cdot P_1 \cdot LFD$ | $2.8 \times 10^{-12}$ | |
| | $F \cdot OP_1 \cdot V_{11} \cdot LFD$ | $2.9 \times 10^{-12}$ | |
| | $F \cdot P_2 \cdot V_{12} \cdot LFD$ | $5.0 \times 10^{-12}$ | |
| | $F \cdot P_2 \cdot P_1 \cdot LFD$ | $2.0 \times 10^{-11}$ | |
| | $F \cdot P_2 \cdot V_{11} \cdot LFD$ | $5.0 \times 10^{-12}$ | |
| | $F \cdot CCP \cdot LFD$ | $3.0 \times 10^{-11}$ | |
| | | $\sum_i = 8.4 \times 10^{-11}$ | |

*Table 7.9 Economic Consequences of Fire Scenarios*

| Scenario Number | Economic Consequence |
|---|---|
| 1 | \$1,000,000 |
| 2 | \$92,000,000 |
| 3 | \$210,000,000 |

7) Risk Calculation and Evaluation

Using values from Table 7.9, we can calculate the risk associated with each scenario. These risks are shown in Table 7.10.

*Table 7.10 Risk Associated With Each Scenario*

| Scenario Number | Scenario Risk |
|---|---|
| 1 | $7.1 \times 10^{-4} \times \$1,000,000 = \$710$ |
| 2 | $3.7 \times 10^{-6} \times \$92,000,000 = \$340$ |
| 3 | $8.4 \times 10^{-11} \times \$210,000,000 = \$0.017$ |

Since this analysis shows that risk due to fire is rather low, uncertainty analysis is not very important. However, one of the methods described in Section 6.2 could be used to estimate the uncertainty associated with each component and the fire-initiating event if necessary. The uncertainties should be propagated through the cut sets of each scenario to obtain the uncertainty associated with the frequency estimation of each scenario. The uncertainty associated with the consequence estimates can also be obtained. When uncertainty associated with the consequence values are combined with the scenario frequencies and their uncertainty, the uncertainty associated with the estimated risk can be calculated. Although this is not a necessary step in risk analysis, it is reasonable to make an estimate of the uncertainties when risk values are high.

Figure 7.9 shows the risk profile based on the values in Table 7.10.

## BIBLIOGRAPHY

1. Dezfuli, H. and M. Modarres (1984). "A Truncation Methodology for Evaluation of Large Fault Trees," *IEEE Transactions on Reliability*, vol. R-33, 4, p.p.325-328.

2. Farmer, F. (1967). "Containment and Siting of Nuclear Power Plants," Proc. of a Symp. on Contain. and Siting of Nucl. Power Plants, Int. Atomic Energy Org., Vienna, Austria.
3. Litai, D. (1980). "A Risk Comparison Methodology for the Assessment of Acceptable Risk," Ph.D Thesis, Dept. of Nucl. Eng., Mass. Inst. Tech., Cambridge, MA.
4. NUREG/CR-4550, (1990). "Analysis of Core Damage Frequency From Internal Events," vol.1, U.S. Nuclear Regulatory Commission, Washington, D.C.
5. Reactor Safety Study (1975). "Reactor Safety Study - An Assessment of Accident Risks in U.S. Commercial Nuclear Power Plants," WASH-1400, U.S. Nuclear Regulatory Commission, Washington, D.C.
6. Rowe, W.D. (1977). *An Anatomy of Risk*, Weily, New York.
7. Paulos, J.A. (1991). Temple University Report, Philadelphia.
8. U.S. Nuclear Regulatory Commission (1986). "Safety Goals for The Operation of Nuclear Power Plants: Policy Statement," Fed, Regist., 51 (149), Washington, D.C.
9. Wilson, R. (1979). "Analyzing the Daily Risks of Life", *Technology Review*, vol. 81, No.4, p.p.41-46, Cambridge, MA.

# Appendix A: Statistical Tables

*Table A.1 Cumulative Standard Normal Distribution* *

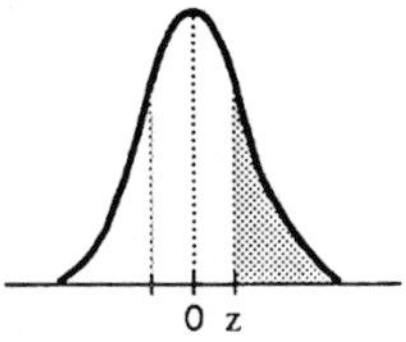


| z | 0.00 | 0.01 | 0.02 | 0.03 | 0.04 | 0.05 | 0.06 | 0.07 | 0.08 | 0.09 |
|---|---|---|---|---|---|---|---|---|---|---|
| 0.0 | 0.5000 | 0.4960 | 0.4920 | 0.4880 | 0.4840 | 0.4801 | 0.4761 | 0.4721 | 0.4681 | 0.4641 |
| 0.1 | 0.4602 | 0.4562 | 0.4522 | 0.4483 | 0.4443 | 0.4404 | 0.4364 | 0.4325 | 0.4286 | 0.4247 |
| 0.2 | 0.4207 | 0.4168 | 0.4129 | 0.4090 | 0.4052 | 0.4013 | 0.3974 | 0.3936 | 0.3897 | 0.3859 |
| 0.3 | 0.3821 | 0.3783 | 0.3745 | 0.3707 | 0.3669 | 0.3632 | 0.3594 | 0.3557 | 0.3520 | 0.3483 |
| 0.4 | 0.3446 | 0.3409 | 0.3372 | 0.3336 | 0.3300 | 0.3264 | 0.3228 | 0.3192 | 0.3156 | 0.3121 |
| 0.5 | 0.3085 | 0.3050 | 0.3015 | 0.2981 | 0.2946 | 0.2912 | 0.2877 | 0.2843 | 0.2810 | 0.2776 |
| 0.6 | 0.2743 | 0.2709 | 0.2676 | 0.2643 | 0.2611 | 0.2578 | 0.2546 | 0.2514 | 0.2483 | 0.2451 |
| 0.7 | 0.2420 | 0.2389 | 0.2358 | 0.2327 | 0.2296 | 0.2266 | 0.2236 | 0.2206 | 0.2177 | 0.2148 |
| 0.8 | 0.2119 | 0.2090 | 0.2061 | 0.2033 | 0.2005 | 0.1977 | 0.1949 | 0.1922 | 0.1894 | 0.1867 |
| 0.9 | 0.1841 | 0.1814 | 0.1788 | 0.1762 | 0.1736 | 0.1711 | 0.1685 | 0.1660 | 0.1635 | 0.1611 |
| 1.0 | 0.1587 | 0.1562 | 0.1539 | 0.1515 | 0.1492 | 0.1469 | 0.1446 | 0.1423 | 0.1401 | 0.1379 |
| 1.1 | 0.1357 | 0.1335 | 0.1314 | 0.1292 | 0.1271 | 0.1251 | 0.1230 | 0.1210 | 0.1190 | 0.1170 |
| 1.2 | 0.1151 | 0.1131 | 0.1112 | 0.1093 | 0.1075 | 0.1056 | 0.1038 | 0.1020 | 0.1003 | 0.0985 |
| 1.3 | 0.0968 | 0.0951 | 0.0934 | 0.0918 | 0.0901 | 0.0885 | 0.0869 | 0.0853 | 0.0838 | 0.0823 |
| 1.4 | 0.0808 | 0.0793 | 0.0778 | 0.0764 | 0.0749 | 0.0735 | 0.0721 | 0.0708 | 0.0694 | 0.0681 |
| 1.5 | 0.0668 | 0.0655 | 0.0643 | 0.0630 | 0.0618 | 0.0606 | 0.0594 | 0.0582 | 0.0571 | 0.0559 |
| 1.6 | 0.0548 | 0.0537 | 0.0526 | 0.0516 | 0.0505 | 0.0495 | 0.0485 | 0.0475 | 0.0465 | 0.0455 |
| 1.7 | 0.0446 | 0.0436 | 0.0427 | 0.0418 | 0.0409 | 0.0401 | 0.0392 | 0.0384 | 0.0375 | 0.0367 |
| 1.8 | 0.0359 | 0.0351 | 0.0344 | 0.0336 | 0.0329 | 0.0322 | 0.0314 | 0.0307 | 0.0301 | 0.0294 |
| 1.9 | 0.0287 | 0.0281 | 0.0274 | 0.0268 | 0.0262 | 0.0256 | 0.0250 | 0.0244 | 0.0239 | 0.0233 |
| 2.0 | 0.0228 | 0.0222 | 0.0217 | 0.0212 | 0.0207 | 0.0202 | 0.0197 | 0.0192 | 0.0188 | 0.0183 |
| 2.1 | 0.0179 | 0.0174 | 0.0170 | 0.0166 | 0.0162 | 0.0158 | 0.0154 | 0.0150 | 0.0146 | 0.0143 |
| 2.2 | 0.0139 | 0.0136 | 0.0132 | 0.0129 | 0.0125 | 0.0122 | 0.0119 | 0.0116 | 0.0113 | 0.0110 |
| 2.3 | 0.0107 | 0.0104 | 0.0102 | 0.0099 | 0.0096 | 0.0094 | 0.0091 | 0.0089 | 0.0087 | 0.0084 |
| 2.4 | 0.0082 | 0.0080 | 0.0078 | 0.0075 | 0.0073 | 0.0071 | 0.0069 | 0.0068 | 0.0066 | 0.0064 |
| 2.5 | 0.0062 | 0.0060 | 0.0059 | 0.0057 | 0.0055 | 0.0054 | 0.0052 | 0.0051 | 0 0049 | 0.0048 |
| 2.6 | 0.0047 | 0.0045 | 0.0044 | 0.0043 | 0.0041 | 0.0040 | 0.0039 | 0.0038 | 0.0037 | 0.0036 |
| 2.7 | 0.0035 | 0.0034 | 0.0033 | 0.0032 | 0.0031 | 0.0030 | 0.0029 | 0.0028 | 0.0027 | 0.0026 |
| 2.8 | 0.0026 | 0.0025 | 0.0024 | 0.0023 | 0.0023 | 0.0022 | 0.0021 | 0.0021 | 0.0020 | 0.0019 |
| 2.9 | 0.0019 | 0.0018 | 0.0018 | 0.0017 | 0.0016 | 0.0016 | 0.0015 | 0.0015 | 0.0014 | 0.0014 |
| 3.0 | 0.0013 | 0.0013 | 0.0013 | 0.0012 | 0.0012 | 0.0011 | 0.0011 | 0.0011 | 0.0010 | 0.0010 |
| 3.1 | 0.0010 | 0.0009 | 0.0009 | 0.0009 | 0.0008 | 0.0008 | 0.0008 | 0.0008 | 0.0007 | 0.0007 |
| 3.2 | 0.0007 | 0.0007 | 0.0006 | 0.0006 | 0.0006 | 0.0006 | 0.0006 | 0.0005 | 0.0005 | 0.0005 |
| 3.3 | 0.0005 | 0.0005 | 0.0005 | 0.0004 | 0.0004 | 0.0004 | 0.0004 | 0.0004 | 0.0004 | 0.0003 |
| 3.4 | 0.0003 | 0.0003 | 0.0003 | 0.0003 | 0.0003 | 0.0003 | 0.0003 | 0.0003 | 0.0003 | 0.0002 |
| 3.5 | 0.0002 | 0.0002 | 0.0002 | 0.0002 | 0.0002 | 0.0002 | 0.0002 | 0.0002 | 0.0002 | 0.0002 |

* Adapted from Table 1 of Pearson, E. S., and H. O. Hartley, Eds.: *Biometrika Tables for Statisticians,* Vol. 1, 3d ed. Cambridge Univ. Press, Cambridge, U.K., 1966. Used by permission.

*Table A.2 Percentiles of the t Distribution* *

| df | $t_{.60}$ | $t_{.70}$ | $t_{.80}$ | $t_{.90}$ | $t_{.95}$ | $t_{.975}$ | $t_{.99}$ | $t_{.995}$ |
|---|---|---|---|---|---|---|---|---|
| 1 | .325 | .727 | 1.376 | 3.078 | 6.314 | 12.706 | 31.821 | 63.657 |
| 2 | .289 | .617 | 1.061 | 1.886 | 2.920 | 4.303 | 6.965 | 9.925 |
| 3 | .277 | .584 | .978 | 1.638 | 2.353 | 3.182 | 4.541 | 5.841 |
| 4 | .271 | .569 | .941 | 1.533 | 2.132 | 2.776 | 3.747 | 4.604 |
| 5 | .267 | .559 | .920 | 1.476 | 2.015 | 2.571 | 3.365 | 4.032 |
| 6 | .265 | .553 | .906 | 1.440 | 1.943 | 2.447 | 3.143 | 3.707 |
| 7 | .263 | .549 | .896 | 1.415 | 1.895 | 2.365 | 2.998 | 3.499 |
| 8 | .262 | .546 | .889 | 1.397 | 1.860 | 2.306 | 2.896 | 3.355 |
| 9 | .261 | .543 | .883 | 1.383 | 1.833 | 2.262 | 2.821 | 3.250 |
| 10 | .260 | .542 | .879 | 1.372 | 1.812 | 2.228 | 2.764 | 3.169 |
| 11 | .260 | .540 | .876 | 1.363 | 1.796 | 2.201 | 2.718 | 3.106 |
| 12 | .259 | .539 | .873 | 1.356 | 1.782 | 2.179 | 2.681 | 3.055 |
| 13 | .259 | .538 | .870 | 1.350 | 1.771 | 2.160 | 2.650 | 3.012 |
| 14 | .258 | .537 | .868 | 1.345 | 1.761 | 2.145 | 2.624 | 2.977 |
| 15 | .258 | .536 | .866 | 1.341 | 1.753 | 2.131 | 2.602 | 2.947 |
| 16 | .258 | .535 | .865 | 1.337 | 1.746 | 2.120 | 2.583 | 2.921 |
| 17 | .257 | .534 | .863 | 1.333 | 1.740 | 2.110 | 2.567 | 2.898 |
| 18 | .257 | .534 | .862 | 1.330 | 1.734 | 2.101 | 2.552 | 2.878 |
| 19 | .257 | .533 | .861 | 1.328 | 1.729 | 2.093 | 2.539 | 2.861 |
| 20 | .257 | .533 | .860 | 1.325 | 1.725 | 2.086 | 2.528 | 2.845 |
| 21 | .257 | .532 | .859 | 1.323 | 1.721 | 2.080 | 2.518 | 2.831 |
| 22 | .256 | .532 | .858 | 1.321 | 1.717 | 2.074 | 2.508 | 2.819 |
| 23 | .256 | .532 | .858 | 1.319 | 1.714 | 2.069 | 2.500 | 2.807 |
| 24 | .256 | .531 | .857 | 1.318 | 1.711 | 2.064 | 2.492 | 2.797 |
| 25 | .256 | .531 | .856 | 1.316 | 1.708 | 2.060 | 2.485 | 2.787 |
| 26 | .256 | .531 | .856 | 1.315 | 1.706 | 2.056 | 2.479 | 2.779 |
| 27 | .256 | .531 | .855 | 1.314 | 1.703 | 2.052 | 2.473 | 2.771 |
| 28 | .256 | .530 | .855 | 1.313 | 1.701 | 2.048 | 2.467 | 2.763 |
| 29 | .256 | .530 | .854 | 1.311 | 1.699 | 2.045 | 2.462 | 2.756 |
| 30 | .256 | .530 | .854 | 1.310 | 1.697 | 2.042 | 2.457 | 2.750 |
| 40 | .255 | .529 | .851 | 1.303 | 1.684 | 2.021 | 2.423 | 2.704 |
| 60 | .254 | .527 | .848 | 1.296 | 1.671 | 2.000 | 2.390 | 2.660 |
| 120 | .254 | .526 | .845 | 1.289 | 1.658 | 1.980 | 2.358 | 2.617 |
| ∞ | .253 | .524 | .842 | 1.282 | 1.645 | 1.960 | 2.326 | 2.576 |
| df | $-t_{.40}$ | $-t_{.30}$ | $-t_{.20}$ | $-t_{.10}$ | $-t_{.05}$ | $-t_{.025}$ | $-t_{.01}$ | $-t_{.005}$ |

When the table is read from the foot, the tabled values are to be prefixed with a negative sign. Interpolation should be performed using the reciprocals of the degrees of freedom.

* The data of this table are taken from Table III of Fischer and Yates: *Statistical Tables for Biological, Agricultural and Medical Research*, published by Longman Group U.K., Ltd., London (previously published by Oliver & Boyd, Ltd., Edinburgh and by permission of the author and publishers. From *Introduction to Statistical Analysis*, 2d ed., by W. J. Dixon and F. J. Massey, Jr. Copyright, 1957. McGraw-Hill Book Company. ) Used by permission.

# *Table A.3 Percentiles of the $\chi^2$ Distribution* *

| df | Per Cent | | | | | | | | | |
|---|---|---|---|---|---|---|---|---|---|---|
| | .5 | 1 | 2.5 | 5 | 10 | 90 | 95 | 97.5 | 99 | 99.5 |
| 1 | .000039 | .00016 | .00098 | .0039 | .0158 | 2.71 | 3.84 | 5.02 | 6.63 | 7.88 |
| 2 | .0100 | .0201 | .0506 | .1026 | .2107 | 4.61 | 5.99 | 7.38 | 9.21 | 10.60 |
| 3 | .0717 | .115 | .216 | .352 | .584 | 6.25 | 7.81 | 9.35 | 11.34 | 12.84 |
| 4 | .207 | .297 | .484 | .711 | 1.064 | 7.78 | 9.49 | 11.14 | 13.28 | 14.86 |
| 5 | .412 | .554 | .831 | 1.15 | 1.61 | 9.24 | 11.07 | 12.83 | 15.09 | 16.75 |
| 6 | .676 | .872 | 1.24 | 1.64 | 2.20 | 10.64 | 12.59 | 14.45 | 16.81 | 18.55 |
| 7 | .989 | 1.24 | 1.69 | 2.17 | 2.83 | 12.02 | 14.07 | 16.01 | 18.48 | 20.28 |
| 8 | 1.34 | 1.65 | 2.18 | 2.73 | 3.49 | 13.36 | 15.51 | 17.53 | 20.09 | 21.96 |
| 9 | 1.73 | 2.09 | 2.70 | 3.33 | 4.17 | 14.68 | 16.92 | 19.02 | 21.67 | 23.59 |
| 10 | 2.16 | 2.56 | 3.25 | 3.94 | 4.87 | 15.99 | 18.31 | 20.48 | 23.21 | 25.19 |
| 11 | 2.60 | 3.05 | 3.82 | 4.57 | 5.58 | 17.28 | 19.68 | 21.92 | 24.73 | 26.76 |
| 12 | 3.07 | 3.57 | 4.40 | 5.23 | 6.30 | 18.55 | 21.03 | 23.34 | 26.22 | 28.30 |
| 13 | 3.57 | 4.11 | 5.01 | 5.89 | 7.04 | 19.81 | 22.36 | 24.74 | 27.69 | 29.82 |
| 14 | 4.07 | 4.66 | 5.63 | 6.57 | 7.79 | 21.06 | 23.68 | 26.12 | 29.14 | 31.32 |
| 15 | 4.60 | 5.23 | 6.26 | 7.26 | 8.55 | 22.31 | 25.00 | 27.49 | 30.58 | 32.80 |
| 16 | 5.14 | 5.81 | 6.91 | 7.96 | 9.31 | 23.54 | 26.30 | 28.85 | 32.00 | 34.27 |
| 18 | 6.26 | 7.01 | 8.23 | 9.39 | 10.86 | 25.99 | 28.87 | 31.53 | 34.81 | 37.16 |
| 20 | 7.43 | 8.26 | 9.59 | 10.85 | 12.44 | 28.41 | 31.41 | 34.17 | 37.57 | 40.00 |
| 24 | 9.89 | 10.86 | 12.40 | 13.85 | 15.66 | 33.20 | 36.42 | 39.36 | 42.98 | 45.56 |
| 30 | 13.79 | 14.95 | 16.79 | 18.49 | 20.60 | 40.26 | 43.77 | 46.98 | 50.89 | 53.67 |
| 40 | 20.71 | 22.16 | 24.43 | 26.51 | 29.05 | 51.81 | 55.76 | 59.34 | 63.69 | 66.77 |
| 60 | 35.53 | 37.48 | 40.48 | 43.19 | 46.46 | 74.40 | 79.08 | 83.30 | 88.38 | 91.95 |
| 120 | 83.85 | 86.92 | 91.58 | 95.70 | 100.62 | 140.23 | 146.57 | 152.21 | 158.95 | 163.64 |

For large values of degrees of freedom the approximate formula

$$\chi_\alpha^2 = n\left(1 - \frac{2}{9n} + z_\alpha\sqrt{\frac{2}{9n}}\right)^3$$

where $z_\alpha$ is the normal deviate and $n$ is the number of degrees of freedom, may be used. For example $\chi_{.99}^2 = 60[1 - .00370 + 2.326(.06086)]^3 = 60(1.1379)^3 = 88.4$ for the 99th percentile for 60 degrees of freedom.

* From *Introduction to Statistical Analysis,* 2d ed., by W. J. Dixon and F. J. Massey, Jr. Copyright, 1957. McGraw-Hill Book Company. Used by permission.

*Table A.4 Critical Values $D_n^{(\gamma)}$ for the Kolmogorov Goodness-of-Fit Test* *

| | γ | | | | |
|---|---|---|---|---|---|
| $n$ | 0.20 | 0.15 | 0.10 | 0.05 | 0.01 |
| 1 | 0.900 | 0.925 | 0.950 | 0.975 | 0.995 |
| 2 | 0.684 | 0.726 | 0.776 | 0.842 | 0.929 |
| 3 | 0.565 | 0.597 | 0.642 | 0.708 | 0.828 |
| 4 | 0.494 | 0.525 | 0.564 | 0.624 | 0.733 |
| 5 | 0.446 | 0.474 | 0.510 | 0.565 | 0.669 |
| 6 | 0.410 | 0.436 | 0.470 | 0.521 | 0.618 |
| 7 | 0.381 | 0.405 | 0.438 | 0.486 | 0.577 |
| 8 | 0.358 | 0.381 | 0.411 | 0.457 | 0.543 |
| 9 | 0.339 | 0.360 | 0.388 | 0.432 | 0.514 |
| 10 | 0.322 | 0.342 | 0.368 | 0.410 | 0.490 |
| 11 | 0.307 | 0.326 | 0.352 | 0.391 | 0.468 |
| 12 | 0.295 | 0.313 | 0.338 | 0.375 | 0.450 |
| 13 | 0.284 | 0.302 | 0.325 | 0.361 | 0.433 |
| 14 | 0.274 | 0.292 | 0.314 | 0.349 | 0.418 |
| 15 | 0.266 | 0.283 | 0.304 | 0.338 | 0.404 |
| 16 | 0.258 | 0.274 | 0.295 | 0.328 | 0.392 |
| 17 | 0.250 | 0.266 | 0.286 | 0.318 | 0.381 |
| 18 | 0.244 | 0.259 | 0.278 | 0.309 | 0.371 |
| 19 | 0.237 | 0.252 | 0.272 | 0.301 | 0.363 |
| 20 | 0.231 | 0.246 | 0.264 | 0.294 | 0.356 |
| 25 | 0.21 | 0.22 | 0.24 | 0.27 | 0.32 |
| 30 | 0.19 | 0.20 | 0.22 | 0.24 | 0.29 |
| 35 | 0.18 | 0.19 | 0.21 | 0.23 | 0.27 |
| $>35$ | $1.07/\sqrt{n}$ | $1.14/\sqrt{n}$ | $1.22/\sqrt{n}$ | $1.36/\sqrt{n}$ | $1.63/\sqrt{n}$ |

* With permission from F. J. Massey (1951). The Kolmogorov-Smirnov Test for Goodness of Fit, *Journal of the American Statistical Association*, Vol. 46, p. 70.

*Table A.5 95th Percentile Values for the F Distribution* *

| $f_1$ / $f_2$ | 1 | 2 | 3 | 4 | 5 | 6 | 7 | 8 | 9 | 10 | 12 | 15 | 20 | 24 | 30 | 40 | 60 | 120 | ∞ |
|---|---|---|---|---|---|---|---|---|---|---|---|---|---|---|---|---|---|---|---|
| 1 | 161 | 200 | 216 | 225 | 230 | 234 | 237 | 239 | 241 | 242 | 244 | 246 | 248 | 249 | 250 | 251 | 252 | 253 | 254 |
| 2 | 18.5 | 19.0 | 19.2 | 19.2 | 19.3 | 19.3 | 19.4 | 19.4 | 19.4 | 19.4 | 19.4 | 19.4 | 19.4 | 19.5 | 19.5 | 19.5 | 19.5 | 19.5 | 19.5 |
| 3 | 10.1 | 9.55 | 9.28 | 9.12 | 9.01 | 8.94 | 8.89 | 8.85 | 8.81 | 8.79 | 8.74 | 8.70 | 8.66 | 8.64 | 8.62 | 8.59 | 8.57 | 8.55 | 8.53 |
| 4 | 7.71 | 6.94 | 6.59 | 6.39 | 6.26 | 6.16 | 6.09 | 6.04 | 6.00 | 5.96 | 5.91 | 5.86 | 5.80 | 5.77 | 5.75 | 5.72 | 5.69 | 5.66 | 5.63 |
| 5 | 6.61 | 5.79 | 5.41 | 5.19 | 5.05 | 4.95 | 4.88 | 4.82 | 4.77 | 4.74 | 4.68 | 4.62 | 4.56 | 4.53 | 4.50 | 4.46 | 4.43 | 4.40 | 4.37 |
| 6 | 5.99 | 5.14 | 4.76 | 4.53 | 4.39 | 4.28 | 4.21 | 4.15 | 4.10 | 4.06 | 4.00 | 3.94 | 3.87 | 3.84 | 3.81 | 3.77 | 3.74 | 3.70 | 3.67 |
| 7 | 5.59 | 4.74 | 4.35 | 4.12 | 3.97 | 3.87 | 3.79 | 3.73 | 3.68 | 3.64 | 3.57 | 3.51 | 3.44 | 3.41 | 3.38 | 3.34 | 3.30 | 3.27 | 3.23 |
| 8 | 5.32 | 4.46 | 4.07 | 3.84 | 3.69 | 3.58 | 3.50 | 3.44 | 3.39 | 3.35 | 3.28 | 3.22 | 3.15 | 3.12 | 3.08 | 3.04 | 3.01 | 2.97 | 2.93 |
| 9 | 5.12 | 4.26 | 3.86 | 3.63 | 3.48 | 3.37 | 3.29 | 3.23 | 3.18 | 3.14 | 3.07 | 3.01 | 2.94 | 2.90 | 2.86 | 2.83 | 2.79 | 2.75 | 2.71 |
| 10 | 4.96 | 4.10 | 3.71 | 3.48 | 3.33 | 3.22 | 3.14 | 3.07 | 3.02 | 2.98 | 2.91 | 2.85 | 2.77 | 2.74 | 2.70 | 2.66 | 2.62 | 2.58 | 2.54 |
| 11 | 4.84 | 3.98 | 3.59 | 3.36 | 3.20 | 3.09 | 3.01 | 2.95 | 2.90 | 2.85 | 2.79 | 2.72 | 2.65 | 2.61 | 2.57 | 2.53 | 2.49 | 2.45 | 2.40 |
| 12 | 4.75 | 3.89 | 3.49 | 3.26 | 3.11 | 3.00 | 2.91 | 2.85 | 2.80 | 2.75 | 2.69 | 2.62 | 2.54 | 2.51 | 2.47 | 2.43 | 2.38 | 2.34 | 2.30 |
| 13 | 4.67 | 3.81 | 3.41 | 3.18 | 3.03 | 2.92 | 2.83 | 2.77 | 2.71 | 2.67 | 2.60 | 2.53 | 2.46 | 2.42 | 2.38 | 2.34 | 2.30 | 2.25 | 2.21 |
| 14 | 4.60 | 3.74 | 3.34 | 3.11 | 2.96 | 2.85 | 2.76 | 2.70 | 2.65 | 2.60 | 2.53 | 2.46 | 2.39 | 2.35 | 2.31 | 2.27 | 2.22 | 2.18 | 2.13 |
| 15 | 4.54 | 3.68 | 3.29 | 3.06 | 2.90 | 2.79 | 2.71 | 2.64 | 2.59 | 2.54 | 2.48 | 2.40 | 2.33 | 2.29 | 2.25 | 2.20 | 2.16 | 2.11 | 2.07 |
| 16 | 4.49 | 3.63 | 3.24 | 3.01 | 2.85 | 2.74 | 2.66 | 2.59 | 2.54 | 2.49 | 2.42 | 2.35 | 2.28 | 2.24 | 2.19 | 2.15 | 2.11 | 2.06 | 2.01 |
| 17 | 4.45 | 3.59 | 3.20 | 2.96 | 2.81 | 2.70 | 2.61 | 2.55 | 2.49 | 2.45 | 2.38 | 2.31 | 2.23 | 2.19 | 2.15 | 2.10 | 2.06 | 2.01 | 1.96 |
| 18 | 4.41 | 3.55 | 3.16 | 2.93 | 2.77 | 2.66 | 2.58 | 2.51 | 2.46 | 2.41 | 2.34 | 2.27 | 2.19 | 2.15 | 2.11 | 2.06 | 2.02 | 1.97 | 1.92 |
| 19 | 4.38 | 3.52 | 3.13 | 2.90 | 2.74 | 2.63 | 2.54 | 2.48 | 2.42 | 2.38 | 2.31 | 2.23 | 2.16 | 2.11 | 2.07 | 2.03 | 1.98 | 1.93 | 1.88 |
| 20 | 4.35 | 3.49 | 3.10 | 2.87 | 2.71 | 2.60 | 2.51 | 2.45 | 2.39 | 2.35 | 2.28 | 2.20 | 2.12 | 2.08 | 2.04 | 1.99 | 1.95 | 1.90 | 1.84 |
| 21 | 4.32 | 3.47 | 3.07 | 2.84 | 2.68 | 2.57 | 2.49 | 2.42 | 2.37 | 2.32 | 2.25 | 2.18 | 2.10 | 2.05 | 2.01 | 1.96 | 1.92 | 1.87 | 1.81 |
| 22 | 4.30 | 3.44 | 3.05 | 2.82 | 2.66 | 2.55 | 2.46 | 2.40 | 2.34 | 2.30 | 2.23 | 2.15 | 2.07 | 2.03 | 1.98 | 1.94 | 1.89 | 1.84 | 1.78 |
| 23 | 4.28 | 3.42 | 3.03 | 2.80 | 2.64 | 2.53 | 2.44 | 2.37 | 2.32 | 2.27 | 2:20 | 2.13 | 2.05 | 2.01 | 1.96 | 1.91 | 1.86 | 1.81 | 1.76 |
| 24 | 4.26 | 3.40 | 3.01 | 2.78 | 2.62 | 2.51 | 2.42 | 2.36 | 2.30 | 2.25 | 2.18 | 2.11 | 2.03 | 1.98 | 1.94 | 1.89 | 1.84 | 1.79 | 1.73 |
| 25 | 4.24 | 3.39 | 2.99 | 2.76 | 2.60 | 2.49 | 2.40 | 2.34 | 2.28 | 2.24 | 2.16 | 2.09 | 2.01 | 1.96 | 1.92 | 1.87 | 1.82 | 1.77 | 1.71 |
| 26 | 4.23 | 3.37 | 2.98 | 2.74 | 2.59 | 2.47 | 2.39 | 2.32 | 2.27 | 2.22 | 2.15 | 2.07 | 1.99 | 1.95 | 1.90 | 1.85 | 1.80 | 1.75 | 1.69 |
| 27 | 4.21 | 3.35 | 2.96 | 2.73 | 2.57 | 2.46 | 2.37 | 2.31 | 2.25 | 2.20 | 2.13 | 2.06 | 1.97 | 1.93 | 1.88 | 1.84 | 1.79 | 1.73 | 1.67 |
| 28 | 4.20 | 3.34 | 2.95 | 2.71 | 2.56 | 2.45 | 2.36 | 2.29 | 2.24 | 2.19 | 2.12 | 2.04 | 1.96 | 1.91 | 1.87 | 1.82 | 1.77 | 1.71 | 1.65 |
| 29 | 4.18 | 3.33 | 2.93 | 2.70 | 2.55 | 2.43 | 2.35 | 2.28 | 2.22 | 2.18 | 2.10 | 2.03 | 1.94 | 1.90 | 1.85 | 1.81 | 1.75 | 1.70 | 1.64 |
| 30 | 4.17 | 3.32 | 2.92 | 2.69 | 2.53 | 2.42 | 2.33 | 2.27 | 2.21 | 2.16 | 2.09 | 2.01 | 1.93 | 1.89 | 1.84 | 1.79 | 1.74 | 1.68 | 1.62 |
| 40 | 4.08 | 3.23 | 2.84 | 2.61 | 2.45 | 2.34 | 2.25 | 2.18 | 2.12 | 2.08 | 2.00 | 1.92 | 1.84 | 1.79 | 1.74 | 1.69 | 1.64 | 1.58 | 1.51 |
| 60 | 4.00 | 3.15 | 2.76 | 2.53 | 2.37 | 2.25 | 2.17 | 2.10 | 2.04 | 1.99 | 1.92 | 1.84 | 1.75 | 1.70 | 1.65 | 1.59 | 1.53 | 1.47 | 1.39 |
| 120 | 3.92 | 3.07 | 2.68 | 2.45 | 2.29 | 2.18 | 2.09 | 2.02 | 1.96 | 1.91 | 1.83 | 1.75 | 1.66 | 1.61 | 1.55 | 1.50 | 1.43 | 1.35 | 1.25 |
| ∞ | 3.84 | 3.00 | 2.60 | 2.37 | 2.21 | 2.10 | 2.01 | 1.94 | 1.88 | 1.83 | 1.75 | 1.67 | 1.57 | 1.52 | 1.46 | 1.39 | 1.32 | 1.22 | 1.00 |

$f_1$ = degrees of freedom in numerator

$f_2$ = degrees of freedom in denominator

* E. S. Pearson and H. O. Hartley, *Biometrika Tables for Statisticians*, Vol. 2 (1972), Table 5, p. 178. Used by permission.

# Appendix B

**GENERIC FAILURE DATA**

*Table B.1 Generic Failure Data for Mechanical Items*

| Component Failure Mode | Range from Other Sources | Suggested Mean Value | Lognormal Error Factor* |
|---|---|---|---|
| Air Operated Valves | | | |
| Failure to Operate | 3E–4/D to 2E–2/D | 2E–3/D | 3 |
| Failure Due to Plugging | 2E–5/D to 1E–4/D, 1E–7/yr | 1E–7/hr | 3 |
| Unavailability Due to Test and Maintenance | 6E–5/D to 6E–3/D | 8E–4/D | 10 |
| Spurious Closure | – | 1E–7/hr | 3 |
| Spurious Open | – | 5E–7/hr | 10 |
| Pressure Regulator Valve | | | |
| Failure to Open | – | 2E–3/D | 3 |
| Motor Operated Valves | | | |
| Failure to Operate | 1E–3/D to 9E–3/D | 3E–3/D | 10 |
| Failure Due to Plugging | 2E–5/D to 1E–4/D | 1E–7/hr | 3 |
| Unavailability Due to Test and Maintenance | 6E–5/D to 6E–3/D | 8E–4/D | 10 |
| Failure to Remain Closed | – | 5E–7/hr | 10 |
| Failure to Remain Open | – | 1E–7/hr | 3 |
| Solenoid Operated Valves | | | |
| Failure to Operate | 1E–3/D to 2E–2/D | 2E–3/D | 3 |
| Failure Due to Plugging | 2E–5/D to 1E–4/D, 1E–7/yr | 1E–7/hr | 3 |
| Unavailability Due to Test and Maintenance | 6E–5/D to 6E–3/D | 8E–4/D | 10 |
| Hydraulic Operated Valves | | | |
| Failure to Operate | 3E–4/D to 2E–2/D | 2E–3/D | 3 |
| Failure Due to Plugging | 2E–5/D to 1E–4/D, 1E–7/yr | 1E–7/hr | 3 |
| Unavailability Due to Test and Maintenance | 6E–5/D to 6E–3/D | 8E–4/D | 10 |

* Defined as $EF = \frac{P_U}{m} = \frac{m}{P_L}$, where $P_U$ and $P_L$ are upper and lower 95 percentile of lognormal distribution and $m$ is its median.

*Table B.1 Generic Failure Data for Mechanical Items (Continued)*

| Component Failure Mode | Range from Other Sources | Suggested Mean Value | Lognormal Error Factor* |
|---|---|---|---|
| Explosive Operated Valves | | | |
| Failure to Operate | 1E–3/D to 9E–3/D | 3E–3/D | 3 |
| Failure Due to Plugging | 2E–5/D to 1E–4/D, 1E-7/yr | 1E–7/hr | 3 |
| Unavailability Due to Test and Maintenance | 6E–5/D to 6E–3/D | 8E–4/D | 10 |
| Manual Valve | | | |
| Failure Due to Plugging | 2E–5/D to 1E–4/D, 1E-7/yr | 1E–7/hr | 3 |
| Unavailability Due to Test and Maintenance | 6E–5/D to 6E–3/D, | 8E–4/D | 10 |
| Failure to Open | – | 1E–4/D | 3 |
| Failure to Remain Closed | – | 1E–4/D | 3 |
| Check Value | | | |
| Failure to Open | 6E-5/D to 1.2E-4/D | 1E–4/D | 3 |
| Failure to Close | – | 1E–3/hr | 3 |
| Safety Relief Valves (SRVs) - BWR | | | |
| Failure to Open for Pressure Relief | – | 1E–5/D | 3 |
| Failure to Open on Actuation | – | 1E–2/D | 3 |
| Failure to Reclose on Pressure Relief | – | 3.9E–6/hr | 10 |
| Relief Valve (Not SRV or PORV) | | | |
| Spurious Open | – | 3.9E–6/hr | 10 |
| Power Operated Relief Valves (PORVs) - PWR | | | |
| Failure to Open on Actuation | – | 2E–3/D | 3 |

* Defined as $EF = \frac{P_U}{m} = \frac{m}{P_L}$, where $P_U$ and $P_L$ are upper and lower 95 percentile of lognormal distribution and $m$ is its median.

*Table B.1 Generic Failure Data for Mechanical Items (Continued)*

| Component Failure Mode | Range from Other Sources | Suggested Mean Value | Lognormal Error Factor* |
|---|---|---|---|
| Failure to Open for Pressure Relief | – | 3E–4/D | 10 |
| Failure to Reclose | – | 2E–3/D | 3 |
| <u>Motor Driven Pump</u> | | | |
| Failure to Start | 5E–4/D to 1E–4/D | 3E–3/D | 10 |
| Failure to Run | 1E–6/hr to 1E–3/hr | 3E–5/hr | 10 |
| Unavailability Due to Test and Maintenance | 1E–4/D to 1E–2/D | 2E–3/D | 10 |
| <u>Turbine Driven Pump</u> | | | |
| Failure to Start | 5E–3/D to 9E–2/D | 3E–2/D | 10 |
| Failure to Run | 8E–6/hr to 1E–3/hr | 5E–3/hr | 10 |
| Unavailability Due to Test and Maintenance | 3E–3/D to 4E–2/D | 1E–2/D | 10 |
| <u>Diesel Driven Pump</u> | | | |
| Failure to Start | 1E–3/D to 1E–2/D | 3E–2/D | 3 |
| Failure to Run | 2E–5/hr to 1E–3/hr | 8E–4/hr | 10 |
| Unavailability Due to Test and Maintenance | – | 1E–2/D | 10 |
| <u>Heat Exchanger</u> | | | |
| Failure Due to Blockage | – | 5.7E-6/hr | 10 |
| Failure Due to Rupture (Leakage) | – | 3E-6/hr | 10 |
| Unavailability Due to Test and Maintenance | – | 3E-5/hr | 110 |
| <u>AC Electric Power</u> | | | |
| <u>Diesel Generator (DG) Hardware Failure</u> | | | |
| Failure to Start | 8E–3/D to 1E–3/D | 3E–2/D | 3 |
| Failure to Run | 2E–4/hr to 3E–3/hr, | 2E–3/hr | 10 |
| <u>DG Test and Maintenance Unavailability</u> | Neg 1 to 4E-2/D | 6E-3/D | 10 |

* Defined as $EF = \frac{P_U}{m} = \frac{m}{P_L}$, where $P_U$ and $P_L$ are upper and lower 95 percentile of lognormal distribution and $m$ is its median.

*Table B.1 Generic Failure Data for Mechanical Items (Continued)*

| Component Failure Mode | Range from Other Sources | Suggested Mean Value | Lognormal Error Factor* |
|---|---|---|---|
| Loss of Offsite Power Other than Initiator | | 2E–4 | 3 |
| AC Bus Hardware Failure | 1E-8/hr to 4E-6/hr | 1E–7/hr | 5 |
| Circuit Breaker | | | |
| Spurious Open | – | 1E–6/hr | 3 |
| Fail to Transfer | – | 3E–3/D | 10 |
| Time Delay Relay | | | |
| Fail to Transfer | – | 3E–4/hr | 10 |
| Transformer | | | |
| Short or Open | – | 2E–6/hr | 10 |
| DC Electric Power | | | |
| Hardware Failure | 6E-10/hr to 1E-4/hr | | |
| Bus | – | 1E–7/hr | 5 |
| Battery | – | 1E–6/hr | 3 |
| Charger | – | 1E–6/hr | 3 |
| Inverter | – | 1E–4/hr | 3 |
| Test and Maintenance Unavailability | | | |
| Battery | – | 1E–3/D | 10 |
| Bus | – | 8E–6/hr | 10 |
| Charger | – | 3E–4/D | 10 |
| Inverter | – | 1E–3/D | 10 |
| Orifice | | | |
| Failure Due to Plugging | – | 3E–4/D | 3 |
| Strainer | | | |
| Failure Due to Plugging | – | 3E–5/hr | 10 |
| Sump | | | |
| Failure Due to Plugging | – | 5E–5/D | 100 |

* Defined as $EF = \frac{P_U}{m} = \frac{m}{P_L}$, where $P_U$ and $P_L$ are upper and lower 95 percentile of lognormal distribution and $m$ is its median.

*Table B.1 Generic Failure Data for Mechanical Items (Continued)*

| Component Failure Mode | Range from Other Sources | Suggested Mean Value | Lognormal Error Factor* |
|---|---|---|---|
| <u>Cooling Coil</u> | | | |
| Failure to Operate | – | 1E–6/hr | 3 |
| <u>Transmitter</u> | | | |
| Failure to Operate | – | 1E–6/hr | 3 |
| <u>Fan (HVAC)</u> | | | |
| Failure to Start | – | 3E–4/D | 3 |
| Failure to Run | – | 1E–5/hr | 3 |
| Unavailability Due to Test and Maintenance | – | 2E–3/D | 10 |
| <u>Instrumentation (Includes Sensor, Transmitter and Process Switch)</u> | | | |
| Failure to Operate | – | 3E–6/hr | 10 |
| <u>Temperature Switch</u> | | | |
| Failure to Transfer | – | 1E–4/D | 3 |
| <u>Transfer Switch</u> | | | |
| Failure to Transfer | – | 1E–3/D | 3 |
| <u>Instrument Air Compressor</u> | | | |
| Failure to Start | – | 8E–2/D | 3 |
| Failure to Run | – | 2E–4/hr | 10 |
| Unavailability Due to Test and Maintenance | –<br>– | 2E–3/D | 10 |
| <u>Flow Controller</u> | | | |
| Failure to Operate | – | 1E–4/D | 3 |
| <u>Cooling tower Fan</u> | | | |
| Failure to Start | – | 4E–3/D | 3 |
| Failure to Run | – | 7E–6/hr | 10 |
| Unavailability Due to Test and Maintenance | – | 2E–3/D | 10 |
| <u>Damper</u> | | | |
| Failure to Open | – | 3E–3/D | 10 |

* Defined as $EF = \frac{P_U}{m} = \frac{m}{P_L}$, where $P_U$ and $P_L$ are upper and lower 95 percentile of lognormal distribution and $m$ is its median.

# Appendix C

## COMPUTER PROGRAMS FOR RELIABILITY RISK ANALYSES

*Table C.1 PC-Based Codes for Logical (Boolean Based) Analysis*

| Code name | Primary functions | How to obtain more information |
|---|---|---|
| CAFTA+/ETA-II | • Full screen fault tree* editor<br>• Top-event cut set generator<br>• Cut set screening editor<br>• Interacts with other mainframe codes<br>• Event tree editor<br>• Integrates fault trees and event trees<br>• Cut set generator<br>• Cut set screening editor<br>• Database for failure data<br>• Cut set quantification | Science Applications International Corp.<br>5150 El Camino Real<br>Suite C-31<br>Los Altos, CA 94022 |
| IRRAS | • Full screen fault tree editor<br>• Top-event cut set generator<br>• Interacts with mainframe<br>• Cut set generator<br>• Event tree editor<br>• Cut set and event tree sequence quantification<br>• Database for failure date<br>• Integrates fault trees and event trees | U.S. Nuclear Regulatory Commission<br>Office of Nucl. Reg. Research<br>Washington, DC 20555 |
| PRA_Workstation | • Works under Microsoft WINDOWS<br>• Fault tree editor<br>• Failure database editor<br>• Combining Boolean logic (e.g., merging fault trees)<br>• Interacts with other major codes | NUS Corp.<br>545 Shoup Avenue<br>Idaho Falls, ID 83401 |

* When a code can handle a fault tree logic, other logical representation such as trees can also be handled by these codes.

*Table C.1 PC-Based Codes for Logical (Boolean Based) Analysis (continued)*

| Code name | Primary functions | How to obtain more information |
|---|---|---|
| SETS | • Boolean equation reduction<br>• Handles complex Boolean equations or fault trees<br>• Logically combines (merges) fault trees<br>• Quantifies fault trees or Boolean equations | |
| RISK-MAN | • Event tree editor<br>• Data base editor<br>• Fault tree editor<br>• Cut set generator<br>• Handles and combine large event trees<br>• Calculate event tree sequence probabilities<br>• Bayesian analysis | PLG, Inc.<br>2260 University Drive<br>Newport Beach, CA 92660 |
| MLD-Code | • Graphically constructs MLD<br>• Propagates effect of failures in the MLD<br>• Generates end-states<br>• Failure data editor<br>• Quantifies end-states<br>• Rank important states | University of Maryland<br>Center of reliability Analysis<br>College Park, MD 20742 |
| MPLD-Software | • Works under Microsoft WINDOWS<br>• Graphically constructs MLD<br>• Propagates effects of failure in the MLD | Scientech, Inc.<br>11821 Parklawn Dr.<br>Suite 100<br>Rockville, MD 20852 |

*Table C.2 Capabilities of PC-Based Software*

| Software Name | Time-Dependent Availability | Reliability Block Diagram | Markov Analysis | Data Trend Analysis | Dependent Failure Analysis | FMEA FMECA | Uncertainty Analysis | Importance Analysis | Human Reliability Analysis | Address |
|---|---|---|---|---|---|---|---|---|---|---|
| FRANTIC ABC | √ | | | | | | | | | Applied Biomathematics<br>100 North County Rd., Bld. B<br>Setauket, NY 11733 |
| CAFTA+ | | | | | | | √ | √ | | See Table B.1 |
| IRRAS | | | | | √ | | √ | √ | | See Table B.1 |
| RISK-MAN | | | | | √ | | √ | √ | | See Table B.1 |
| TEMAC | | | | | | | √ | √ | | Sandia National<br>Laboratories<br>Albuquerque, NM 87185 |
| MARKOV1 | | | √ | | | | | | | Decision System Associate<br>746 Crompton<br>Redwood City, CA 94061 |
| RAMS-CALS | | | | | | √ | | | | Management Sciences Inc.<br>6022 Constitution Ave., NE<br>Albuquerque, NM 87110 |
| RBDA | | √ | | | | | | | | Science Application Int. Corp.<br>5150 El Camino Real<br>Suite C-31<br>Los Altos, CA 94022 |
| BRAT | | | | | | | √ | | | The Craig Mark Company.<br>P.O. Box 192<br>Del Mar, CA 92014 |
| SETS | | | | | √ | | | √ | | See Table B.1 |
| SLIM-MAUD | | | | | | | | | √ | |

# INDEX

## A



## B



## C



## D

## P



## Q



## R



## S

# T



# U

## V



## W

# About the Author

**M. MODARRES** is a Professor of Reliability and of Nuclear Engineering and a cofounder of the Reliability Engineering Program at the University of Maryland, College Park. A consultant to government, industry, and international organizations, he is the author or coauthor of more than 80 professional papers and a member of the American Nuclear Society and the Institute of Electrical and Electronics Engineers. Dr. Modarres received the B.S. degree (1975) in mechanical engineering from Tehran Polytechnic Institute, Iran, and the M.S. degree (1977) in mechanical engineering and Ph.D. degree (1979) in nuclear engineering from the Massachusetts Institute of Technology, Cambridge.